SUPERNOVAE AND SUPERNOVA REMNANTS

ASTROPHYSICS AND SPACE SCIENCE LIBRARY

A SERIES OF BOOKS ON THE RECENT DEVELOPMENTS OF SPACE SCIENCE AND OF GENERAL GEOPHYSICS AND ASTROPHYSICS PUBLISHED IN CONNECTION WITH THE JOURNAL SPACE SCIENCE REVIEWS

VOLUME 45

SUPERNOVAE AND SUPERNOVA REMNANTS

PROCEEDINGS OF THE INTERNATIONAL CONFERENCE ON SUPERNOVAE HELD IN LECCE, ITALY, MAY 7–11, 1973

Edited by

CRISTIANO BATALLI COSMOVICI

Astrophysics and Space Science Group, Institute of Physics,
University of Lecce, Lecce 73100, Italy

Assisted by

EMILIA D'ANNA and ALFREDO BORGHESI

D. REIDEL PUBLISHING COMPANY

DORDRECHT-HOLLAND / BOSTON-U.S.A.

Library of Congress Catalog Card Number 73–91428

ISBN-13:978-94-010-2168-5 e-ISBN-13:978-94-010-2166-1
DOI: 10.1007/978-94-010-2166-1

Published by D. Reidel Publishing Company,
P.O. Box 17, Dordrecht, Holland

Sold and distributed in the U.S.A., Canada and Mexico
by D. Reidel Publishing Company, Inc.
306 Dartmouth Street, Boston,
Mass. 02116, U.S.A.

Softcover reprint of the hardcover 1st edition 1974

TABLE OF CONTENTS

* Contributor who delivered the paper at the conference.

PART III / SPECTRA OF SUPERNOVAE AND THEIR INTERPRETATION

PART IV / STATISTICS OF SUPERNOVAE

PART V / SUPERNOVA REMNANTS

PART VI / THEORIES ON SUPERNOVAE AND SUPERNOVA REMNANTS

PREFACE

> This conference is a tribute to those astronomers who pioneered the investigation of this subject such a short time ago and who carried it through to its present state.
>
> (H. Arp, *Concluding Remarks of the Conference*)

A previous conference, covering mainly the observational aspects of Supernovae, was held at the Haute Provence Observatory in September 1963.

In the following ten years this field of research has considerably increased; it seemed, therefore, the right time to organize an international conference on Supernovae taking into account that in the meantime important discoveries, such as the Pulsars, had been made, and new techniques of observation were available. This book contains the proceedings of this conference held at Porto Cesareo (Lecce), Italy, during the period May 7 through 11, 1973. About one hundred participants from eighteen countries attended the conference.

It was also the first attempt to hold an international conference in the Salento, the southernmost region of Apulia, in whose capital, Lecce, the newly founded Faculty of Sciences of the University of Lecce is located.

The program of the conference included the results and techniques of Supernova surveys, photometric and spectral studies, statistics of Supernovae, Supernova Remnants, and finally, theories on Supernovae and Supernova Remnants.

The review and contributed papers were of an excellent scientific standard, thus assuring the complete success of the conference. Among the people who contributed to the organizational success I would like first to thank Prof. Francesco Bertola for suggesting this conference and for continuous help in preparing it, and all the other members of the Scientific Committee for their valuable suggestions and advice.

The Director of the Institute of Physics, Prof. Saverio Mongelli was most helpful in providing financial support and all necessities for the conference.

Dr Emilia D'Anna, general secretary of the conference, was an excellent and untiring collaborator in its organization, and in the preparation of the proceedings.

Dr Alfredo Borghesi and Dr Livio Ruggiero were especially cooperative in providing excellent residential and conference facilities.

My special thanks are also due to Prof. Guido Pizzella, to the Rector of the University of Lecce, Prof. Giuseppe Codacci-Pisanelli, to the President of the Faculty of Sciences, Prof. Ida Gasparini-Cattaneo, and to my wife, Beatrice, for their moral support.

The conference would not have been possible without the financial support of the following institutions:

Consiglio Nazionale delle Ricerche, Ministero della Pubblica Istruzione, Ministero per gli Affari Esteri, Regione Puglia, Provincia di Lecce, Comune di Lecce, Comune di Monteroni, Comune di Nardò, Banca del Salento, Ente Provinciale per il Turismo, Banca Vallone e Venturi.

Lecce, September 1973 CRISTIANO BATALLI COSMOVICI

SCIENTIFIC COMMITTEE

H. ARP, Hale Observatories, Pasadena
F. BERTOLA, Università di Lecce
E. R. MUSTEL, Academy of Sciences, Moscow
L. ROSINO, Osservatorio Astronomico, Padova
B. ROSSI, MIT, Cambridge, Mass.
L. WOLTJER, Columbia University, New York
F. ZWICKY, Caltech, Pasadena

ORGANIZING COMMITTEE

(Università di Lecce)

A. BORGHESI, S. MONGELLI, L. RUGGIERO, and G. SOLIANI

Chairman: C. BATALLI COSMOVICI
General Secretary: E. D'ANNA

LIST OF PARTICIPANTS

Adams, D. J., Dept. of Astronomy, University of Leicester, Leicester, Great Britain
Agrinier, B., Centre d'Études Nucléaires de Saclay, 91-Gif-Sur-Yvette, France
Aiello, S., Istituto di Fisica, Università di Firenze, Firenze, Italy
Ardeberg, A., European Southern Observatory, Santiago, Chile
Arp, H., Hale Observatories, Pasadena, Calif., U.S.A.
Barbieri, C., Osservatorio Astrofisico, Asiago, Italy
Barbon, R., Osservatorio Astrofisico, Asiago, Italy
Belli, B. M., Lab. di Astrofisica CNR, Frascati, Italy
Bertola, F., Istituto di Fisica, Università di Lecce, Lecce, Italy
Biermann, L., Max-Planck-Institut für Physik und Astrophysik, München, Germany
Bollea, A., Laboratorio del Plasma nello Spazio, Frascati, Italy
Bonetti, A., Università di Firenze, Firenze, Italy
Bonoli, F., Osservatorio Astronomico Universitario, Bologna, Italy
Borghesi, A., Istituto di Fisica, Università di Lecce, Lecce, Italy
Branch, D., Norman, Okla., U.S.A.
Broglia, P., Osservatorio Astronomico, Merate, Italy
Buchler, J. R., Belfer Graduate School of Science, Yeshiva University, New York, U.S.A.
Caloi, V., Laboratorio di Astrofisica, CNR, Frascati, Italy
Capaccioli, M., Osservatorio Astronomico, Padova, Italy
Casini, C. B., Osservatorio Astronomico, Milano-Brera, Italy
Cavaliere, A., Laboratorio di Astrofisica, CNR, Frascati, Italy
Celnikier, L. M., Section d'Astrophysique 92, Observatoire de Paris, Meudon, France
Chiosi, C., Osservatorio Astronomico, Padova, Italy
Chiosi, E., Osservatorio Astronomico, Padova, Italy
Ciatti, F., Osservatorio Astrofisico, Asiago, Italy
Cline, T. L., Nasa Goddard Space Flight Center, Greenbelt,Md., U.S.A.
Conconi, P., Osservatorio Astronomico, Merate, Italy
Cosmovici, C. B., Istituto di Fisica, Università di Lecce, Lecce, Italy
Dallaporta N., Osservatorio Astronomico, Padova, Italy
D'Anna, E., Istituto di Fisica, Università di Lecce, Lecce, Italy
De Bruyn, A. G., Sterrewacht te Leiden, Leiden, Holland
De Groot, M., European Southern Observatory, Santiago, Chile
Detre, L., Konkoly Observatory, Szabadsaghegy, Budapest, Hungary
Dickel, J. R., University of Illinois, Dept. of Astronomy, Urbana, Ill., U.S.A.
D'Odorico, A., Osservatorio Astrofisico, Asiago, Italy
Duin, R. M., Sterrewacht te Leiden, Leiden, Holland

Ferrari, A., Istituto di Fisica dell'Università, Torino, Italy
Friedjung, M., Institut d'Astrophysique, Paris, France
Gallino, R., Istituto di Fisica dell'Università di Torino, Torino, Italy
Gordon, C., IBM Thomas J. Watson Research Lab., Yorktown Heights, N.Y., U.S.A.
Gorenstein, P., Center for Astrophysics, Smithsonian Astrophys. Observ., Cambridge, Mass., U.S.A.
Goss, W. M., Sterrenkundig Lab. Kapteyn, Rijksuniversiteit te Groningen, Groningen, Holland
Gull, S. F., Cavendish Laboratory, Cambridge, Great Britian
Hassan, S. M., Helwan Astronomical Observatory, Cairo, Egypt
Heidmann, J., Section d'Astrophysique, Observatoire de Paris, Meudon, France
Henbest, S. N., Cavendish Laboratory, Cambridge, Great Britain
Immirzi, G., Istituto di Fisica, Università di Lecce, Lecce, Italy
Jelley, J. V., AERE, Nuclear Physics Division, Harwell, Great Britain
Johansson, K., Astronomical Observatory, Uppsala University, Uppsala, Sweden
Köbke, K., Institut für Kernphysik, Technische Hochschule Darmstadt, Darmstadt, Germany
Leggieri, G., Istituto di Fisica, Università di Lecce, Lecce, Italy
Lopez-Garcia, A., Facultad de Ciencias, Universidad de Valencia, Observatorio Astronómico, Valencia, Spain
Lovas, M., Konkoly Observatory, Szabadsaghegy, Budapest, Hungary
Madore, B. F., Dept. of Astronomy, University of Toronto, Toronto, Ontario, Canada
Mansfield, V., Center for Radiophysics and Space Research, Cornell University, Ithaca, N.Y., U.S.A.
Massaro, E., Istituto di Fisica, Università di Palermo, Palermo, Italy
McCarthy, M. F., Specola Vaticana, Castel Gandolfo, Vatican
Meikle, W. P. S., Dept. of Natural Philosophy, University of Glasgow, Great Britain
Mongelli, S., Istituto di Fisica, Università di Lecce, Lecce, Italy
Moore, E., New Mexico Institute of Mining and Technology, Socorro, N.M., U.S.A.
Palumbo, G. G. C., Laboratorio TE.S.R.E. CNR, Bologna, Italy
Pizzella, G., Istituto di Fisica, Università di Lecce, Lecce, Italy
Pizzichini, G., Laboratorio TE.S.R.E. CNR, Bologna, Italy
Porter, N. A., Belfield Physics Department, University College, Dublin, Ireland
Reeves, H., Commissariat à l'Énergie Atomique, Centre d'Étude Nucleaires de Saclay, Gif-sur-Yvette, Saclay, France
Robba, N. R., Istituto di Fisica, Università di Palermo, Palermo, Italy
Romano, G., Osservatorio Astronomico, Padova, Italy
Rosino, L., Osservatorio Astronomico, Padova, Italy
Rossi, B., Dept. of Physics, Massachusetts Institute of Technology, Cambridge, Mass., U.S.A.
Ruggiero, L., Istituto di Fisica, Università di Lecce, Lecce, Italy
Sartori, L., Dept. of Physics, University of Nebraska, Lincoln, Neb., U.S.A.
Scarsi, L., Istituto di Fisica, Università di Palermo, Palermo, Italy

Searle, L., Hale Observatories, Pasadena, Calif., U.S.A.
Setti, G., Istituto di Fisica, Università di Bologna, Bologna, Italy
Soliani, G., Istituto di Fisica, Università di Lecce, Lecce, Italy
Spinelli, G., Istituto di Matematica, Università di Milano, Milano, Italy
Stephenson, F. R., Dept. of Geophysics and Planetary Physics, University of Newcastle Upon Tyne, Newcastle Upon Tyne, Great Britain
Taffara, S., Osservatorio Astronomico, Padova, Italy
Tammann, G. A., Astronomisch-Meterologische Anstalt der Universität Basel, Binningen, Switzerland
Treves, A., Istituto di Scienze Fisiche, Università di Milano, Milano, Italy
Wickett, A. J., Dept. of Nuclear Physics, University of Oxford, Oxford, Great Britain
Wild, P., Astronomisches Institut der Universität Bern, Bern, Switzerland
Wilson, A. S., Sterrewacht te Leiden, Leiden, Holland
Woltjer, L., Dept. of Astronomy, Columbia University, N.Y., U.S.A.
Wood, R., Royal Greenwich Observatory, Herstmonceux Castle, Hailsham, Great Britain
Zarnecki, J. C., Dept. of Physics, University College London, Mullard Space Science Laboratory, Great Britain
Zwicky, F., California Institute of Technology, Pasadena, Calif., U.S.A.

Participants at the International Conference on Supernovae
Club Azzurro – Porto Cesareo (Lecce)
May 7–11, 1973

(0) Mrs Taffara
(1) Biermann
(2) Mrs Romano
(3) Mrs Rosino
(4) Heidmann
(5) D'Odorico
(6) Mrs Mansfield
(7) Mansfield
(8) Leggieri
(9) Aiello
(10) Mongelli
(11) Cosmovici
(12) Villa
(13) Wickett
(14) Bonoli
(15) Woltjer
(16) Mrs Biermann
(17) Taffara
(18) Miss Pizzichini
(19) Rossi
(20) Mrs Chiosi
(21) Bertola
(22) Mrs Casini
(23) Gull
(24) Miss D'Anna
(25) Soliani
(26) Cline
(27) Friedjung
(28) Reeves
(29) Miss Scordari
(30) Bollea
(31) Porter
(32) Romano
(33) Zwicky
(34) Searle
(35) Arp
(36) Lovas
(37) Sartori
(38) Dallaporta
(39) Rosino
(40) Ardeberg
(41) Gorenstein
(42) Jelley
(43) Wilson
(44) Barbon
(45) Capaccioli
(46) Detre
(47) Gallino
(48) Ferrari
(49) Lopez-Garcia
(50) Mrs Lopez-Carcia
(51) Ciatti
(52) Zarnecki
(53) Chiosi
(54) Meikle
(55) Henbest
(56) Miss Caloi
(57) Duin
(58) Ruggiero
(59) Broglia
(60) Treves
(61) Mrs Belli
(62) Mrs Zwicky
(63) Moore
(64) Palumbo
(65) De Groot
(66) Conconi
(67) Spinelli
(68) Mrs Gordon
(69) Mrs Immirzi
(70) Buchler
(71) Dickel
(72) Wood
(73) Goss
(74) Johansson
(75) Hassan
(76) Köbke
(77) Tammann
(78) McCarthy
(79) Wild
(80) Stephenson
(81) Mrs Cosmovici
(82) Branch
(83) Adams
(84) Bonetti
(85) Robba
(86) Borghesi

REVIEW OF THE RESEARCH ON SUPERNOVAE

F. ZWICKY
Hale Observatories, Pasadena, California Institute of Technology, Calif., U.S.A.

1. The Original Scenario

In the 1928 April issue of Harper's Magazine there appeared an article by George Ellery Hale, then Director of the Mount Wilson Observatory, which was entitled 'The Possibilities of Large Telescopes'. The article starts out with the following sentence,

> "Like buried treasures, the outposts of the universe have beckoned to the adventurous since immemorial times."

It has been related that Mr Rockefeller read only this first sentence of the article and immediately called up Hale, offering to support any of his big plans financially. A few months later the International Education Board of the Rockefeller Foundation awarded the California Institute of Technology in Pasadena six million dollars for the construction of a new Astrophysical Observatory and Laboratory, to be conducted in close cooperation with the Mount Wilson Observatory of the Carnegie Institution of Washington. Hale's plans for the new observatory envisaged the building of a 200-in. reflecting telescope and all necessary auxiliary apparatus and devices.

At the California Institute of Technology, there were at that time five of us who had previously dabbled with various problems of theoretical, observational and instrumental astronomy, namely R. C. Tolman, Professor of Physical Chemistry, Dr J. J. Johnson, who had been on many solar eclipse expeditions, Dr John Strong, working on various phases of applied astronomy and on infrared radiation, Dr Sinclair Smith, who had helped perfecting many instruments at the Mount Wilson Observatory and who had considerable experience as an observer, and especially in the evaluation of the performance characteristics of various telescopes and the effects upon them of atmospheric seeing, and finally myself. As developments went, the five of us subsequently worked together most efficiently for many years on a great variety of problems. Three of the men having passed away, there remain only Professor John Strong and myself cooperating more intensively than ever on problems in astronomy, as well as in morphological research, he and I acting respectively as the Vice President and the President of the International Society for Morphological Research.

The temptation, in 1928, for the five of us above mentioned faculty members at CIT to switch our allegiance from physics to astrophysics and astronomy was of course very great, and we were soon engaged working on various aspects of the new observatory, as well as making plans for future observational projects. We were accompanied on our way by Hale's advice,

> "Do not make any mean plans."

C. B. Cosmovici (ed.), Supernovae and Supernova, Remnants, 1–16.

With this license, and Hale's confidence in us, we started on projects which fairly stunned the Babylonians among the staid astronomical fraternity, but which, to our own amazement turned out to be successful beyond all expectation.

Personally I felt that, although Hale had established astrophysics as a new discipline, astronomers in general simply did not know enough physics to apply its fundamental principles to their science and that they did not at all realize that beyond measuring positions, motions, apparent luminosities and spectra of conspicuous stars and galaxies with admirable precision, there awaited us an unknown buried multitude of hidden treasures, that is, new cosmic bodies and phenomena which could only be divined through systematically directed intuition and subsequent tenacious search with proper instruments. Actually to this very day the morphological method of directed intuition has been studied and applied by only a very few. This is not the place to systematically explain this method. To those who eventually intend to make use of the morphological approach, however, it may be of interest if we sketch briefly how, with the method of directed intuition, specifically the field of supernovae was opened up and how the search for and the research on cosmic implosions and explosions was planned and conducted during the past four decades.

Morphological thought attempts to visualize all possible solutions of any given problem and all possible explanations for any set of facts which are not immediately and uniquely interpretable. To explore and to evaluate all aspects of a given situation, however, is only possible in limited cases in which all of the items involved, as well as the interactions between them, are clearly known. In practical cases we must limit ourselves to what Professor John Strong has proposed to call MODEST MORPHOLOGY.

Thus, when I started occupying myself with problems in astronomy I was not satisfied, for instance, with accepting the theory of the expansion of the universe as the only possible explanation of the large redshifts in the spectra of distant galaxies and, in several papers (*Proc. Nat. Acad. Sci* **15** (1929) 773–779 and *Phys. Rev.* **34** (1929) 1623–1624) I studied other alternatives. Among these the possibility, nay even the certainty of the existence of a gravitational drag on light, which lowers the frequency of quanta while they travel through intergalactic space has remained foremost in my mind.

My principal interest, however, was directed toward predicting the existence of new cosmic bodies and phenomena and then use optimally suited instruments and procedure for their discovery. After tenaciously implementing this approach for the past forty years I feel satisfied that it has proved most fruitful and that it has been successfully tested in a sufficient number of cases to merit incorporation in the curriculae of instruction of all institutions of learning. Among the scientists, who came to the same conclusion, Dr P. Chamaraux of Paris Observatory in Meudon, in a letter of 1972 to me states this conclusion most succintly by writing about the discovery of compact galaxies, "Je trouve en effet très remarquable que vous ayez prévu l'existence de ces objects avant de les *découvrir*, puis de les *étudier*. Cela me fait beaucoup penser à la découverte de Neptune par Le Verrier, excepté que Le Verrier n'a pas lui-même identifié la planète."

Concerning the prediction of the existence of new cosmic bodies and phenomena, I started in 1928 (*Proc. Nat. Acad.* **14** (1928) 592–597) from a few elementary principles or socalled pegs of knowledge that promised to direct the intuition into the proper channels.

(a) Obviously, because of atractive forces acting between the elementary particles and bodies of microscopic and cosmic sizes, matter in general has a tendency to agglomerate and to compact.

(b) The tendency toward compaction is counteracted by preexisting kinetic energy possessed in different degree by the various bodies in the universe.

(c) Furthermore, compaction cannot proceed unilaterally since, in the process, energy will be released which automatically will result in some of the matter involved being ejected with high speeds of escape. Thus, at the same time that compact bodies are formed, matter is also being dispersed into space, interplanetary, interstellar or intergalactic.

(d) The process of compaction of matter can take place in many ways, slowly, by accretion for instance, or by fast implosion. This point is exceedingly important, since some of the most serious mistakes have been made by prejudices of various investigators concerning the kinetics of the compacting processes. One of these 'Denkfehler' was largely responsible for the face that my original theory of the existence of neutron stars and their role as the compact remnants of some supernovae was not accepted for more than three decades.

Contemplating the above stated aspects of possible avenues to discoveries to be made at our new observatory on Palomar Mountain, it struck me in particular that the study of cosmic implosions and explosions and their inevitable results of compaction and of dispersion of matter had not been given enough attention in the past. In fact, Henry Norris Russell, who spent some time every year at the Mount Wilson Observatory as an Associate of the Carnegie institution tried to impress on me time and again that implosive processes were freaks and played no role whatever in the evolution of matter in the universe. Eddington in 1930, when I lectured in Cambridge, England on cooperative phenomena and the physics of crystals told me the same thing.

Disregarding these wise councils, Dr Walter Baade and I focussed our attention on one of the best known phenomena in the realm of fast cosmic reactions and engaged in a study of novae, that is of stellar implosions and explosions which within a few hours or days result in outbursts that increase the luminosity of the stars involved by factors up to the order of one million. A careful perusal of the literature, however, revealed that, in addition to what we later on proposed to call common novae, such as the well known Nova Persei of 1901, Nova Aquilae of 1918 and others, some outbursts had been observed and reported on in the past which indicated the occurrence of a much brighter class of novae. These, in 1933, Baade and I proposed to call supernovae.

There were in particular two types of observations pointing to the existence of supernovae, thousands of times as bright as common novae, or billions of times as bright as the Sun, at maximum luminosity.

In the first place, since the discovery in 1885 of a temporary star near the nucleus of

Messier 31 by Hartwig at the Dorpat Observatory, about a dozen stars had flared up in the line of sight of distant galaxies, nearly equalling the apparent brightness of these galaxies. Since no similar one time events had been observed in the enormously greater regions of the sky which are not covered by the brighter galaxies, Baade and I concluded that the mentioned flareups must have been caused by exploding stars that were members of the respective galaxies themselves.

Secondly, the socalled Tycho star of 1572 for a few days became as bright as Venus. If it had been an ordinary nova it would have had to be within a few dozen light years of the Earth. Both its stellar remnants and the expanding gas clouds would have been easy to locate, since Tycho had given a very good position of the nova. But no remnant of any kind could be found to the limit of the 100-in. telescope on Mt. Wilson. From this we could conclude that Tycho's star must have been so distant that its absolute brightness at maximum must have been hundreds of millions of times that of the Sun or even greater.

2. Predictions about Supernovae and Associated Phenomena

Using the principles of morphological thought sketched above, as well as the few scanty facts just mentioned as pegs to direct the intuition, I visualized the following consequences.

(1) There occur implosions in highly concentrated cosmic agglomerations of matter of stellar dimensions, which lead to explosive outbursts emitting at maximum visible light equalling that of hundreds of millions of suns or more.

(2) In large galaxies like Messier 31 and the Milky Way, such outbursts which Baade and I called supernovae, occur at the average rate of the order of about one in a thousand years.

(3) The total visible radiation emitted by a supernova is greater than 10^{48} ergs and therefore equals that emitted by the Sun over a period of ten million years.

(4) Velocities of the gas clouds ejected from supernovae were estimated to be of the order of from 15000 km s^{-1} to 30000 km s^{-1}.

These and other conclusions Baade and I published in a fundamental paper in the *Proc. Nat. Acad. Sci.* **20** (1934) 254–259.

Pushing the application of the morphological method of directed intuition to the limit I ventured a series of predictions, which I presented first to my students at the California Institute of Technology in an ad hoc course on Astrophysics in 1933 and then to my colleagues in a big physics seminar in November 1933. This seminar was reported on at great length in the Los Angeles Times of December of 8, 1933 (see excerpts reproduced in F. Zwicky, '*Catalogue of Selected Compact Galaxies and of Post-Eruptive Galaxies*, Gümligen 1971), as well as in the *Science News Letter*, Washington, D.C. of January 20, 1934. Its editor, my old friend, Mr Watson Davis concluded his account of the new theory, tying supernovae to cosmic rays and neutron stars with the dictum "Scientists are not inclined to accept the new theory until it shows its ability to withstand unfriendly criticism."

It must be left to the readers to decide how far the unfriendly criticism of the past forty years has succeeded to demonstrate that the conclusions and the predictions made in November 1933 are incorrect. Some additional conclusions and predictions were as follows.

(5) Supernovae are caused by the collapse of stars or clouds of gas and dust into neutron stars. These consist of matter which is electrically neutral in regions of nuclear dimensions (10^{-12} cm) already, as compared with ordinary matter which is neutral only in regions of dimensions greater than Bohr's length (10^{-8} cm). The average density of neutron stars is of the order of 10^{14} g cm^{-3} and their diameters are of the order of 10 km.

On the basis of the neutron star hypothesis the magnitude and the speed of the flareups of supernovae became understandable, as well as the absence of any long lived and easily observable stellar remnants.

The fact that this theory, among astronomers, found practically no favor for more than thirty years probably was due to the general but mistaken belief that white dwarfs represent the final stages of the evolution of stars.

(6) From the neutron star theory it followed that great parts of the gravitational energy released would be radiated into space as ultraviolet light, X-rays and γ-rays. Emission of radiowaves was to be expected from the gas clouds ejected at high speed after they had become tenuous enough.

(7) Two types of magnetic fields would be generated, that is very intense ones near the neutron stars and very extended ones in the tenuous plasmas, the latter because of ion streams in the expanding shells of gas.

(8) Enormous electric potential differences would be built up between charged parts of the expanding gas clouds and between them and the central neutron star. According to my estimates these potential differences could reach values of the order of 10^{19} V. Upon the collapse of the electric fields, at a later stage, after the radiation which sustained them had escaped, electrons, protons, nuclei and ions of *all* atoms present would be accelerated to high speeds and thus escape into interstellar and intergalactic space, making their appearance as *cosmic rays* (except those that became trapped in the interstellar magnetic fields or suffered too many collisions with all sorts of cosmic matter). I therefore predicted (*Proc. Nat. Acad. Sci.* **25** (1939) 338–344) that *all* nuclei would be found represented as constituents of the cosmic rays and that their maximum energies would be as high as 10^{19} eV. It must be remembered that at that time the highest energies observed in the so-called Auger showers were at most 10^{14} eV and it was firmly believed that cosmic rays contained only protons. My above predictions have been confirmed in the meantime.

(9) Cosmic rays in the 1930's were also believed to come only from Milky Way sources and nobody of any stature in physics gave the supernova cosmic rays origin any credence, although, postulating this origin I had demonstrated that, taking into account the proper 'redshift' for the cosmic ray particles, my theory predicted the proper value of about 10^{-3} ergs cm^{-2} s^{-1} for the total intensity of the cosmic rays impinging on the top of the earth's atmosphere.

There were a number of additonal predictions, which cannot be described here and which still await confirmation. It should, however, be mentioned, that some secondary conclusions regarding peculiar phenomena associated with supernova outbursts can in all probability be assumed to be in agreement with observations made during the past twenty years. Among these I mentioned the supernova theory of the origin of runaway stars, of expanding star associations and the existence of pygmy stars.

Also, the possibility of antimatter being present in the universe in equal amounts as ordinary matter may be considered as eliminated, because of confirmation of the tests which I proposed in my Letter of 1935 entitled 'Where Can Negative Protons Be Found' (*Phys. Rev.* **48** (1935) 169).

My old friend David Hilbert of the University of Gőttingen, who visited us every year in Zürich under the pretext of finding Swiss physicians capable of curing some of his imagined ailments repeatedly had advised us then young men to be brief in our scientific communications. He claimed in fact that the ideal would be the presentation of one's life's work in ten minutes or on no more than ten printed pages. I feel that Hilbert would have been pleased to read about the original predictions made of the significance of supernovae for the evolution of matter in the universe.

3. The Search for Supernovae

To prove that Baade and I were right concerning the previously unrecognized existence of supernovae I bought a Wollensack $3\frac{1}{4}$ in. lens camera and from 1934 to 1936 started photographing the rich Virgo cluster of galaxies from the roof of the Robinson astrophysics building of the California Institute of Technology, accompanied by the hilarious laughter of most professional astronomers and my colleagues at Caltech. Luck was not with me, since no supernova appeared on my plates during that time, although I had expected to find two or three, having calculated from meager historical data that in the galaxies within the reach of my camera supernovae would make their appearance on the average about once in a thousand years in any one of the galaxies.

During my fruitless search in Pasadena I heard that Bernhard Schmidt in Hamburg had invented and successfully constructed a very powerful type of wide angle telescope, which is now famous and bears his name. I visited Schmidt in Hamburg in 1935 and, upon my return managed to persuade our chief, Dr R. A. Millikan that, with a fair size Schmidt telescope chances were great to find a few supernovae every year. He agreed, stating laughingly that my project had scientific sex appeal. He also helped me persuade Dr Hale to allocate $ 25000 of the Rockefeller grant to build an 18-in. Schmidt telescope, on the pretext to the Rockefeller Foundation that it was needed as a scout instrument for the 200-in. telescope.

With the expert help of Drs J. A. Anderson, Sinclair Smith, Russell Porter and superb mechanics like Albert Brower we built the 18-in. Schmidt telescope within a year, and I put it in operation on the night of September 5, 1936, spending thereafter twenty-one nights in a row to photograph the comparison fields for the projected supernova search and to show the professional astronomers what a determined physicist could do. Beating the 'tar' out of the sky I found my first supernova in the spiral galaxy NGC 4157 in March 1937. On August 26, 1937 I discovered my second supernova in the dwarf spiral galaxy IC 4182, which has so far remained as the brightest one discovered in this century. A third excellent one I found on September 9, 1937, at the end of the same observing period, in NGC 1003. The investigation of the light curves and the spectra of these objects by Baade, Minkowski and myself firmly established the base for all future searches and investigations on supernovae. (See F. Zwicky, *PASP* **49** (1937) 204 – W. Baade and F. Zwicky, *Astrophys. J.* **88**, (1938) 411 and R. Minkowski, *Astrophys. J.* **89** (1939) 143).

After that, Dr J. J. Johnson and I kept rolling at a steady pace, discovering about 4 supernovae every year. In parenthesis I call attention to the curious fact that the various aspects of the supernova search have brought forth statistical improbabilities which are quite incredible and which are examples for the conviction held by all seasoned statisticians that the impossible happens far more often than is generally expected. In my forthcoming book on supernovae I shall enlarge on a dozen or more of these most improbable aspects and here only casually mention some of them, in order to warn both those who have come up with unrealistic and ill defined results on the frequency of supernovae, as well as the amateurs in the calculus of probabilities who sell the myths of extrasensory perception.

Among the first improbable events, there stands out the fact that the brightness of the second of my supernovae has not been equalled by any of the 350 objects subsequently discovered and, furthermore, that it was six magnitudes brighter than the whole galaxy IC 4182 in which it appeared. Also, the first twelve supernovae which I found were all of the type I. We almost concluded that there was only one type as far as light curves and spectra are concerned, until Dr Johnson found his first two supernovae, that is Palomar's numbers 13 and 14, in NGC 5907 and 4725, which were of an entirely different type II. Several of my collaborators continued to find types II, III, IV and V, while I never discovered any of these, until my 36th turned out to be of the type II.

Within 2 years, from 1936 to 1938, we had established the Schmidt telescope as one of the most powerful innovations in astronomy, having also started work on the six volume catalogue of galaxies and clusters of galaxies which was completed after 30 years in 1968 and by having discovered the previously unknown dwarf galaxies, luminous intergalactic matter, many clusters of galaxies and other cosmic objects. These successes induced Dr Hale to ask the Rockefeller Foundation for an additional $ 450000 for the construction of the 48-in. Schmidt telescope, which I put in operation on the night of January 31, 1949. Thereafter it was used exclusively for the project of a total sky survey and it became available for the other projects, including the search for supernovae, only in 1958. This search, with the 18-in. Schmidt, had also been interrupted because of World War II. While working on the six volume catalogue of galaxies, some of my collaborators, among them in particular P. Wild and H. S. Gates, looked for and found a few supernovae in the 1950's. A new large scale search with the 48-in.Schmidt was initiated in 1959 when Milton L. Humason joined my group after having retired from the staff of the Mount Wilson Observatory. Things really started to move on an international scale after I induced the International Astronomical Union to establish a Committee for the Supernova Research at its general assembly in Berkeley in 1961. I have served as the Chairmen of this committee since its inception.

Thanks to the joint efforts of about 15 observatories all over the world the number of bona fide supernovae discovered since 1936 has now risen to about 380, of which 270 have been discovered by the group of a dozen collaborators working with me at the California Institute of Technology. I myself found my hundredth supernova about 2 months ago, a milestone (parsec stone) which, I think, will entitle me to write a book on the subject.

It may be useful to sketch here briefly the different methods that have been used to search for supernovae. There is of course the possibility of sighting some of the nearby galaxies with field glasses or small telescopes, but, so far only a few supernovae have been discovered this way.

With the 18-in. Schmidt at Palomar we use films taken at different epochs and then superpose two of them, viewing the films under small power binoculars. This has proved most efficient. As alternatives we built first a blink machine, but the comparison of the films proved too time consuming and, secondly, we started making positives. Superposing respectively a positive and a negative of films obtained at sufficiently different dates and projecting the ensemble, we hoped to discover supernovae as white spots appearing on the projection screen. This method again is not only time consuming and costly but was defeated by the presence of too many defects.

Discussing the problem with Dr V. Zworykin of the RCA, we decided that a most efficient search for supernovae could be organized, if one could interest hundreds or thousands of amateur astronomers, and perhaps even some professionals, to regularly watch fields scanned by powerful Schmidt telescopes on their television screens. Each amateur would have been responsible for a few limited fields of the sky, watching them on live television and comparing the images of galaxies appearing in them with basic illuminated transparencies of the same fields mounted on their apparatus. Actually, Dr Zworykin suggested at that time in 1939, that the RCA might cooperate, free of charge, to put this scheme into action. Our ideas, however, could not be realized, firstly because we did not succeed in mobilizing the amateur astronomers and secondly because of the advent of the Second World War.

As already mentioned, the 48-in. Schmidt telescope became available for the supernova search only towards the end of the 1950's It has been used successfully for this purpose ever since. Unfortunately, the observatory council of the California Institute of Technology has never allowed us to use films with the 48-in. Schmidt, although this was proved feasible and would have facilitated the search for new objects, as well as rendered it much cheaper. With the use of plates it is of course necessary to compare them by blinking or by some similar methods, which are all much more time consuming than the comparison of films by superposition and viewing them with binoculars.

Recently again, various proposals have been made to speed up and to improve the search for supernovae. Some of the tasks facing us are to catch supernovae on the rise, to get early spectra and to find objects which appear in the luminous disks and cores of various galaxies, such as those of the elliptical types – the SO's, Sa's and compacts.

Some improvements will be possible with Mr Paul Wild's twin Schmidt telescope at the Zimmerwald Observatory of the University of Berne in Switzerland. Simultaneous photographs of the same field will eliminate the disturbing uncertainties introduced by photographic defects and will also speed up the certainty of discovery.

Dr S. Colgate at Socorro, New Mexico started building elaborate equipment for quick scanning of all brighter galaxies and evaluating the signals received through a computer in order to insure quick early discovery. In my estimate, however, this scheme could not have worked, even if the mechanics and electronics of the equipment had been perfect. In fact, every astronomer knows that on fast scanning the disturbances in the atmosphere will begin to interfere in a disastrous manner, since, within fractions of seconds the apparent brightness of stellar images and of luminous spots may vary by several magnitudes. Confirming my doubts about the easy feasibility of the Colgate scheme, no supernova seems to have been discovered as yet at Socorro, in spite of inordinately high expenditures by the U.S. National Science Foundation. It should further be emphasized that photographic records obtained in the supernova search at Palomar, now continuously from 1936, are of immense value for various other purposes. This 'memory' of the photographic plates is actually very difficult to match by any other scheme known to present technology.

Dr J. A. Hynek at Correlitos, New Mexico has in some limited way come back to the original idea explored by Dr V. Zworykin and myself in 1939 of using a television search for supernovae, while at the same time photographing the screen to retain a lasting integral record of the fields viewed. This approach has resulted in the discovery of a few supernovae during the past few years but cannot as yet be rated as being an overwhelming success.

Finally, continuing the efforts of Professor R. W. Wood of Johns Hopkins and myself, more than thirty years ago Professor John Strong of the Four College Observatory at Amherst and I promoted the construction of large laminar and echelette objective gratings. (*Appl. Opt.* **8** (1969) 1021). Three 45 × 45 cm transparent objective gratings of excellent quality are now available, one of which I have tested with the Zimmerwald Schmidt telescope near Berne. It is intended to promote the construction of such gratings 1 m × 1 m in size or greater. Large Schmidt telescopes equipped with such gratings would make possible the immediate and certain discovery of supernovae by their spectra, on single plates, without the necessity of blinking. Spectrum, redshifts and so on would be available on the discovery plates, which also would contain information on flare stars, outbursts of compact galaxies and, before all the tantalizing question of the possible occurrence of truly intergalactic supernovae.

I propose therefore, as one of the foremost projects to be considered to plan for the construction of very large transparent objective gratings either of the laminar or of the echelette type. Such gratings, mounted on the large Schmidt telescopes would, I am certain, enable us to make enormous inroads on the various problems confronting us not only in our researches on supernovae, but on many of the most important aspects of the large scale, as well as the small scale distribution of matter in the universe. I touched upon this subject in an article on 'Objektivprismen und Objektivgitter' (*Die Sterne* **43** (1967) 89–92) which, as I hear, helped accelerate the decision of the authorities in charge of the 52-in. Schmidt Telescope at Tautenburg to construct a prismatic Schmidt plate for this instrument. This plate apparently has been successfully tested and it is to be hoped that it will also be made available for the search for and the research on supernovae.

In view of future efforts it is of some interest to touch upon the estimated costs of the search for supernovae. As for the 270 supernovae found at Palomar between 1936 and the Spring of 1973 I estimate that the total expenditure including the construction of the 18-in. Schmidt telescope, plates and films bought, salaries for assistents paid, etc., was less than $ 150000, that is, on the average our supernovae 'cost' less than $ 550 a piece. This of course does not include the cost of the 48-in. Schmidt telescope and it also disregards the fact that the 18-in. Schmidt was used for many other projects.

4. The Search for Old Supernovae

This search concerns the perusal of old and very old historical records, as well as belated survey of photographic records obtained by ourselves or our contemporaries.

Professional astronomers as well as men in all walks of life through the ages were interested in celestial phenomena and recorded them in many ways, in writings, paintings, inscriptions and construction of various monuments and artifacts. As far as supernovae are concerned, there are no indications so far for any ancient recordings of supernovae in extragalactic systems, although such outbursts could certainly have been seen if occurring in the Magellanic Clouds or possibly in the Great Nebula of Andromeda. On the other hand we have the writings of several civilizations on galactic supernovae, like the Chinese 'guest star' of 1054 whose remnant, the Crab Nebula with its central probable neutron star has been extensively studied in our time. It is also probable that some of these galactic supernovae have been depicted on smooth cliffs and canyon walls by cave dwellers in various parts of the Earth.

Secondly, past systematic surveys of the sky, at the present represent a rich source for the discovery of supernovae. Although these are not 'alive', much statistical information can be gained by examining these surveys. The systematic Palomar Sky Survey made with the 48-in. Schmidt Telescope appeared particularly inviting for such an examination. It is of course to be regretted, that the observatory council of the California Institute of Technology, on the short-sighted advice of the director and some of the Babylonians of the Mount Wilson Observatory staff did not allow us to explore the plates while they were being taken, such that the opportunity was lost to catch at least one hundred supernovae in action. Nevertheless, in my state of retirement, and not being permitted to use the Palomar Schmidt telescopes any more. I decided that, for various purposes it would be useful to find out how many supernovae are buried in Palomar Sky Survey Plates. Assuming a frequency of about one supernova per 300 years for an average assembly of galaxies in the range of indicative absolute photographic magnitudes between -15.0 and -20.0, I estimated that on the survey plates about one hundred supernovae will be found, that is, about fifty at the actual interception of the light curves of apparent photographic magnitudes brighter than $m_p = 19.0$ and about fifty in the range $19.0 < m_p < 20.5$. For the set of the comparison plates, the same estimate holds true, so that as a grand total I expect to find about 200 supernovae. So far, I have located some seventy supernovae, the distribution of which, in apparent magnitudes and magnitudes of the parent galaxies is shown in the Figures 1 and 2. From the fraction of plates so far blinked, the above number of supernovae found on them very closely confirms the prediction made.

It should be noted that in addition to the great Palomar Sky Survey other wholesale surveys have been made which remain to be searched for supernovae. Among these, most notably are the surveys for counts of galaxies to the 19th magnitude by Shane and Wirtanen at the Lick Observatory, the 48-in. Schmidt Palomar Survey of Luyten for proper motion stars, various surveys at Palomar and other observatories for blue stellar objects, proper motion stars and variable stars and so on. In all of the plate material available I feel certain that another 500 to 1000 images of supernovae remain

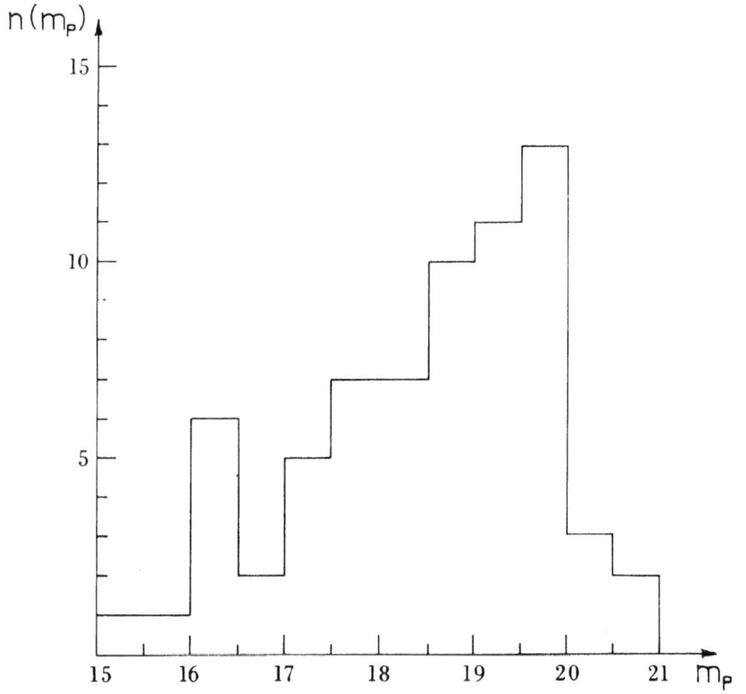

Fig. 1. $n(m_p)$ gives the number of supernovae, as a function of their apparent photographic magnitudes m_p at interception that were so far found by Zwicky on the 48-in. Schmidt Palomar Sky Survey plates and a set of comparison plates.

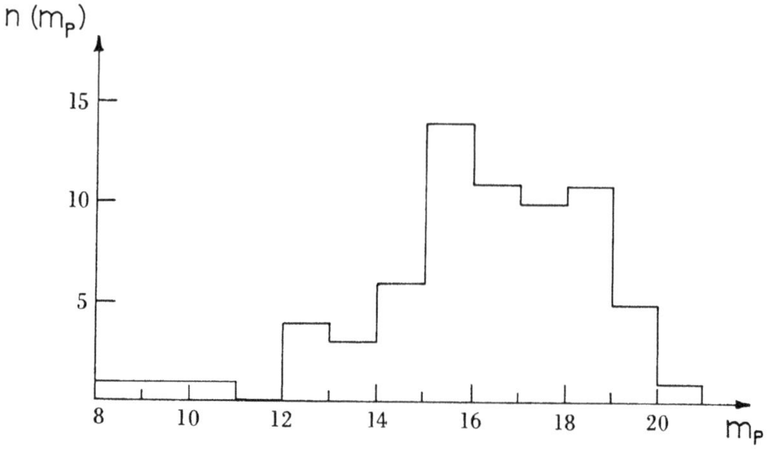

Fig. 2. $n(m_p)$ represents the distribution, as a function of their apparent photographic magnitudes m_p, of the parent galaxies in which the supernovae included in Figure 1 have appeared.

buried which it would be worthwhile for a few retired astronomers to unearth. At the same time no doubt many additional confirmations for supernovae already known will be found, furnishing us possibly even with some fragmentary data of their light curves.

5. Types of Supernovae

In addition to the five types of supernovae previously known ('Supernovae', in *Stars and Stellar Systems*, Vol. 8, University of Chicago Press, 1965) no new types have been definitely identified. During my search of the Palomar Sky Survey Plates, however, I have found images of supernovae much bluer than any previously known. Because of this fact, and on general principles to be discussed elsewhere, we may expect the existence of additional supernovae, different from those known so far.

From the investigations of the past few years two most important results have emerged.

In the first place, as I stressed it in many previous communications, supernovae of the type I are of very uniform characteristics as far as absolute luminosities, light curves and spectra are concerned, as long as they are located far away from dense agglomerations of various types of matter in the respective parent galaxies. Supernovae of the type I therefore almost certainly have their origin in the implosion and explosion of unique and well defined cosmic bodies. Supernovae of this type are therefore the only types of objects so far known, which we shall eventually be able to use as standards for the determination of cosmic distances to the limit of vision of large telescopes. It must be emphasized again and again that determinations of the Lundmark-Hubble redshift constant through the use of characteristics of H II regions, of globular clusters, variable stars and supposedly luminosities or sizes of galaxies are worthless except as very crude and unreliable estimates.

In the second place, important progress has been made in the interpretation of the spectra of supernovae of the type I, this especially because of the discovery of the bright object in NGC 5253 and the careful analysis of a sequence of its spectra by Professor Oke and collaborators. The results which will be reported on at this conference by Dr Searle, I think, are in accordance with my long held opinion that the origin of supernova of the type I is to be associated with the catastrophic implosion of what I symbolically proposed to call an 'iron white dwarf', that is a white dwarf containing only heavier nuclei, at the exlusion of hydrogen. Naturally, as a result of a violent collapse of such an iron white dwarf into a neutron star, some processes of fission will take place and small proportions of hydrogen will be regenerated giving rise to the appearance of the respective emissions of light, such as the Balmer and the Lyman lines.

6. The Frequency of Occurrence of Supernovae

Recently an unfortunate confusion has arisen concerning the frequency of occurrence of supernovae. Investigators, who are not themselves actively engaged in the search have derived most debatable results because they did not and could not have actually exactly defined what they are talking about. Any attempts at the present time to tie supernovae to the problems of the evolution of stars in galaxies, or to manipulate the numbers of supernovae observed by introducing losses caused by the photographically burnt out disks of near and distant galaxies, to introduce corrections due to supposed losses because of the inclinations of spirals and other 'flat' system, etc. etc. cannot be taken seriously. It will therefore be useful to state again, how I have proposed to treat the problem, using exclusively principles of *dimensionless and phenomenological morphology* (see F. Zwicky, *Morphological Astronomy*, Springer Verlag, Berlin, 1957) and abstaining entirely from invoking the principles of *absolute morphology*.

To start with, supernovae, as I proposed to define them, form a very varied assembly of objects, since I fixed them by the single requirement that their indicative luminosity at maximum be greater than that corresponding to an *indicative absolute photographic magnitude* M_p (indicative) $= -11.0$. Note that the indicative absolute magnitude (according to F. Zwicky and M. L. Humason, *Astrophys. J.* **132** (1968) 628), in the sense of phenomenological morphology is defined operationally by the observed apparent magnitude of the object combined with the observed redshift, thus in no way prejudging what the real absolute magnitude might be. The limiting value -11.0 was arbitrarily chosen, because no common nova brighter than the absolute photographic magnitude -11.0 has ever been observed. In my terminology, therefore, *dwarf novae* reach real absolute photographic magnitudes between zero and minus 5, those of *common novae* lie in the range $-5.0 > M_p > -11.0$, while *supernovae* are brighter than the indicative absolute magnitude -11.0, leaving the question open what this limit will become, once we actually know the exact value of the universal redshift constant (as differing from the here adopted *indicative value* of 100 km s^{-1} and million parsecs), as well as the exact distances of all nearby galaxies.

In recent years a secondary difficulty has arisen concerning the above definition of supernovae. Indeed, tremendous temporary outbursts have been observed in some of the Zwicky compact galaxies, almost certainly equalling the absolute rises of luminosities in supernova outbursts and probably often far surpassing them. Since this phenomenon so far, observationally, involves the whole of the respective compact galaxies, any further localization within them is not at the present possible. I, therefore, propose not to include them in the master list of supernovae, but to keep their records separate, until we know more about the phenomena involved.

True intergalactic supernovae, which do not appear on visibly luminous bridges between galaxies no doubt exist, but have not so far been identified with any certainty. They are therefore not considered in the following. Furthermore, indicatively absolutely very faint supernovae, such as η Carinae of 1843 in our own Milky Way and SN 1954j in NGC 2403, in the past, could readily be found only in a few thousand of the nearest extra-galactic systems. These, however, are mostly so thinly spread over the whole sky that a close survey is not easily possible and has certainly not been carried out. Since only a few objects of this type are known so far, the determinations of the frequency of occurrence of supernovae given by the writer during the past 35 years and, following him by many other observers, refer essentially to the numbers of supernovae observed brighter than about the indicative absolute photographic magnitude -14.0 and having appeared in galaxies brighter than the absolute indicative magnitude $M_p = -15.0$.

For these supernovae the frequencies of occurrence per galaxy are on the average as follows:

(1) The frequency is of the order of one in 300 years as derived by Zwicky already 30 years ago.

(2) For galaxies of different indicative luminosity the frequency is roughly proportional to this luminosity, but the proportionality factor is not the same for galaxies of different types and material contents.

(3) According to the statement (2) the frequency within a group of galaxies of the same type should, for instance, on the average be one in five years for galaxies of M_p (indicative) $= -20.0$ if it were one in 500 years for galaxies of M_p (ind) -15.0. Actually, so far, the brighter galaxies, per unit indicative absolute luminosity do not seem to produce supernovae quite as often as the fainter galaxies in our samples. The number of supernovae known for the different luminosity classes, however, is not yet large enough to establish this result with certainty.

(4) As to the relative productivity of different types of galaxies it is most enlightening to compare the frequencies in different classes of clusters of galaxies. For instance, the Leo A and Leo B clusters with, respectively, symbolic velocities of recession of the order of $V_s \sim 10000$ km s^{-1} and 12000 km s^{-1}, produced per galaxy in the range of indicative absolute photographic magnitudes -20.0 to -15.0 in the observing period from 1960–1973 at least twice as many supernovae as the Coma cluster, $V_s \sim 6500$ km s^{-1}, even if we make no correction for the supernovae being visible a longer time in the latter cluster than in the former two.

(5) Finally if we make the statistics for the appearance of n supernovae in an as-

sembly of N galaxies it is found that, from the data available, single and double appearances check with the theoretical expectations, while triple, quadruple and quintuple appearances cannot thus be accounted for and thus point to the existence of fast producers of supernovae among otherwise seemingly equal type galaxies, for instance Sc's of the same indicative absolute magnitude.

7. Absolute Luminosities of Supernovae of the Type I and their Possible Use as Distance Indicators

Observations of supernovae of the type I that occurred in any given cluster of galaxies, for instance in the Coma cluster, indicate that their absolute photographic luminosity at maximum appears to be of a quite definite magnitude, with only very small deviations. This is a most fortuitous circumstance because, so far we know of no other distinct cosmic objects that could reliably be used as standards for the determination of distances on a truly cosmic scale.

One of our principle tasks therefore will be not only to make certain of the result just mentioned, but also to determine the absolute luminosity of supernovae of type I which flare up in unobscured regions of nearby galaxies. This should preferably be done by one of the methods of dimensionless or phenomenological morphology which I have described in other places (for instance in *Morphological Astronomy*, Springer Verlag 1957). As a first case the bright supernova No. Z25 of 1937 in IC 4182 might well serve for this purpose, since the 'disk' of the expanding clouds of ejected gases possibly now has reached an apparent diameter of $\frac{2}{5}''$.

For the present, however, several somewhat more indirect approaches must be used which, in contradiction of persistently advertised values of about 50 km s^{-1} per 10^6 pc for the Lundmark-Hubble redshift constant indicate that a value of about 150 km s^{-1} per million parsecs is more realistic. One of the approaches leading to this result is, quite rudimentarily stated, as follows.

Adopting for the time being the same starting point as our opposition, we assume the distance modulus of M31 to be 24.2. With an apparent photographic magnitude $m_p = 7.2$ for S-Andromedae, this supernova is thus assigned an absolute magnitude $M_p = -17.0$.

Since the supernova No. Z25 in IC 4182 at maximum reached $m_p = 8.0$, the distance modulus of its Sc dwarf parent galaxy becomes equal to $8.0 + 17.0 = 25.0$, a value which would appear to be in good agreement with the following characteristics. IC 4182 has an apparent photographic magnitude $m_p = 14.0$, that is six magnitudes fainter than the supernova which appeared in it. It is therefore a dwarf galaxy, in agreement with the result, derived above that its distance modulus is equal to 25.0 and its absolute photographic magnitude equal to $M_p = -11.0$. In such a system, as expected, we do not find any pronounced H II regions. Furthermore, the brightest (blue) stars, with $m_p = 20.5$ and the distance modulus derived above must be assigned an absolute photographic magnitude $M_p = -4.5$, again in agreement with expectations. Because of the interference of night sky bands I never succeeded in obtaining a readable spectrum of the very faint nucleus. A single spectrum showing a weak $\lambda 3727$ in emission indicates a symbolic velocity of recession of 280 km s^{-1}. But this value remains to be confirmed through the more certain determination of the redshift of stellar absorption lines.

With an average absolute photographic magnitude $M_p = -17.0$ for supernovae of the type I and an apparent magnitude of 15.7 for these objects in the Coma cluster we obtain a distance modulus 32.7 for this cluster, or a distance of 35×10^6 pc. Since the average symbolic velocity of recession of the parent galaxies of the supernovae found in the Coma cluster is about 6500 km s^{-1} it follows that our preliminary value of the redshift constant is about 185 km s^{-1} per million parsecs.

Several analogous approaches can be used and lead to most probable values between 150 and 200 km s^{-1} per million parsecs for the Lundmark-Hubble redshift constant.

DISCUSSION

H. Arp: Do you have enough information on spectra and light decay curves for the variable compact galaxies to establish a SN class? What are you suggesting about the variable galaxies by this reference to a SN class.

F. Zwicky: No spectrum available during any flashy outbreak.

D. Branch: Is it fair to say that your value of the Hubble constant depends directly on the assumed absolute magnitude of the brightest stars in IC 4182?

F. Zwicky: This is only one of the arguments.

D. Branch: There are several lines of evidence that IC 4182 is a dwarf, with brightest stars of absolute magnitude about −5. For instance, the supernova was several (6) magnitudes brighter than the galaxy.

F. Zwicky: Also, nucleus is weak, there are no H II regions and $V_s \sim 280$ km s^{-1}, if we can trust determination of this velocity from a single spot emitting $\lambda 3727$.

A. Treves: Which is the rate of supernova birth in our Galaxy?

F. Zwicky: I do not know.

PART I

RESULTS AND TECHNIQUES OF SUPERNOVA SURVEYS

THE ASIAGO SUPERNOVA SEARCH

L. ROSINO and G. DI TULLIO
Asiago Astrophysical Observatory, University of Padova, Italy

The systematic search of supernovae began at Asiago on 1959, when a Schmidt telescope 40–50 cm became available. Previously, an occasional survey, which continued for several years, was made by Romano who found two supernovae 1957 *b* in NGC 4374 and 1961 *h* in NGC 4561 using for the survey two small astrographs. When the Schmidt was ready, the supernova survey entered in the plan of cooperative search organized by Prof. Zwicky.

The Schmidt telescope employed for the survey has a focal length of 1 m. The photographs are taken on circular films having a free diameter of 11 cm. Since the scale is 1 mm = 200″ the photographs cover a circular field of 6°.1 (diam) and 29.3 sq. degrees. The plate limit is about 17.5 and can be reached with 103a-0 without filter in 5 min, while with films Royal Panchro or Tri-X Kodak the limit is attained in 15–20 min, depending on the brightness of the sky. Unfortunately, for the expansion of the small city of Asiago and the steep increase of public illumination, the sky has become very bright, so that longer exposures without filters are now impossible. However a new dependence of the Asiago Observatory has been made on Mount Ekar, about 5 km from the old Observatory at an altitude of 1350 m, far from the lights. We plan to move there the Schmidt as soon as possible.

Although the plate limit is about 17.5, a supernova can be discovered in our plates only if its brightness is larger than 15.5. Occasionally some fainter supernova has been discovered, but in general a supernova weaker than 15.5 is simply overlooked. Supernovae are also easily lost when they appear in the high luminous centers of condensed galaxies, in consequence of the large focal ratio of the telescope (F:2.5). From time to time, but not systematically, photographs of the fields under survey are taken with the larger Schmidt telescope of 67-92-210 cm, which easily reaches magnitude 18.5. However, this telescope is mostly employed for the photographic study of supernovae, rather than for the searching, which is more quick on the films taken with the 40 cm Schmidt. Although we dispose of a very good blink comparator, we prefer to adopt for the discovery of supernovae the simple method of superposing two thin films of the same field, slightly displaced. The method has been found very rapid and as effective as the blinkage.

We meet however one very serious difficulty on the survey: this is due to the peculiar weather of Asiago with most of the clear nights concentrated from October to March, while the period from April to July is very poor. Unfortunately, this period is the best for the observation of high galactic latitude fields, where is higher the probability of finding supernovae. We shall see later that many supernovae which could have been found at Asiago were lost because they appeared in June or July.

C. B. Cosmovici (ed.), Supernovae and Supernova Remnants, 19–27.

We began in 1959 the systematic search of supernovae in 10 fields (β Per, χ UMa, γUMa, 6 Com, 3CVn, 11Com, BD+6°2599, BD+15°2469, 30ϱ Vir, BD+29°2348). The survey brought to the discovery of two SN (1960*r* in NGC 4382; 1961*h* in NGC 4564); This proved that the survey could be effective and therefore the search was extended to other field. At the Novae and Supernovae Colloquium in Haute Provence, in 1964, a list was given of 40 fields which were surveyed at Asiago. Presently the fields under control are 74, distributed all over the sky, mostly at high galactic latitudes. However, in the following considerations I shall mostly refer to 36 fields (Table I), which were maintained under a more strict control.

TABLE I

Galaxies and Supernovae in 36 fields

Field	Plates	Years	Months	F	N_g	SN(15)	SN_A
Cet IV	110	10	4	11	10	0	0
GC 1103	84	9	4	15	26	0	0
Psc I	145	12	4	10	13	1	1
Psc II	135	11	5	12	3	0	0
Cet II	111	11	4	12	9	0	0
And I	266	14	6	10	17	4	1
Cet I	138	12	4	10	13	1	0
Cet III	104	11	4	15	12	1	1
β Per	96	15	6	30	9	1	0
Eri III	89	11	3	11	11	0	0
Eri II	134	11	4	10	15	2	2
Eri I	114	12	3	9	21	0	0
9i UMa	149	10	8	16	7	0	1
GC13265	103	10	8	23	4	1	0
M81–82	125	13	8	25	9	0	0
40 LMi	128	11	4	10	6	0	0
Leo I	140	11	5	12	18	1	0
β UMa	188	11	8	14	12	1	0
Leo II	118	11	5	14	10	0	0
Leo III	107	11	5	15	17	0	0
χ UMa	236	14	8	14	12	1	1
γ UMa	241	14	8	14	20	1	0
6 Com	180	14	5	12	35	2	2
γ Com	140	11	5	12	22	1	1
2 CVn	164	11	3	10	10	0	0
3 CVn	229	14	8	15	20	1	0
11 Com	179	14	5	12	13	1	1
BD + 6 2599	160	14	5	11	37	3	1
BD + 15 2469	217	14	5	10	48	4	2
ψ Vir	92	12	3	12	30	1	0
ϱ Vir	160	14	4	10	25	0	0
37Vir	99	12	4	14	14	1	0
BD + 29 2348	109	14	6	23	17	1	0
K Boo	112	11	8	23	6	1	0
110 Vir	99	11	4	13	15	0	0
Peg I	134	11	6	15	19	2	2
	5135	11.7	5.25	14	585	33	(16) 13

In the successive columns Table I gives:

(1) Name of the field.

(2) Number of photographs from the beginning of the survey.

(3) Number of years covered by the survey.

(4) Number of months per year during which the field is controlled.

(5) Frequency of the observations (mean interval, in days, between two consecutive plates, during the effective period of observation).

(6) Number of galaxies brighter than 14 in each field.

(7) Number of SN brighter than 15 found from the beginning of the survey in each field.

(8) Number of SN discovered at Asiago within these fields. The total number is 16, but only 13 have been discovered in the survey with the 40 cm Schmidt.

The average interval between two consecutive plates in the effective period of observation is 14 days. But as one can see by looking to column 5 this interval ranges from 9 to 30 days, with a strong maximum between 10 and 14 days. Considering that a supernova looses about half a magnitude or more within 14 days from maximum and that the telescope does not allow, in general, the discovery of objects fainter than 15.5, we see that only supernovae which have been, at maximum, brighter than 15 can be found, unless the discovery plate has been taken just within two or three days from maximum. On the other hand, since supernovae are mostly weaker than the parent galaxy by 1.5 mag, we have considered, as supernovae productor only galaxies brighter than 14 mag, in total *585* in 36 fields. A very rough guess on the frequency of supernovae in these galaxies can be obtained in this way: in the average the 585 galaxies brighter than 14 have been kept under control 11.7 yr, but only for 5.25 months each year. So the time survey in the average, is 5.12 yr. During this period 33 SN brighter than 15 have been found in these field at Asiago or elsewhere, with a rate of 33:5.12 = = 6.45 SN/y in 585 galaxies. The frequency is therefore of (585:6.45) 1 SN per galaxy every 91 years. This however implies that all of the supernovae brighter than 15 have been found, in the fields under control, during the effective period of observation. If some SN have been overlooked the frequency increases. For instance, considering only the 13 SN discovered at Asiago, it is apparent that the frequency becomes lower, about 1 SN per galaxy every 230 years.

We are well conscious, that, in consequence of the many assumptions made, these figures give only an order of magnitude and are purely indicative. We believe however that also a more precise calibration will not give a completely different result and that the frequency of SN in galaxies brighter than 14, up to mag 15, is of the order of 1 SN every 100 years.

Table II gives the list of the 21 supernovae (including the two supernovae discovered by Romano) which have hitherto been found at Asiago. The supernovae are evenly distributed between magnitudes 10 to 15. Their mean magnitude (excluding those fainter than 15) is 13.2. Supernovae have not been found, even by other observers, in some of the fields which nevertheless are rich in galaxies, as Eri II and Eri III, M81–82, Leo II and Leo III.

TABLE II

Supernovae discovered at Asiago

SN	m_{SN}	Type	Gal. NGC	m_{gal}	Type	Field	Disc.
1957 b	12.5	I	4374	10.8	E1	ϱ Vir	Rom
1960 r	12.0	I	4382	10.2	SO	11 Com	Ros
1961 h	11.2	I	4564	12.2	E6	ϱ Vir	Rom
1962 L	13.9	I pec	1073	12.5	SBc	III Cet	Ros
1962 p	14.5		1654	14.2	E	II Eri	Ros
1963 d	15.0	I	4146	13.8	SBb	γ Com	Ros
1964 f	14.0	I	4303	10.9	Sc	BD + 6 2599	Ros
1965 i	13.5	I	4753	11.7	SOp	γ Vir	Ros
1966 a	15.5		anon	14.0	S	9i UMa	Ros
1966 n	15.0		anon	16.0		II Eri	Ros
1969	13.9		6946	10.5	Sc	6946	Ros
1969 c	13.7	I	3811	13.0	S	χ UMa	Ros
1969 L	12.8	II	1058	11.8	Sc	And I	Ros
1970 j	14.4	I	7619	12.7	E3	Peg I	Ros
1970	13.7		anon			γ Com	Rom
1971 g	13.5	I	4165	14.7	Sc	6 Com	Ros
1971 p	15.8	I	7319	14.8	SBb	NGC 7331	Ros
1971 s	15.6	II	493	13.0	Sc	Psc I	Pig
1971	17.0		anon			M 33	Bian
1972	14.2	I	7634	13.5	E:	Peg I	Pig
1972	15.8	II	4254	10.2	Sc	6 Com	Ros

Table II shows that most of the SN discovered at Asiago are type I (12 to 15); only three are type II.

Table III lists the SN brighter than 15.5 which have appeared in fields under Asiago survey and have *not* been discovered at Asiago. Some of the supernovae were lost simply because other observers, who were surveying the same fields, found the supernova some day before the star could be observed on the Asiago material. However, it can be seen that many of the supernovae in Table III were found during the period April to July, when the sky conditions at Asiago are very poor and the interval between two consecutive plates becomes higher than its mean value. By examining Table III, together with Table I, one gets the impression that perhaps, in future, it would be convenient to reduce the number of fields under control and increase the time of survey in the remaining fields, particularly in the months of poor weather.

As we said at the beginning, some of the fields are also surveyed, occasionally, with the 67–92 cm Schmidt. We intend in future to improve the survey by using more extensively the large Schmidt and give more time for the searching of supernovae. Also some galaxies with a high supernova production (NGC 6946 (5), NGC 5236 (4), NGC 2276, 3184, 4303, 4321, 5447 (all with three SN)) will be kept under a more strict control.

The Asiago supernova survey is not limited to the finding of supernovae. The program is more ambitious and includes the spectroscopic and photographic study, in *B*, *V* and,

TABLE III

SN in Asiago fields *not* discovered ($M < 15.5$)

SN	Gal.	Type	Field	M	Date	Disc.
1960 f	4496	SBc	BD + 6°2599	11.6	April 16, 1960	Humason
1960 h	4096	Sc	3 CVn	14.5	June 1960	Wild
1961 i	4303	Sc	BD + 6 2599	13.0	June 1961	Humason
1962 a	Anon	SO	BD + 29 2348	15.5	Jan 25, 1962	Zwicky
1963 j	3913	S	γ UMa	13.7	May 26, 1963	Zwicky
1963 L	Anon.	SBc	And I	15.0	June 1963	Reaves
1963 p	1084	Sc	Cet I	14.0	Sep. 28 1963	Wild
1963 s	Anon.	S	Psc I	15.0	Dec. 1963	Chavira
1964 e	Anon.	SBcp	γ UMa	12.9	March 2 1964	Lovas
1965 h	4666	Sc	γ Vir	14.0	May 1965	Chavira
1966 b	4688	SBc	37 Vir	15.0	Feb. 1966	Yakimov
1966 e	4189	Sc	6 Com	14.8	July 1966	Wild
1966 g	521	SBb	Psc I	15.5	Aug. 1966	Reaves
1967 c	3389	Sc	Leo I	13.0	Feb. 28 1967	Lovas
1967 h	4254	Sc	6 Com	14.6	July 1967	Zwicky
1968 a	1275	E	β Per	15.5	Jan. 1968	Lovas
1968 d	6946	Sc	6946	13.5	Feb. 1968	Wild
1968 i	4981	Sc	ψ Vir	13.5	April 1968	Lovas
1968 u	4183	Sc	2 CVn	14.5	Oct. 1968	Dunlap
1970 a	3476	Sc	BD + 15 2469	14.0	Feb. 1970	Grisunov
1970 b	Anon.	Irr	Leo I	15	March 1970	Kowal
1970 g	5457	Sc	K Boo	11.0	July 1970	Lovas
1970 L	2968	Ip	GC 132651	15.5	Oct. 1970	Wild

when possible, in U, of all of the SN which are within the possibilities of the Asiago telescopes, besides the supernovae discovered at Asiago.

The photographic study of the SN is made, as soon as the announcement of the discovery is given, mostly with the 67–92 cm Schmidt telescope but sometimes also with the 122 cm reflector. The spectra are normally taken with this instrument by means of prism or grating spectrographs giving a dispersion from 130 Å mm^{-1} at Hγ to 280 Å mm^{-1}. The use of image intensifiers allow the observation of the spectra even when the supernova has weakened of two or three magnitudes, and also in the infrared. The new telescope of 182 cm of Mount Ekar will offer very soon the possibility of extending and improving the spectroscopic researches on the supernovae.

Table IV gives the list of 30 supernovae studied at Asiago during the last 15 years, mostly by Bertola, and more recently by Barbon, Ciatti and the writer, for which the light curves or spectra or both have been obtained and discussed. Other seven supernovae have been recently examined at Asiago. Ciatti will later refer on these stars. So a total number of nearly 40 supernovae have been hitherto studied at Asiago on a very homogeneous material, with the same instruments and methods of observation and reduction.

Among the results of this extensive study of supernovae, we should like to quote the following points:

TABLE IV

Supernovae observed at Asiago for spectra and light curves

SN	Gal	M_{sn}	Type	Author	Notes	Bibl.
1957 a	2841	14.0	IIp	Bertola (1964)	Light curve	A.J. 69,3.
1957 b	4374	12.5	I	Bertola (1964)	Light curve	A.J. 69,3.
1960 f	4496	11.6	I	Rosino and Bertola (1961) Bertola (1964)	Spectrum, Light curve	As.Co.166 A.J. 69,236.
1960 r	4382	12.0	I	Bertola (1962, 1964)	Spectrum, Light curve	As.Co.128; A.J.69,236.
1961 f	3003	13.1	IV	Bertola (1965)	Light curve	As.Co.174
1961 h	4564	11.2	I	Bertola (1964a)	Light curve	A.J.69,236.
1961 p	Anon.	14.3	I	Bertola (1962; 1964)	Light curve	As.Co.128; A.J.69,236.
1961 u	3938	13.7	II	Bertola (1963)	Spectrum and Light curve	As.Co.135
1961 v	1058	12.2	V	Bertola (1964; 1965) Bertola and Arp (1970)	Spectrum and Light curve	Ann.Ap.27,319; As.Co.17 PASP 82,488.
1962 j	6835	13.6	I	Bertola (1965)		As.Co.171
1962 L	1073	13.9	I pec	Bertola (1964)	Spectrum and light curve	Ann.Ap.27,319.
1963 d	4146	15.8	I	Bertola (1965)	Light curve	As.Co.182
1963 i	4178	14.3	I	Bertola (1965)	Spectrum, Light curve	As.Co.182
1963 j	3913	13.7	I	Bertola (1965)	Spectrum, Light curve	As.Co.182
1963 o)	5905	15.95		Bertola *et al.* (1965)	Light curve	As.Co.174
1963 p	1084	14.0	I	Bertola *et al.* (1965)	Light curve	As.Co.174
1964 L	3938	13.3	I pec	Bertola *et al.* (1965)	Light curve	As.Co.174
1965 i	4753	13.5	I	Ciatti and Barbon (1971)	Spectrum, Light curve	As.Co.254
1965 L	3631	16.0	II ?	Ciatti and Barbon (1971)	Light curve	As.Co.254
1966 a	Anon.	15.5	I	Ciatti and Barbon (1971)	Light curve	As.Co.254
1966 n	Anon.	15.0	I	Ciatti and Barbon (1971)	Light curve	As.Co.254
1968 e	2713	13.5	I	Ciatti and Barbon (1971)	Light curve	As.Co.254
1969 b	3566	16.1	II	Ciatti and Barbon (1971)	Light curve	As.Co.254
1969 c	3811	13.7	I	Bertola and Ciatti (1971)	Spectrum, Light curve	As.Co.253
1969 L	1058	12.8	II	Ciatti *et al.* (1971)	Spectrum, Light curve	As.Co.255
1970 j	7619	14.4	I	Barbon *et al.* (1973)	Light curve	As.Co.282
1970 L	2968	12.5 ?	I ?	Barbon *et al.* (1973)	Light curve	As.Co.282
1970 g	5457	11.5	II	Barbon *et al.* (1973)	Spectrum, Light curve	A. & A. 29, 57
1971 g	4145	13.5	I	Barbon *et al.* (1973)	Light curve	As.Co.282
1971 i	5055	11.6	I	Barbon *et al.* (1973)	Spectrum, Light curve	As.Co.282
1971 L	6384	13.6	I	Barbon *et al.* (1973)	Light curve	As.Co.282
1972	3147	14.9	I	Barbon *et al.* (1973)	Light curve	A. & A. 29, 57
1972 r	2841	12.5 ?	I ?	Barbon *et al.* (1973)	Light curve	A. & A. 29, 57
1972	4254	12.5 ?	II	Barbon *et al.* (1973)	Light curve	A. & A. 29, 57

(a) The fact that in some type I supernova (as e.g. in SN 1962 L in NGC 1073) the peculiar absorption feature at 6150 is absent. When this characteristic feature is missing, also the typical progressive redshift of the violet emission bands fails to be observed (Bertola, 1964a; Bertola *et al.*, 1965).

(b) The presence of a progressive redshift of the emission bands in the type II SN 1961n in NGC 3938 (Bertola, 1973). This shift has been later oserved in other type II supernovae and seems to be a characteristic of these stars (Ciatti *et al.*, 1971).

(c) The evidence of the presence in the light curve of type I supernovae, after the early steep decline and before the linear final decline, of a sort of hump (Bertola,

1964b), which seems to be a characteristic feature of all type I SN and which has been recently confirmed by Barbon *et al.* (1973) by their study of the composite curve of supernovae type I.

(d) The first discovery of the peculiar spectral characters of the SN 1961v in

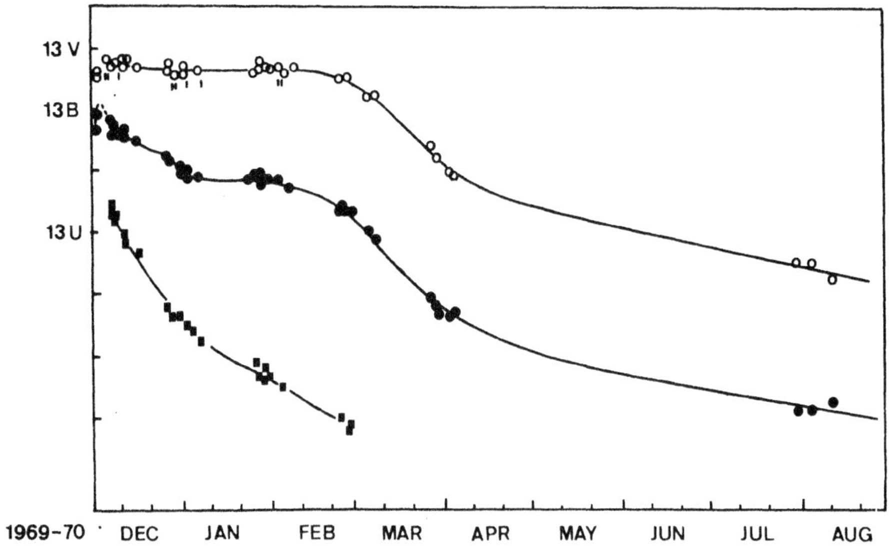

Fig. 1. Light curves of the type II SN 1969 L in NGC 1058.

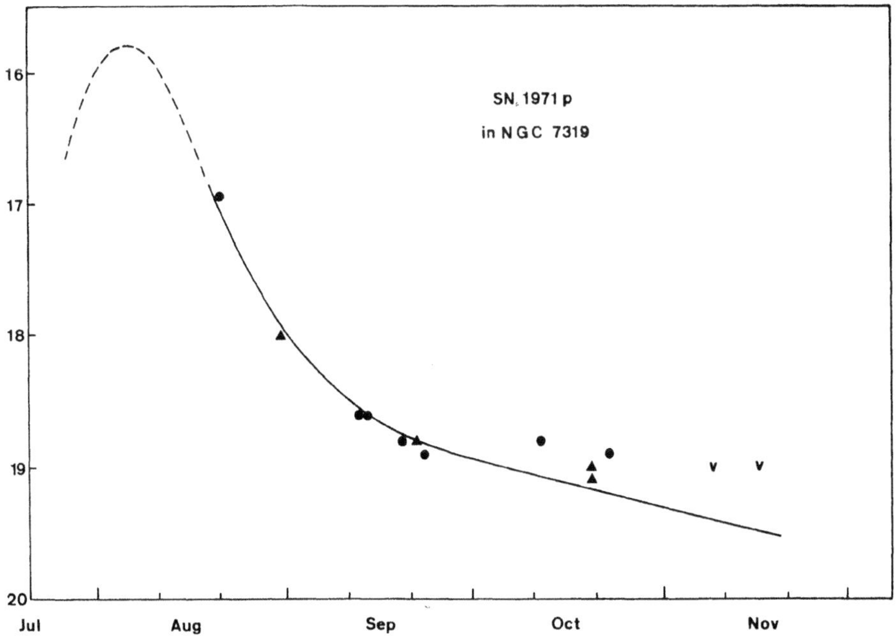

Fig. 2. Light curve of SN in NGC 7319.

NGC 1058, (Bertola, 1964a), and the study of its light curve. Dr Zwicky classified this supernova as representative of the new type V.

Finally, as a conclusion of the present review on the supernova work at Asiago, we should like to present (a) The light curves in *UBV* of the type II SN 1969L in NGC 1058 (Ciatti *et al.*, 1971) determined over a period of more than one year (Figure 1). Their are strictly correlated with the spectral evolution and seem to be typical of at least one group of type II Supernovae. (b) The light curves of two other Supernovae which have motives of great interest: SN 1971p in NGC 7319 (Figure 2), is one of the

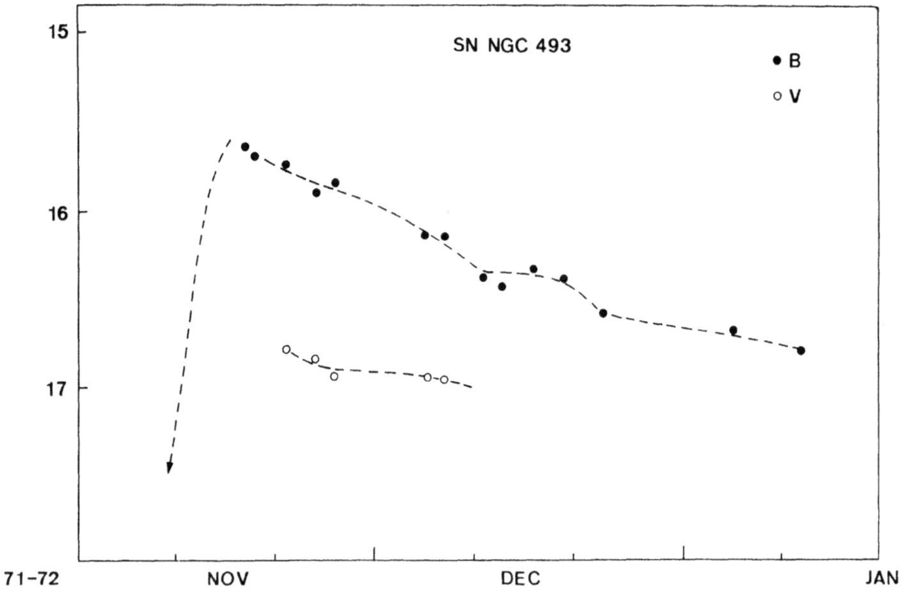

Fig. 3. Light curves of SN in NGC 493.

components of the Stephan Quintet recently studied by Arp. In our opinion the supernova is a normal type I; in this case the distance of the galaxy is in agreement with its redshift. The other Supernova of which we present the light curve is SN 1971 in NGC 493 discovered by Luisa Pigatto which was observed by Sprott *et al.* in the X-ray range. The SN is very likely of type II (Figure 3). No X-rays have been detected and an upper limit of 5×10^{43} erg s^{-1} has been determined for the X-ray energy flux.

References

Barbon, R., Ciatti, F., and Rosino, L.: 1973, *Astron. Astrophys.* **25**, 241.
Bertola, F.: 1963, *As. Co.* 135.
Bertola, F.: 1964a, *Ann. Astrophys.* **27**, 319.
Bertola, F.: 1964b, *Astron. J.* **69**, 236.
Bertola, F., Mammano, A., and Perinotto, M.: 1965, *As. Co.* 174.
Ciatti, F., Rosino, L., and Bertola, F.: 1971, *As. Co.* 255.

DISCUSSION

H. Arp: Do each SN marked type I pec marked so solely because of the absence of the absorption feature in the spectrum?

L. Rosino: Yes.

F. Bertola: When the absorption feature is missing the usual feature in the blue do not progress in redshift as in normal type I spectra.

THE SUPERNOVA SEARCH AT ZIMMERWALD

P. WILD
Astronomisches Institut der Universität Bern, Switzerland

In our observatory at Zimmerwald, six miles to the south of Bern, at an altitude of 900 m, we have a 24-in. Cassegrain and a 16-in. Schmidt camera (40/60/104 cm). The fields of the Schmidt has a diameter of $6\frac{2}{3}$ deg; the scale is 198″ mm^{-1}. We use film, since it is easier to handle than plates (especially so for rapid searches), and we buy the fastest panchromatic emulsions generally available (at the moment Kodak Tri X), the special emulsions for astronomy being by far too expensive for our small budget.

Of the four thousand exposures made since 1959, 1800 were on the supernova search proper, which encompasses 71 fields. The films are examined by direct superposition in a simple binocular viewer, allowing faster work than a blink comparator. Normally the whole field is searched through, not only the galaxies; so there is a second crop, not negligible, in other interesting variables, in comets and new minor planets. These objects often lead me astray; this has added hundreds of exposures outside the regular search, but rich enough in galaxies to be combed through, too. We discovered in Zimmerwald 24 supernovae; this can be considered a fair success, with our modest equipment and under not too favorable skies. On the average, we can count on 55 clear moonless nights a year, of which about 35 are spent for the supernova search.

The rather irregular way of our search makes it quite difficult to derive a 'true frequency' of supernovae. Even the regular fields I photographed with highly unequal frequencies and under that diverse conditions that I would practically have to re-examine each film, e.g. for its limiting magnitude; and even then I could hardly obtain anything more certain than what you find already in the literature. On the other hand it seems to be the special strength of our sample of supernovae, that, by chance, the various types and peculiarities are well represented. I remind you especially of the two type V supernovae in NGC 1058 and 3631, of the type IV in NGC 3003, and of the two unusually faint supernovae in NGC 2841. (The second of these appeared last December, very close to an 18th-magnitude star, of which I should like to see the spectrum, too). A few weeks before this conference I found that a star 40″ west of the center of M 81, always of magnitude about 17 during the last 13 years, had brightened by one magnitude at some recent date. To be judged from a photograph in *Astrophys. J.* **92** (1940) 22, it had been of similar brightness at least once before, probably in the 1930's. Now (June 1973) it has fallen back to its normal state. It would certainly be interesting to know whether this variable belongs to M 81 or to our Galaxy. Along this line I would also mention the four eruptive variables of yet undecided type and unknown distance, which I found far away from perceptible galaxies. (Prof. Rosino, who also discovered some similar objects, believes that at least some of them are U Geminorum stars in our Galaxy.)

One of the supernovae discovered at Zimmerwald has given me the strong impres-

C. B. Cosmovici (ed.), Supernovae and Supernova Remnants, 29–32.

sion that its explosion might have a peculiar cause: supernova 1966e in NGC 4189 flared up exactly on the rim of a strange circular gap in the spiral arm (with an emission nebulosity in the center), on which Kowal had drawn attention before (*Astron. J.* **69** (1964) 757). I find it hard to believe that the coincidence is purely accidental, but would rather suggest that the round gap was produced by a disturbance of very high energy and that that event has now triggered the observed supernova. In proximity of similar, though less pronounced lacunae lay the supernovae 1937c in IC 4182 and 1965m in NGC 7606. All these cases might be recent analoga to the supernova remnants N 49 and N 63A in the Large Magellanic Cloud. These were found by Westerlund and Mathewson to lie embedded in a huge hydrogen ring, the origin of which they attributed to the explosion of a super-supernova (*Montly Notices Roy. Astron. Soc.* **131** (1966) 371).

Continuing the search, we plan to use more often a set of twin Schmidts of ten inches opening and focal ratio f/1.6, covering a field of 11 degrees down to magnitude 15½ only. We could then search a wider part of the sky for bright supernovae, and we would practically eliminate the great nuisance of film defects.

It is a pity that we have no larger telescope yet available in Switzerland to follow up the supernovae found at Zimmerwald, and especially no spectrographic equipment. We are obliged to many colleagues throughout the world for taking over where we had to leave 'our' stars. Cordial thanks are due above all to Prof. L. Rosino and to several other astronomers at Asiago, who repeatedly and graciously helped us to check on suspicious objects, when our weather deteriotated.

List of Supernova Search Fields: Zimmerwald (Switzerland)

α_{1950}	δ_{1950}	No. of good exposures since 1959	α_{1950}	δ_{1950}	No. of good exposures since 1959
0^h19^m	−23°52′	7	11^h50^m	+44°30′	6
0 35	+ 0 18	27	11 52	+53 55	13
0 36½	−11 20	16	11 58	+52 50	40
0 37½	−23 15	17	11 53	−18 00	20
0 40	+40 25	33	12 04	+46 32	65
$0^h42½^m$	+48°00′	22	12^h18^m	+15°52′	41
0 53	− 7 25	20	12 20	+ 2 10	12
1 13	+ 2 30	32	12 20	+39 10	20
1 30	+15 00	5	12 20	+63 30	12
1 44	+30 00	46	12 24	+ 7 25	35
2^h08^m	+34°00′	23	12^h28^m	+31°10′	19
2 30	+39 50	40	12 35	+12 05	8
2 40	+37 10	49	12 38	+12 55	35
2 34	− 0 45	52	12 36	+ 2 05	16
2 42½	− 9 00	25	12 47½	+ 2 40	9
3^h12^m	−22°05′	26	12^h43^m	+27°15′	15
3 33	−21 40	20	12 44	− 8 56	4
4 04	+66 30	15	12 55	− 7 35	29
4 40	− 4 50	16	12 49	−11 00	6
6 10	−22 55	17	13 08	+39 50	30

(Continued List of Supernova Search Fields

7^h11^m	$+83°05'$	28	13^h19^m	$-18°30'$	19
7 22	+66 30	24	13 45	+47 12	12
8 30	+76 10	27	14 10	+56 50	33
9 00	+51 10	29	14 28	+ 5 00	30
9 34	+70 25	40	14 53	+ 1 55	25
9^h44^m	$+ 2°50'$	27	15^h21^m	$+57°20'$	51
9 45	+32 20	62	15 34	+15 00	21
10 20	+22 10	40	17 50	+73 10	28
10 25	+43 30	27	19 42	−12 50	32
10 47	+14 35	27	20 35	+62 50	47
11^h16^m	$+15°48'$	23	22^h06^m	$+31°07'$	6
11 15	+56 15	32	22 35	+34 10	21
11 24	+56 18	44	22 57½	+13 36	21
11 27½	+54 35	6	23 10	+10 00	20
11 46	+46 20	21	23 24	− 6 50	26
			23 54	− 3 20	5

List of Supernovae discovered at Zimmerwald

Designation	Galaxy (NGC)	Type	Brightest magnitude	Type	References & remarks
1957 a	2841	S b	14.0	II p	(1), (2)
1960 h	4096	S c	<14.5		(3)
1961 f	3003	S bc	13.1	IV	(4), (5)
1961 l	3221	S m	<17.5		
1961 p	Anon.	S c	14.3	I	(2), (6)
1961 u	3938	S c	13.7	II	(5), (7) (19)
1961 v	1058	S c	12.2	V	(5), (8) to (13)
1962 j	6835	SBa	13.6	I	(14), (15)
1963 j	3913	S c	13.7	I	(15) to (18)
1963 o	5905	SBb	15.9		(4)
1963 p	1084	S c	14.0	I	(4), (18)
1964 a	3631	S c	<17.0	V	
1964 l	3938	S c	13.3	I p	(4), (19)
1965 l	3631	S c	<16.0	(II)	(20)
1965 m	7606	S b	<16.0		
1966 e	4189	S c	14.8	(II)	
1966 j	3198	SBc	<13.2	II	(21), (22)
1968 d	6946	S c	<13.5	II	(23)
1968 x	4939	S bc	<16.0		
1969 b	3556	S c	<14.2	(II)	(20), (24)
1970 c	Anon.	S c	<16.0		
1970 l	2968	I o	<15.5		(25)
1972 r	2841	S b	<16		
1973	4939	S bc	<16		

Remarks and References (not intended to be complete):
(1) discovered by Schürer, with 10-in. Schmidt. (2) Bertola, *Contrib. Asiago* 153 (=*Astron. J.* **69**, 236). (3) also discovered by Humason. (4) Bertola, Mammano and Perinotto, *Contrib. Asiago*, 174. (5) Zwicky, *Stars and Stellar Systems*, Vol. VIII (Stellar Structure), p. 367. (6) Bertola, *Contrib. Asiago* 128. (7) Bertola, *Contrib. Asiago* 135. (8) Bertola, *Contrib. Asiago* 142. (9) Bertola, *Contrib. Asiago* 159. (10) Bertola, *Contrib. Asiago* 171. (11) Zwicky, *Astrophys. J.* **139**, 514. (12) Bertola and Arp, *PASP* **82**, 894. (13) Branch and Greenstein, Astrophys. *J.* **167**, 89. (14) also discovered by Bertaud. (15) Bertola, *Contrib. Asiago* 182. (16) also discovered by Zwicky. (17) Chincarini and Margoni, *Contrib. Asiago* 149. (18) Zwicky, *Ann. Astrophys.* **37**, 300. (19) Bertaud, *Astronomie* **79**, 133. (20) Ciatti and Barbon, *Contrib. Asiago* 254. (21) Chalonge and Burnichon, *Contrib. IAP (Paris)* **B 349** (=*J.O.* **51**, 5). (22) Chincarini and Perinotto, *Contrib. Asiago* 205. (23) also discovered by Dunlap. (24) also discovered by Balázs. (25) on the luminous bridge to NGC 2970 (type E).

Eruptive variables (U Gem Stars or Intergalactic Supernovae?) found at Zimmerwald.

α_{1950}	δ_{1950}	Maximum	
		Magnitude	Date
$15^h07^m28^s$	$+55°39'.5$	15	Sept. 1964
13 45 57	−30 53.7	$13^1/_2$–14	March 1970
9 06 19	+44 58.8	$13^1/_2$	March 1972
22 56 13	+10 53.1	$13^1/_2$	Oct. 1972

DISCUSSION

L. Rosino: Can you give me the exact position of the variable peculiar object you found on M 81? I have several recent plates of this galaxy and perhaps some useful data can be derived.

P. Wild: 40″ W and about 6″ S of the center of M 81; mag. about 17.

H. Arp: Why are you interested in the spectrum of the star next to the SN in NGC 2841?

P. Wild: I should like to know if it is a normal foreground star (galactic) or a bright member of NGC 2841.

THE PALOMAR SUPERNOVA SEARCH

W. L. W. SARGENT, L. SEARLE and C. T. KOWAL
Hale Observatories, California Institute of Technology, Carnegie Institution of Washington, Pasadena, Calif., U.S.A.

1. History and Achievements

The Palomar Supernova Search was initiated by F. Zwicky when the 18-in. Schmidt Telescope was put into operation in 1936. At that time only 20 supernovae had been discovered worldwide since 1885 when the first supernova of modern times was found in M31. The introduction of the 18-in. Schmidt lead to a rapid increase in the rate of discovery; a particularly notable event was Zwicky's discovery of the SN 1937c in IC 4182 which led to Baade and Minkowski's delineation of the evolution of the spectrum and light curve characteristic of a type I supernova.

Since 1958 the Palomar Supernova Search has been conducted primarily on the 48-in. Schmidt Telescope. By April 1973 the Palomar observers had discovered 250 of the 378 supernovae discovered worldwide. This includes many supernovae found by F. Zwicky on accepted and rejected plates obtained for the Palomar *Sky Survey*.

At the present time two main programs are under way. The search on the 48-in. Schmidt covers 38 fields containing either nearby groups and clusters of galaxies or distant rich clusters including the Coma and Coronae Borealis clusters. The 18-in. Schmidt Telescope has been used since the beginning of 1971 to survey fields containing 42 Sc I galaxies which, according to Tammann (1970), have an abnormally high supernova rate. Associated with the survey work proper, programs are currently under way to set up UBV photoelectric sequences in each of the 48-in. fields, to determine the redshifts of supernova parent galaxies, and to study the physics of individual supernova outbursts.

2. Methods

Currently three dark nights each month are allocated to the supernova search on the 48-in. Schmidt Telescope. Table I contains a list of the plate centers for the 38 fields observed. Ten inch square 103a D plates with a Wratten 12 filter are used; the exposure times, listed in the last column of Table I range from 6 to 25 min depending on the field being investigated. Even with the short exposures used for the nearer groups and clusters, the images of the brighter galaxies are strongly overexposed at the center. Consequently supernovae are probably missed if they occur near the centers of galaxies, particularly ellipticals. From this point of view it would be advantageous to photograph many of the fields with exposure times as short as one minute; this is now under consideration as a possible addition to the existing program. The plates are blinked against standard plates of the fields as soon as possible after they have been exposed. The work is begun on the mountain and in some cases confirmatory

C. B. Cosmovici (ed.), Supernovae and Supernova Remnants, 33–49.

TABLE I

48-In. supernova fields

Cluster	α (1970)	δ	Guide star	m_v	Exp.
Zw. Gr. C4	$00^h16^m09^s$	+26°59′	AGK2 +29°33	(9.4)	15 min.
Zw. Gr. C3	00 36 58	+29 09	+29 89	4.5	15
M31	00 43 00	+40 31	GC 880	7.5	10
Zw. Gr. C2	01 04 32	+32 01	+31 91	6.6	15
NGC 474	01 18 38	+02 55	+ 2 141	(9.0)	12
Gr. C1 + M33	01 24 43	+32 15	+32 124	(9.4)	15
Cl. 0123 − 0133	01 25 56	−01 21	− 1 128	9.4	12
Cl. 0151 + 3615	01 49 07	+35 42	+35 168	(6.7)	10
NGC 1058	02 36 25	+37 36	+37 304	6.3	10
NGC 1068	02 39 57	+00 25	+ 0 232	6.8	10
Perseus Cl.	03 18 18	+41 47	+41 353	(7.5)	12
Cl. 0510 + 0458	05 13 08	+05 07	+58 607	5.8	20
Cancer Cl.	08 18 36	+20 51	+20 965	5.9	12
M81 Group	09 52 38	+69 03	GC 13606	8.7	5
NGC 3379	10 44 43	+12 54	+13 1070	6.8	8
Dbl. Cl. A + B	11 11 48	+29 43	+29 1186	(8.3)	15
NGC 3627	11 17 33	+13 33	+13 1127	7.1	6
NGC 3656	11 22 56	+54 08	+54 808	(9.1)	12
Cl. 1142 + 2126	11 43 48	+20 03	+20 1288	7.4	15
Ursa Major Cloud	12 00 46	+51 44	+51 868	(8.3)	12
Virgo 4	12 10 11	+15 03	+15 1289	(9.0)	10
Virgo 2	12 10 51	+06 14	+ 6 1505	(8.4)	10
NGC 4258	12 13 46	+47 17	+47 956	(8.5)	6
Virgo 8	12 27 06	+10 31	+10 1541	(10.2)	10
Virgo 7	12 28 02	+06 24	+ 6 1535	(9.2)	6
Virgo 9	12 28 16	+14 49	+14 1292	7.2	6
Virgo 13	12 44 01	+10 44	+10 1566	(9.2)	6
Coma Cl.	12 58 06	+28 14	+28 1287	7.1	12
M101 Group	14 05 14	+55 34	+55 918	(7.5)	10
Boötes Cl.	14 31 05	+31 46	+31 1278	8.6	25
Shane Cl.	15 13 42	+05 03	+ 5 1999	5.4	25
Corona Bor. Cl.	15 21 30	+28 10	+28 1486	(7.5)	25
Hercules Cl.	16 02 29	+17 53	+17 1565	6.9	20
Cl. 1625 + 4006	16 25 21	+40 53	+40 1520	7.0	15
Cl. 1710 + 6401	17 05 56	+64 59	+65 805	(8.0)	25
Pegasus Cl.	23 18 34	+07 52	+ 7 3370	(9.2)	10
Cl. 2322 + 1427	23 19 27	+14 09	+14 2601	(8.6)	25
Cl. 2335 + 2449	23 35 23	+26 45	+26 2807	(8.9)	20

plates of faint supernova suspects are taken during the same observing run.

The work on the 18-in. Schmidt Telescope involves photographing fields containing bright Sc I galaxies. These are listed in Table II. The films used on this telescope cover a circle with a diameter of 8 degrees. An exposure time of three minutes is used with IIa O emulsion and no filter; in general it is not necessary to guide the telescope during these short exposures. The limiting magnitude on the films is about $B = 17.0$. The Sc I galaxies are examined immediately; later the whole field on each film is compared with a standard film of the same field in a binocular microscope with the images slightly

TABLE II

18-in. supernova fields

NGC	Guide star	α (1971)	δ
157	GC 610	$0^h29^m37^s$	$-10°14'35''$
514/628	$AGK_2 + 12°153$	1 26 31	+12 37 54
877	GC 2768	2 16 31	+13 17 48
908	GC 2838	2 19 52	−20 02 51
1084	GC 3291	2 42 30	− 8 02 00
1187/1232	GC 3676	3 02 45	−20 49 48
1385	GC 4285	3 33 38	−25 40 44
IC 342	GC 4463	3 43 15	+67 06 48
2276	$AGK_2 + 85°99$	7 15 44	+85 47 46
2532	GC 11073	8 08 18	+35 32 42
2535	GC 11091	8 08 43	+25 35 55
2805	GC 12838	9 17 52	+65 08 20
3031	GC 13596	9 52 05	+69 35 19
3294	GC 14663	10 38 32	+38 33 18
3367	GC 14814	10 44 53	+14 20 55
3631-42	$AGK_2 + 56°843$	11 17 32	+56 03 36
3646	GC 15608	11 20 18	+18 21 01
3726	GC 15860	11 32 32	+48 43 44
3810	GC 16065	11 40 34	+12 26 51
3893	GC 16137	11 44 32	+47 56 24
3938	GC 16417	11 59 08	+43 46 58
4030	GC 16400	11 58 05	−01 45 25
4088	GC 16537	12 04 39	+49 19 11
4254/4321	GC 16800	12 17 52	+14 42 11
4303-35	GC 16907	12 22 42	+06 07 56
4666	GC 17307	12 42 03	+02 46 46
5194	GC 18519	13 40 36	+46 21 59
5236	----	13 35 24	−29 43 00
5248	GC 18460	13 37 47	+10 39 25
5427	GC 18996	14 02 44	−05 14 33
5457	GC 19031	14 04 20	+56 39 30
6070	GC 21734	16 08 13	+00 53 43
6181	GC 22218	16 30 01	+19 20 51
6217	GC 22301	16 31 53	+77 30 13
6946	GC 28660	20 33 23	+61 38 35
7314	GC 31588	22 36 33	−27 35 42
7678	GC 32666	23 27 39	+22 53 21

offset. Three nights each month are allocated to this program. In fair weather it is possible to photograph all the accessible fields in one night. For this reason a set of lower priority fields, not listed in Table II, are observed on some occasions.

3. Statistics of the 48-in. Supernova Search

The supernova search has been conducted on the 48-in. Schmidt Telescope in practically its present form since 1958*. The completion by Zwicky and his associates of the

* An account of the work for each calendar year has been published annually in the *Publ. Astron. Soc. Pacific* since 1959, volume 71.

'Catalogue of Galaxies and Clusters of Galaxies' (CGCG) has made it possible to obtain reliable information of the distribution of apparent magnitude of the galaxies in each of the supernova fields down to $m_p = 15.7$, the limiting magnitude of the catalogue. In Table III we give the number of galaxies in each magnitude interval and the number of supernovae discussed between 1958 and the end of 1971 for each field. It will be seen

TABLE III

Summary of observations since 1958: magnitude distribution[a]

Field	m_{pg} galaxy ≤11.0 Gal.	SN's	m_{pg} galaxy 11.1–12.0 Gal.	SN's	m_{pg} galaxy 12.1–13.0 Gal.	SN's	m_{pg} galaxy 13.1–14.0 Gal.	SN's	m_{pg} galaxy 14.1–15.0 Gal.	SN's	m_{pg} galaxy 15.1–15.7 Gal.	SN's
Zw.Gr.C4	–	– –	–	– –	–	– –	4	– –	24	1 –	49	1 –
Zw.Gr.C3	–	–	–	– –	–	– –	3	– –	19	– –	33	– –
Zw.Gr.C2	–	– –	–	– –	2	– –	12	1 –	38	– –	49	4 –
NGC 474	–	– –	2	– –	5	1[1]	5	1 –	15	– –	48	–[1]
Zw.Gr.Cl + M33	1	– –	–	– –	2	– –	14	– –	43	1 –	57	1 –
Cl. 0123 – 0133	–	– –	–	– –	–	– –	7	– –	32	1 –	39	– –
Cl. 0151 + 3615	–	– –	–	– –	2	– –	17	– –	33	– –	59	– –
NGC 1058	1	– –	2	–[1]	2	– –	3	–[1]	12	1 –	32	– –
NGC 1068	1	– –	1	– –	3	2[1]	5	– –	24	– –	42	1 –
Perseus	–	– –	–	– –	1	–[1]	1	– –	40	– –	57	–
Cl. 0510 + 0458	–	– –	–	– –	–	– –	1	– –	9	– –	34	– –
Cancer	–	– –	–	– –	–	– –	6	1 –	24	2 –	52	2 –
M81 Group	4	– –	–	– –	–	– –	1	– –	3	– –	17	– –
NGC 3379	5	– –	3	–[1]	2	– –	5	– –	9	–	42	1 –
Dbl. Cl. A & B	–	– –	1	– –	1	– –	1	– –	16	1 –	66	– –
NGC 3627	2	– –	3	– –	1	– –	2	– –	7	– –	36	– –
NGC 3656	2	–[1]	2	– –	3	– –	4	– –	12	– –	25	– –
Cl. 1142 + 2126	–	– –	–	– –	–	– –	12	– –	47	– –	112	3 –
Ursa Major	2	– –	5	– –	2	– –	3	– –	11	– – –	18	– –
Virgo 4	2	2 –	2	– –	6	–[1]	4	– –	24	2[1]	67	– –
Virgo 2	–	1 –	1	– –	3	– –	14	– –	20	– –	35	– –
Virgo 8	1	–[1]	5	– –	15	–[1]	23	– –	25	– –	78	– –
Virgo 7	2	–[1]	3	– –	4	– –	16	1 –	33	– –	64	1 –
Virgo 9	5	– –	10	– –	13	– –	19	–[1]	17	– –	46	– –
Virgo 13	2	– –	3	– –	12	– –	3	– –	10	– –	57	– –
Coma	–	– –	–	– –	2	– –	13	2 –	58	1 –	141	1 –
M101 Group	1	–[1]	2	– –	2	– –	5	– –	5	– –	17	– –
Boötes	–	– –	–	– –	1	– –	2	– –	16	– –	19	– –
Shane Cld.	–	– –	–	– –	2	– –	–	– –	11	– –	103	2 –
Cor. Bor.	–	– –	–	– –	–	– –	–	– –	2	1 –	30	1 –
Hercules	–	– –	–	– –	–	– –	1	– –	32	– –	108	1–
Cl. 1625 + 4006	–	– –	–	– –	–	– –	3	– –	17	– –	67	2 –
Cl. 1710 + 6401	–	– –	–	– –	–	– –	2	– –	4	– –	14	– –
Pegasus	–	– –	–	– –	3	–[1]	6	– –	22	– –	64	1 –
Cl. 2322 + 1427	–	– –	–	– –	–	– –	2	– –	10	– –	44	– –
Cl. 2335 + 2449	–	– –	1	– –	–	– –	1	– –	18	– –	54	2 –
Totals	31	3[4]	46	–[2]	89	3[6]	220	6[2]	742	11[1]	1875	24[1]

[a] The first number in columns 2–7 gives the number of galaxies in each field in the particular magnitude interval. Next is the number of SN discovered at Palomar and, in square brackets, the number discovered at other observatories.

from these data that the Palomar Supernova Search encompasses 3003 galaxies of magnitude 15.7 or brighter, and that 63 supernovae have been discovered in these galaxies in the interval 1958–71. Of these supernovae, 47 were found at Palomar and 16 at other observatories. In Table IV we summarize the actual observational coverage of each field; the last two columns give the number of months in which observations were made as compared with the number of months which elapsed between a particular field being put on the regular program and the end of the period under discussion.

TABLE IV

Summary of observations since 1958 observational coverage

Field	No. of galaxies $m_{pg} \leqslant 15.7$	Supernovae found				Total SN	Months observed	Months since start
		Galaxies $\leqslant 15.7$		All galaxies				
		Pal.	[Other]	Pal.	[Other]			
Zw. Gr. C4	77	2	–	3	–	3	90	169
Zw. Gr. C3	55	–	–	1	–	1	82	168
Zw. Gr. C2	101	5	–	5	[1]	6	84	168
NGC 474	75	2	[2]	5	[2]	7	74	149
Zw. Gr. Cl + M33	117	2	–	2	–	2	17	41
Cl. 0123 – 0133	78	1	–	3	–	3	62	159
Cl. 0151 + 3615	111	–	–	–	–	–	19	52
NGC 1058	52	1	[2]	4	[2]	6	73	123
NGC 1068	76	3	[1]	6	[1]	7	78	159
Perseus	99	–	[1]	3	[1]	4	88	170
Cl. 0510 + 0458	44	–	–	1	–	1	31	73
Cancer	82	5	–	6	–	6	86	169
M81 Group	25	–	–	–	–	–	23	65
NGC 3379	66	1	[1]	2	[2]	4	36	81
Dbl. Cl. A & B	85	1	–	2	–	2	19	40
NGC 3627	51	–	–	–	–	–	31	77
NGC 3656	48	–	[1]	1	[1]	2	16	40
Cl. 1142 – 2126	171	3	–	5	–	5	60	118
Ursa Major	41	–	–	2	–	2	13	36
Virgo 4	105	4	[2]	5	[2]	7	85	156
Virgo 2	73	1	–	1	–	1	26	64
Virgo 8	147	–	[2]	–	[2]	2	28	64
Virgo 7	122	2	[1]	3	[1]	4	80	156
Virgo 9	110	–	[1]	–	[1]	1	85	156
Virgo 13	87	–	–	–	–	–	27	64
Coma	214	4	–	8	–	8	96	166
M101 Group	32	–	[1]	–	[1]	1	21	84
Boötes	38	–	–	–	–	–	9	44
Shane Cld.	116	2	–	3	–	3	82	140
Cor. Bor.	32	2	–	8	–	8	86	141
Hercules	141	1	–	8	–	8	105	168
Cl. 1625 + 4006	87	2	–	4	–	4	67	127
Cl. 1710 + 6401	20	–	–	1	–	1	18	39
Pegasus	95	1	[1]	2	[1]	3	49	84
Cl. 2322 + 1427	56	–	–	1	–	1	44	76
Cl. 2335 + 2449	74	2	–	4	–	4	47	78
Totals	3003	47	[16]	99	[18]	117		

This table also lists the total number of supernovae found in each field, including those which occurred in galaxies fainter than $m_p = 15.7$.

In Table V we give crude estimates of the supernova frequency using the data in Tables IV and V. For each magnitude interval in column one we list in column two the number of supernovae discovered at Palomar and elsewhere, while column two gives the total number of galaxies in all the search fields in that magnitude interval.

TABLE V

Observed frequency of supernovae

Galaxy mag.	No. of SN	No. of galaxies	Galaxy months		Frequency Galaxy-years per Supernovae	
			Lower Limit	Upper limit	Lower limit	Upper limit
$\leqslant$ 11.0	7	31	1418	2974	16.9	31.6
11.1–12.0	2	44	2285	4639	95.2	193.3
12.1–13.0	9	89	7465	9215	69.1	85.3
13.1–14.0	8	220	12147	23952	126.5	249.5
14.1–15.0	12	742	47913	101705	332.7	706.3
15.1–15.7	25	1875	115212	220115	384.0	733.7

Column four gives an estimate of the coverage of the Palomar search of the galaxies in column three. It is the sum over the number of fields of the number of galaxies in a particular magnitude interval in each field multiplied by the number of months in which that field was observed. Column five is similar except that the number of galaxies in each field is multiplied by the total number of months in which the field was in the regular program. The last two columns of Table V give estimates of the supernovae frequency, as a function of the apparent magnitude of the parent galaxy, based on the two estimates of the coverage.

It is an extremely complicated problem to compute the true supernova frequency from the available data. Account has to be taken of the fact that the supernovae which occur in nearby galaxies can be discovered many months after they reach maximum light, while those that occur in distant galaxies must be caught near maximum if they are to be discovered. This is one reason why the last two columns of Table V show a progressively decreasing supernova frequency with increasing faintness of the parent galaxy. (Another reason is that the type of distribution of galaxies changes with apparent magnitude; there appears to be an excess of spirals in general and Sc I galaxies in particular among the apparently bright galaxies.) However it is clear that the main effect in explaining the last two columns of Table V must be that supernovae are missed in the fainter galaxies, probably in their central regions. Support for this conclusion is found in Table VI which gives the radial distribution of supernovae in galaxies of known redshift. Beyond 5 kpc the number of supernovae falls off rapidly with radius; the fall-off farther in is probably due to supernovae being missed in the over-exposed centers of galaxies. Thus our preliminary conclusion is that the supernova

TABLE VI

Radial distribution of supernovae in galaxies ($H = 55$ km s^{-1} Mpc^{-1})

Radial distance (kpc)	<2	2–3.9	4–5.9	6–7.9	9–10.9	11–12.9	13–14.9
No. of supernovae	11	12	19	10	10	6	5
Radial distance (kpc)	15–16.9	17–18.9	19–20.9	21–22.9	23–24.9	>25	
No. of supernovae	4	3	0	2	1	2	

TABLE VII

The color distribution for galaxies

$(B-V)_0$	< +0.2	0.2–0.29	0.3–0.39	0.4–0.49	0.5–0.59	0.6–0.69	0.7–0.79	0.8–0.89	>0.90
No. of galaxies	4	9	38	81	94	81	139	172	32
No. with supernovae[a]			7(3)	13(7)	11(7)	13(5)	3(2)	8(1)	4(0)
Percentage with supernovae			18(8)	16(9)	12(7)	16(6)	2(1.4)	5(0.6)	13(0)
Percentage expected[b]			11.5	9.2	7.0	4.6	2.3	0.2	0.0

[a] In lines 3 and 4 the data in brackets are for supernovae of type II alone.
[b] From the calculations by Searle *et al.* (1973) normalized to 7% at $B-V = 0.5$–0.59.

frequency in galaxies is one every 20 or 30 years, a considerably higher rate than that advocated by Zwicky, but compatible with Tamman's (1970) recent conclusions.*

4. Associated Programs

As an appendix we give Table VIII, an updated version of the *Master List of Supernovae Discovered since 1885* which is maintained by the Palomar Supernova Search and which was last published by Kowal and Sargent (1971). The most striking thing about this list is that, although the rate of discovery of supernovae has risen dramatically in the last few years, there is a remarkable paucity of data on the types of most of these supernovae and on the redshifts and colors of the parent galaxies. These data are

* At the Conference F. Zwicky objected that the reason for the decline in supernova rate with fainter apparent magnitudes is not a matter of observational selection, as we have claimed. Instead, Zwicky proposed that the *apparently* fainter galaxies are also *intrinsically* fainter and, therefore, have a lower supernova frequency than the apparently bright galaxies. A plot of redshift versus apparent magnitude for the parent galaxies, using the data in Table VIII, shows that Zwicky's assertion is not correct. Such a plot is given in Figure 1; the line of slope 5 through the data points shows that there is little change of intrinsic luminosity with redshift for the supernova parent galaxies and that, if anything, the fainter parent galaxies are intrinsically *more* luminous.

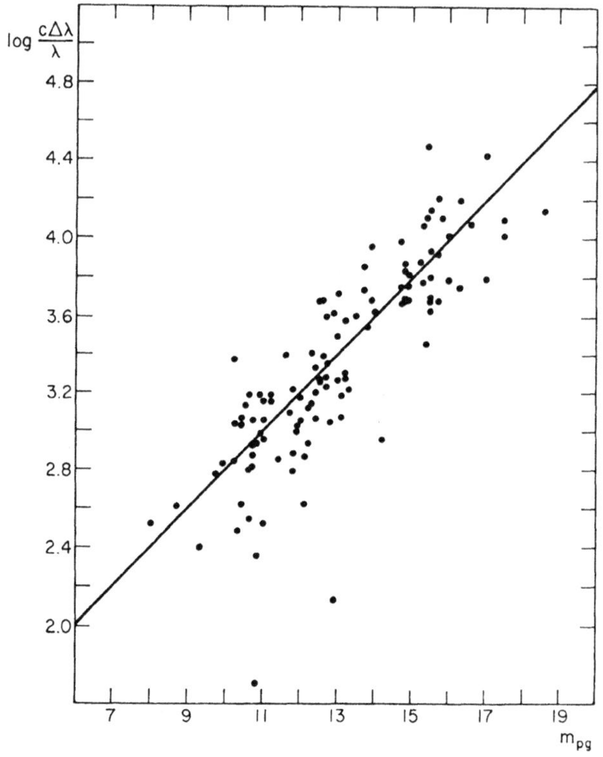

Fig. 1. Magnitudes of supernova parent galaxies vs the logarithms of their redshifts.

essential for a thorough discussion of the stellar population characteristics of supernovae of different types; accordingly the Hale Observatories group is making a systematic effort to obtain these measurements. However, the task is enormous and other observatories are urged to join in the work. In particular, it would be extremely useful if one spectrogram could be obtained of each supernova with the image tube spectrographs now available; this would seem to be the most economical way of determining the type of each supernova.

5. Remarks on the Stellar Populations of Supernovae

Supernovae of type II are found in spiral galaxies but not in ellipticals, while supernovae of type I are found in galaxies of all types – spirals, ellipticals and irregulars. From these facts it has been inferred that supernovae of type II are associated with Baade's Population I and that they are produced by the evolution of massive stars. Similarly, type I supernovae have been associated with Baade's Population II, although the corresponding inference that these objects are produced by stars of low mass (around one solar mass) does not seem popular with theoreticians.

From the theoretical standpoint our main concern is not with the Population assignment of supernovae but with the masses of the stars which become supernovae. One

TABLE VIII

Supernovae discovered since 1885 – by date-designation

ALL MAGNITUDES ARE PHOTOGRAPHIC, UNLESS PRECEDED BY 'B' OR 'V' .
AN ASTERISK AFTER A MAGNITUDE MEANS 'BRIGHTER THAN, OR EQUAL TO' .
A PARENTHESIS FOLLOWING ANY QUANTITY INDICATES AN APPROXIMATION

		1950		-----GALAXY-------			-----SUPERNOVA------					
SN	NGC	R.A.	DEC	TYPE	V	M PG	D1	D2	M	TYPE	T-MAX	NO.
1885A	N 224	0040.0	+4100	S B	-299	4.3	15W	4S	V 5.8)		AUG18	1
1895A	N4424	1224.7	+0942	SBB		13.1	75E	11S	12.5*		MAR	2
1895B	N5253	1337.1	-3124	IRR	403	10.8	16E	23N	8.0	I	JUL 6	3
1901A	N2535	0808.2	+2520	S C	4135	13.5	19E	7N	14.7*		JAN	4
1901B	N4321	1220.4	+1606	S C	1617	10.6	110W	4N	15.6*		MAR	5
1907A	N4674	1243.4	-0823	S B		14.5	10W	11N	13.5)		MAY10	6
1909A	N5457	1401.5	+5435	S C	266	8.7			12.1)	PEC)	FEB)	7
1912A	N2841	0918.5	+5112	S B	631	9.9	50W	20N	13.0*		FEB)	8
1914A	N4321	1220.4	+1606	S C	1617	10.6	24E	111S	15.7*		FE/MA	9
1915A	N4527	1231.6	+0256	S B	1727	12.4	44E	8S	15.5*		MAR)	10
1917A	N6946	2033.8	+5959	S C	80	10.5	37W	105S	14.6*		JUL	11
1919A	N4486	1228.3	+1240	E0	1261	10.4	15W	100N	12.3)	(I)	FEB26	12
1920A	N2608	0832.2	+2839	SBC	2119	13.2	19W	5N	11.8*		JAN)	13
1921A	N4038	1159.3	-1835	S PEC	1650	11.0			.		MAR)	14
1921B	N3184	1015.3	+4140	S C	418	10.4	32E	160S	13.5)		APR)	15
1921C	N3184	1015.3	+4140	S C	418	10.4	79E	236S	11.0		DEC11	16
1923A	N5236	1334.3	-2937	S C	506	8.0	109E	58N	14.0	PEC	MAY	17
1926A	N4303	1219.3	+0445	S C	1671	10.9	11W	69N	14.0)	(II)	MAY 1	18
1926B	N6181	1630.1	+1956	S C	2158	12.7	000	48N	14.8*		JUN)	19
1934A	I4719	1829.0	-5646	S		13.9	6E	13S	13.6)		OCT11	20
1936A	N4273	1217.4	+0537	S C	2302	12.4	000	29N	14.4	II	JAN13	21
1936B	ANON	0118.4	+1526	SBC		15.6			14.0)		SEP	22
1937A	N4157	1208.6	+5047	S C	916	11.9	42E	42N	16.2*	(II)	JAN)	23
1937B	ANON	2207.9	-2256	SBB		14.8	29E	31S	15.3*		AUG)	24
1937C	I4182	1303.5	+3752	S/IRR		14.0	30E	40N	8.4	I	AUG22	25
1937D	N1003	0236.1	+4040	S C	585	12.1	48E	1S	12.8	I	SEP16	26
1937E	N1482	0352.4	-2039	SA P		14.4	24W	51N	15.0*		DEC)	28
1937F	N3184	1015.3	+4140	S C	418	10.4	5E	149S	13.5	(II)	DEC12	27
1938A	ANON	0234.6	+3414	SBC	4800	13.9	8W	28S	15.2)		NOV	29
1939A	N4636	1240.3	+0258	E1	883	11.8	26W	20N	12.2	I	JAN22	30
1939B	N4621	1239.5	+1155	E3	414	11.0	000	53S	11.9	I	MAY 2	31
1939C	N6946	2033.8	+5959	S C	80	10.5	215W	24N	13.0	II	JUL 6	32
1939D	ANON	0055.0	-0516	S B		14.5	9W	11N	16.0*		NOV)	33
1940A	N5907	1514.6	+5630	S B	535	11.4	137E	310S	14.3	II	FEB16	34
1940B	N4725	1248.0	+2546	S(B)B	1114	10.2	95E	118N	12.8	II	MAY 8	35
1940C	I1099	1505.6	+5644	S A		15.0			16.3*		APR)	36
1940D	N4545	1232.4	+6348	S C		13.1	9W	20N	15.0*		JUL	37
1940E	N 253	0045.1	-2534	S C	96	7.0	51W	17S	14.0*	I	NOV	38
1941A	N4559	1233.4	+2814	S C	856	10.7	30W	26N	13.2	II	FEB26	39
1941B	N3254	1026.5	+2945	S B	1228	12.4	10W	34S	15.1*		MAR	40
1941C	N4136	1206.8	+3012	S C	445	12.1	44E	67S	16.8*	II	APR	41
1945A	N5195	1327.9	+4731	PEC	527	10.6	6W	4S	V14.0*	I	FEB)	42
1946A	N3977	1153.5	+5540	S A P	5707	14.7			18.0*		MAY)	43
1946B	N4632	1240.0	+0010	S C		12.6	16E	6N	15.7*	(II)	MAY)	44
1947A	N3177	1013.8	+2122	S B	1220	12.8	17E	40S	16.5*	II	MAR)	45
1948A	N4699	1246.5	-0824	S B	1511	10.5	000	46N	17.0*		MAR)	46
1948B	N6946	2033.8	+5959	S C	80	10.5	222E	60N	14.9*	(II)	JUN)	47
1950A	I4051	1258.5	+2817	E1	4932	14.8	2W	13N	17.7*	I	FEB	48

1950B	N5236	1334,3	-2937	S C	506	8,0	105W	000	14,5*		MAR	4
1950C	N5033	1311,2	+3651	S C	916	10,9	225E	50S	18,2*		MAY)	24
1950D	ANON,	0840,3	+1821	S C		17	8W	12S	16,6*		MAR)	24
1950E	ANON,	1017,5	+1335	S B		15,7	74W	000	17,8*		APR)	25
1950F	ANON	1402,3	+0904	SBC		14,9	11W	1N	16,2*		FEB)	31
1950G	ANON	0941,4	+0935	S C		17,5	3E	2S	18,3*		FEB)	37
1951A	ANON,	2152,7	-0432	SBB		15,6	8E	8S	16,6*		AUG)	25
1951B	ANON	1401,2	+1137	S C		15,1	10W	15N	18,2*		APR)	32
1951C	ANON	1016,5	+0821	E		17,8	12W	8S	18,2*		FEB)	32
1951D	ANON	1109,2	+0358	E		18,5	6E	1S	19,6*		MAY)	32
1951E	ANON	1522,5	+0830	S C		18,4	7W	2S	20,0*		MAY)	33
1951F	ANON	2357,5	-0639	S C		15,4	18E	9S	17,3*		AUG)	34
1951G	ANON	0008,0	-0634	E PEC		17,6	6E	2N	18,8*		AUG)	34
1952A	ANON,	0149,8	+3622	S C		13,9	5W	12N	18,6*		SEP)	24
1952B	ANON,	0923,0	+2940	S C		16	9E	11S	17,8*		JUN)	25
1952C	ANON	1257,6	-0305	IRR		18,1	5W	10S	18,7*		JA/FE	28
1952D	ANON	1114,9	-0248	S C		16,7	11W	15N	19,5*		APR)	28
1952E	ANON	1307,2	-0313	S		18,0	10E	10S	19,8*		APR)	28
1952F	ANON	1225,9	+1337	IRR		18,5	1,5E	2S	16,9*		JAN)	31
1953A	N3561	1108,5	+2859	S B P	8696	15,5	5E	1N	16,0*		APR)	16
1953B	ANON	1619,2	+4014	S C		15,2	20E	6S	17,5*		MAY)	22
1953C	ANON,	1827,4	+4813	S C		14,9	15W	35N	19,0*		JUL)	25
1953D	N3200	1016,2	-1744	S B		12,6	15E	56N	19,5*		MAR)	25
1953E	ANON	2331,2	+2946	S B B		15,1	11E	16S	19,5*		DEC)	26
1953F	ANON	2335,4	+3143	S A		14,1	1E	10N	19,0*		DEC)	26
1953G	ANON	1056,3	+5032	S		16	8E	6N	16,5*		FEB)	27
1953H	ANON	1100,8	+5007			19,5	1E	3S	17,0*		MAR)	27
1954A	N4214	1213,1	+3637	IRR	290	10,3	84E	216S	9,8	I	APR19	5
1954B	N5668	1430,9	+0440	S C	1737	12,7	000	32S	12,3	I	APR29	5
1954C	N5879	1508,4	+5712	S B	876	11,9	14W	11S	14,9	II	OCT 4	5
1954D	ANON	0029,5	+3124	S B		15,1	12W	17S	17,4*		SE/OC	18
1954E	N 753	0154,8	+3540	S C	4766	12,6	42W	25S	18,5*		SEP)	24
1954F	ANON,	0018,5	+2614	S		15,5	6W	N	18,8*		AUG)	24
1954G	ANON,	1232,8	-1858	S C		17,5	13W	5S	16,6*		MAR)	24
1954H	ANON,	1307,9	-0723	IRR		14,5	27E	17S	17,4*		MAY)	24
1954I	ANON	1112,1	-2038	S		17,5	6E	2N	17,3*		JAN)	29
1954J	N2403	0732,0	+6543	S C	136	9,3	36E	100N	16,0*	(V)	OCT)	29
1954K	ANON	0841,5	+6119	S C		16,0	3E	4N	15,0*		NOV)	30
1954L	ANON	1058,5	+1236	S C		15,7	20W	6N	19,3*		FEB)	31
1954M	ANON	1447,2	+1029	S C		17,7	3W	0 NS	17,5*		JUN)	31
1954N	ANON	1443,4	+1023	S C		16,0	2W	6N	18,0*		JUN)	31
1954O	ANON	1417,3	+1334	S C		16,8	15E	2S	17,6*		MAY)	32
1954P	ANON	1345,5	+0936	E		19,2	2W	2S	19,5*		MAY)	32
1954Q	ANON	1351,0	+0958			20,0	12W	8N	18,5*		MAY)	32
1954R	N 210	0038,0	-1409	S(B)B	1768	12,5	86E	80S	15,9*		SEP)	32
1954S	ANON	0033,2	-0847	IRR		18,5	1W	6S	17,4*		SEP)	33
1954T	ANON	2234,7	+0328	S O		16,5	7E	7S	19,0*		JUL)	33
1954U	ANON	2356,3	-0553	IRR		18,7	6W	6N	17,8*		SEP)	34
1954V	ANON	0010,6	-0752	EO		19,0	4W	7N	20,0*		SEP)	34
1954W	ANON	1426,1	+0046	S C		15,2	15E	5N	19,4*		MAY)	36
1954X	ANON	1348,8	+1753	S C		16,7	3E	9S	18,5*		APR)	36
1954Y	ANON	1352,1	+1517	SO		14,1	13W	000	19,4*		APR)	36
1954Z	ANON	0942,6	+0920	S B		13,8	14W	19S	16,0*		FEB)	36
1954AA	N4129	1206,3	-0845	S B		12,9	53W	6N	19,9*		APR)	37
1954AB	ANON	1155,7	-0539	SC/IR		18,9	000	1N	17,5*		APR)	37
1955A	N4157	1208,6	+5047	S B	916	11,9	103E	40N	16,0*		APR	151
1955B	ANON	0105,0	-1330	S	16024	15,7	2W	5N	15,8		OCT 3	53
1955C	N 23	0007,3	+2539	SBA	4568	12,5	10E	10N	V16,0*		OCT	59
1955D	ANON	0049,5	-1641	S PEC		15,1	1E	12N	15,5*		NOV	---
1955E	N4335	1220,6	+5844	E		13,7	19W	7S	16,3*		MAY)	206

955F	ANON	1201,2	+0159	S B		15,0	15E	17N	19,5*		APR)	284
955G	ANON	1123,5	+0346	S A		15,2	8E	10S	19,4		APR)	325
955H	ANON	1108,2	+0312	IRR		19,4	3E	3S	18,8*		APR)	328
955I	ANON	1409,0	+0123	S A		15,7	1E	6S	19,3*		MAY)	363
955J	ANON	1409,2	+0057	S A		15,6	3E	2N	17,9*		MAY)	364
955K	ANON	1420,0	+1518	S A		14,2	3W	7S	18,4*		MAY)	365
955L	ANON	1102,2	+3018	PEC		14,7	8E	14N	16,5*		MAR)	371
956A	N3992	1155,0	+5339	SBB	1059	10,7	67E	9S	12,3	I	MAR18	54
956B	N4782	1252,0	-1216	DBL E	4343	12,9	40W	15S	18,6*		APR)	239
956C	ANON	1240,6	+0405	S		16,7	5E	27S	17,6*		FEB)	269
956D	I 850	1305,2	-0035	S		14,8	12E	6N	19,2*		FEB)	287
956E	ANON	1128,5	-1552	S C		17,5	3E	4S	19,5*		FEB)	291
956F	ANON	1320,5	+0704	IRR		19,5	1E	1S	19,6*		MAR)	333
956G	ANON	1314,6	+0236	S C		17,5	9W	5N	19,3*		MAR)	334
956H	ANON	1318,2	+0232	S C		17,5	2W	11N	19,3*		FEB)	335
956I	ANON	1315,2	-0118	IRR		18,7	4W	000	20,2*		APR)	366
956J	ANON	1308,2	+0018	SC/IR		15,4	9E	5S	19,8*		APR)	367
956K	ANON	1311,2	-0320	S B		15,1	12W	1S	19,6*		MAR)	368
957A	N2841	0918,5	+5112	S B	631	9,9	106W	73N	14,0	IIP	MAR 3	55
957B	N4374	1222,5	+1310	E1	954	10,8	8W	47N	12,5	I	MAY10	56
957C	N1365	0331,8	-3618	SBB	1724	11,2	54W	75N	16,5*		OCT	57
957D	N5236	1334,3	-2937	S C	506	8,0	41W	145N	15,0*		DEC	58
958A	ANON	1558,6	+1952	E		17,5	1W	5S	19,0*		AUG	114
958B	ANON	0234,5	+0107	SO		16,0	1W	7N	18,0*		DEC	80
958C	ANON	1255,4	-3018	S C		17,5	21W	22N	19,6*		APR)	292
958D	ANON	1151,2	-3028	S C		16,0	6W	13S	19,5*		APR)	294
959A	N1350	0329,1	-3347	SBC	1802	11,8	22E	70N	16,0*		JAN	60
959B	N4921	1259,0	+2808	S A	5459	13,7	16E	48S	18,5*	PEC)	FEB)	61
959C	ANON	1308,8	+0340	SBC	2990	15,4	7E	3S	13,6	I	JUN28	62
959D	N7331	2234,8	+3410	S B	794	10,4	32W	13N	13,4	II	JUL 4	63
959E	N4321	1220,4	+1606	S C	1617	10,6	58E	21S	17,5*	I	AU/SE	67
959F	ANON	0247,2	-0044	SBC		15,1	28E	20N	18,5*		DEC	64
960A	ANON	0232,8	+0153	S		16,5	5E	8S	16,0*		JAN	76
960B	ANON	1232,2	+0917	S	12924	15,4	5E	6S	16,0*		FEB	65
960C	ANON	1204,6	+1716	S B	6740	14,9	4W	9N	17,0*		FEB	66
960D	ANON	0817,4	+2102	SBC	5008	15,5	18E	26S	16,4*		MAR	68
960E	ANON	1215,1	+4810	S		17,8	1E	8S	16,5*		MAR)	90
960F	N4496	1229,1	+0413	SBC	1773	13,3	38E	24N	11,6	I	APR16	69
960G	ANON	1128,5	+1840	S C		16,0	14E	6S	17,5*		APR	70
960H	N4096	1203,4	+4745	S C		11,6	67E	114N	14,5*	I	JUN	71
960I	ANON	1224,6	+4833	S(B)B	4627	14,7	5E	11S	18,5*		JUN	72
960J	N4375	1222,5	+2850	S A	9165	13,9	37E	17N	18,5*		JUN	73
960K	ANON	2237,4	+3407	SBB	7386	15,2	10W	6S	19,0*		JUN	74
960L	N7177	2158,3	+1730	S B	1105	12,2	6E	54S	16,0*		AUG	75
960M	N2565	0816,9	+2211	SBB	3684	13,8	13W	34N	15,7*	I	SEP	77
960N	ANON	0823,2	+2137	S C	4349	15,5	11E	12N	16,6*	I	OCT	78
960O	ANON	2333,9	+2739	S		17,0	1W	10N	18,5*		NOV	79
960P	ANON	0103,4	+3108	S		15,5	4E	2N	17,5*	I	NOV	81
960Q	ANON	0133,4	-0545	S		18,5	1E	2S	17,5*		DEC	82
960R	N4382	1222,9	+1828	SO	765	10,2	8E	132S	12,0)	I	DEC19	86
961A	ANON	0241,5	+0034	SBB		16,1	9E	27N	19,0*		JAN	83
961B	I2363	0822,9	+1936	S C		15,0	9E	18N	18,5*		JAN	84
961C	ANON	0212,3	+4053	S C		16,4	3E	8S	18,2*		JAN	85
961D	ANON	1248,3	+2806	EO	7638	14,8	32E	12S	V16,5	I	JAN 6	87
961E	ANON	1515,1	+0514	SO	12900	15,8	10W	2S	17,0*		JAN	88
961F	N3003	0945,6	+3339	S C	1476	12,3	34E	17N	13,1	IV	FEB23	89
961G	ANON	1019,6	+2129	S B	16125	16,3	9W	2S	18,2*		APR	91
961H	N4564	1233,9	+1143	E6	941	12,2	000	5N	11,2	I	MAY 8	92
961I	N4303	1219,3	+0445	S C	1671	10,9	82E	12S	13,0)	III	JUN	93
961J	ANON	1609,8	+2942	S)		19,5	1W	1S	16,0*	III)	JUN)	94

1961K	ANON	1200,8	+1647	S C		14,6	10W	5S	16,3*		MY/JU	9
1961L	N3221	1019,6	+2150	S		14,3	19W	53N	17,5*		JUL)	10
1961M	ANON	0107,2	+3205	E		14,0	16E	37N	17,0*		AUG	9
1961N	I5342	2336,2	+2644	EO		15,4	1E	7S	16,8*		SEP)	9
1961O	ANON	0223,5	+4324	S		17,0	2W	10N	17,0*		SEP)	9
1961P	ANON	0232,5	+3725	S A	3665	13,2	000	36S	14,3	I	SEP12	10
1961Q	N 550	0124,1	+0145	S B		13,6	20W	13N	17,2*		NOV)	10
1961R	ANON	0107,2	+3238	S		15,5	11W	5N	17,0*		OCT	10
1961S	ANON	1018,3	+2158	S		15,8	19E	6S	18,3*		NOV)	10
1961T	ANON	0936,0	+3340	IRR		19,5	3W	5N	17,5*		NOV)	11
1961U	N3938	1150,2	+4423	S C	874	11,0	81E	104N	13,7	II	DEC29	10
1961V	N1058	0240,3	+3708	S C	480	11,8	76E	17N	12,2	V	DEC 5	9
1962A	ANON	1304,3	+2808	SO	6137	16,0	11W	7N	15,6	I	JAN25	10
1962B	ANON	1520,7	+2957	S		14,9	5E	1S	17,0*	I	JAN	10
1962C	ANON	1602,8	+1743	S		18,5	12E	6N	18,0*		JAN	10
1962D	ANON	1030,8	-2739	IRR		17,4	000	2S	16,0*		FEB	10
1962E	ANON	1112,5	+2610	EO	14250	15,5	4W	14S	V17,5*	(I)	FEB	11
1962F	ANON	0814,5	+2150	SBC		14,5	38E	40S	16,5*		MAY	11
1962G	ANON	1526,0	+2911	SO		16,5	4E	4S	19,0*		MAY	11
1962H	I4237	1321,8	-2052	SBB		13,8	19W	3S	13,0)	(II)	JUN	11
1962I	ANON	1300,5	+2747			19,0	8W	7N	17,6)		MAY	11
1962J	N6835	1951,8	-1242	S(B)A	1720	13,0	42W	23S	13,6	I	SEP 1	11
1962K	N1090	0244,0	-0027	SBC		12,8	52W	10N	18,2*		SEP)	11
1962L	N1073	0241,1	+0110	SBC	1874	12,5	10E	77N	B13,9	IPEC	DEC 7	11
1962M	N1313	0317,6	-6640	SBC	254	10,8			11,7*	(II)	NOV	12
1962N	ANON	1519,0	+2631	S		15,6	000	3N	17,0*		DEC	12
1962O	ANON	0321,2	+3951	E	14150	18,6	15E	11N	19,8*		NOV	12
1962P	N1654	0443,3	-0211	E		14,2			14,5*		SEP	14
1962Q	N2276	0710,5	+8551	S C	2391	12,3	34W	11S	16,9*		FE/MA	19
1963A	ANON	1524,9	+2637	S PEC	10400	17,5	10W	1N	18,1*		JAN	12
1963B	ANON	1509,0	+0526	S	11800	15,3	3E	6S	17,0*	I	JAN	12
1963C	ANON	1255,3	+2809	E	6050	15,3	5W	6N	15,7	I	JAN30	12
1963D	N4146	1207,8	+2642	SBB		13,8	19E	27S	15,8	I	JAN20	12
1963E	I1703	0123,8	-0154	S	5688	14,9	2E	11S	16,5*	I	JAN	12
1963F	ANON	1520,3	+2801	S		17,7	1W	7S	V17,6*		MAR	12
1963G	I3112	1215,3	+2618	S C		15,3	7E	17S	15,8*		FEB	12
1963H	ANON	1519,9	+0534	S		15,2	6W	5N	18,5*		MAR	13
1963I	N4178	1210,3	+1108	S C	233	12,9	10W	22S	14,3*	I	MAY	13
1963J	N3913	1148,0	+5537	S	831	14,2	5W	12S	13,7	I	MAY26	13
1963K	N3656	1120,8	+5407	S PEC		13,4	11E	19S	15,0*		JUN	13
1963L	ANON	0244,8	+3720	S(B)C		14,3	21W	27S	15,0*	I	JUN	13
1963M	ANON	1255,5	+2820	SO P		16,3	2W	7S	V15,9)		JUN	13
1963N	N 536	0123,6	+3427	SB		13,2	28W	7S	17,7*	(II)	JUN	13
1963O	N5905	1514,0	+5542	SBC		13,6	60W	9S	16,0*		AUG	13
1963P	N1084	0243,5	-0747	S C	1465	11,2	33E	8S	B14,0	I	SEP28	13
1963Q	I1195	1604,4	+1719	S	29881	15,4	8W	3N	17,0*		SEP	13
1963R	ANON	0233,2	+3544	S		17,5	2W	1S	16,8*		OCT	14
1963S	ANON	0121,0	+0118	S		15,7	5E	5S	15,0*		DEC	14
1963T	ANON	0121,6	+3347	S		17,7	000	000	17,5*		OCT	14
1963U	ANON	0950,3	+3618	SBC		14,9	2W	9N	15,0*	PEC)	DEC	14
1963V	ANON	0232,0	-0618	IRR		16,5	6W	5S	16,0*		SEP	14
1963W	ANON	0227,8	+4344	S		17	8E	3S	17,0*		AUG)	15
1964A	N3631	1118,2	+5327	S C	1087	11,0	74W	92N	17,0*	(V)	FEB	14
1964B	ANON	0314,8	+4011	S		16,3	22W	12S	16,7*	I	FEB	14
1964C	ANON	1603,8	+1735	SO		16,5	7W	5S	17,3*	I	FEB	14
1964D	N4887	1258,1	-1424	SO		14,8	11E	4S	16,5*		FEB	14
1964E	ANON	1156,6	+5259	SBC P		14,5	83W	44S	12,9*	I	MAR 2	15
1964F	N4303	1219,3	+0445	S C	1671	10,9	28W	1S	14,0*	I	JUN	15
1964G	ANON	1621,9	+3920	S	8365	15,7	3E	11N	16,0*		MY/JU	15
1964H	N7292	2226,1	+3003	S/IRR	934	13,1	28W	9S	14,8*	(II)	JUN	15

1964I	ANON	1621,4	+4120	SBB		18,0	17E	18N	18,0*		SEP	156
1964J	ANON	0033,6	-1010	SC/IR	4866	14,9	4E	10S	17,0*		OCT	157
1964K	ANON	2312,2	+0726	S C	4648	15,7	1W	8S	18,0*		NOV	158
1964L	N3938	1150,2	+4423	S C	874	11,0	3W	31N	13,3	I P)	DEC 8	159
1964M	ANON	1024,5	+2042	S	5763	16,3	14W	5S	18,0*		DEC	162
1964N	ANON	0223,6	+2936	S		15,7	24W	000	16,0*		SEP)	195
1965A	N4410	1223,9	+0917	S PEC		13,6	17E	10N	16,0*		JAN)	160
1965B	ANON	1248,3	-1405	S B		14,2	6W	8N	16,0*		JAN)	161
1965C	ANON	1140,7	+1848	S		16,3	1E	8S	18,0*		JAN	163
1965D	ANON	1037,1	-2739	S		14,5	17E	29N	14,0*		FEB	164
1965E	ANON	0947,2	+3439	S C		16,9	36W	5N	16,4*		MAR	165
1965F	ANON	1608,4	+1911	S(B)A		16,6	11W	2N	18,0*		MAR	166
1965G	N4162	1209,3	+2424	S C	2546	12,6	19E	23S	14,0*		MAR	167
1965H	N4666	1242,6	-0010	S C	1645	12,0	22W	29S	14,0*	II	MAY	168
1965I	N4753	1249,8	-0056	SO P	1364	11,7	98W	68N	13,5*	I	JUN	170
1965J	N1310	0319,1	-3719	S C P		14,8	14E	7S	18,0*		AUG	171
1965K	ANON	0226,4	+3115	S A P		14,5	18W	8S	16,0*		SEP	172
1965L	N3631	1118,2	+5327	S C	1087	11,0	38W	58N	16,0*		SEP	173
1965M	N7606	2316,5	-0846	S C	2341	11,6	30W	50N	16,0*		OCT	174
1965N	N3074	0956,7	+3538	S C		14,8	6E	9N	15,8*		DEC	175
1965O	ANON	1159,8	+5012	NOT CONFIRMED								197
1965P	N2599	0829,2	+2244	S A		13,4	13E	5N	15,7*		MAR	210
1966A	ANON	0911,6	+4707	S	4212	14,0	3E	10S	15,5*		JAN	176
1966B	N4688	1245,2	+0436	SBC		14,5	24W	38S	B15,0*		JA/FE	177
1966C	ANON	1627,5	+4106	S	10334	16,0	12E	4N	16,8*		JAN	178
1966D	ANON	1133,6	+2048	E	6533	15,5	12W	8S	19,5*		JUN	179
1966E	N4189	1211,3	+1342	S C	2044	12,7	20W	36N	14,8)	(II)	JUL	180
1966F	N4453	1226,2	+0647	S(B)C		15,4	7E	6S	V17,5*		JUL	181
1966G	N 521	0122,0	+0128	SBB		12,9	60E	13N	15,5*		AUG	182
1966H	ANON	0118,0	+0310	S C		17,0	6E	1S	19,0*		SEP	183
1966I	ANON	0031,8	+3007	S C	6062	17,0	1W	000	16,0)		SEP15	184
1966J	N3198	1016,9	+4549	S C	649	10,7	100W	1658	13,C*	I	NOV)	186
1966K	ANON	1115,6	+2833	S	9808	14,7	26W	15N	17,8*		DEC	187
1966L	ANON	0108,3	+3335	S C		15,5	7W	3S	17,6*		DEC	188
1966M	ANON	0234,0	+3756	S B		16,5	1E	5S	18,5*		DEC	189
1966N	ANON	0434,3	-0308			16,0	19E	17N	15,0*		OCT	198
1966O		1113,5	+2938	INTER-GALACTIC - NATURE UNCERTAIN					15,5*		FEB)	---
1967A	ANON	0113,9	+0245	IRR)		18,5	5E	000	17,0*		JAN)	190
1967B	ANON	1115,8	+0408	IRR)		18,5	3E	000	14,5*		JAN	191
1967C	N3389	1045,8	+1248	S C	1276	12,0	43W	44N	13,0	I	FEB28	192
1967D	ANON	1151,7	+2018	S B		15,6	12W	9N	19,5*		FEB	193
1967E	ANON	0223,2	+4238	IRR)	26544	17	18W	6N	18,0*		FE/MA	194
1967F	DOES NOT EXIST								.			199
1967G	ANON	1603,9	+1828			18	1E	5N	19,5*		MAY	200
1967H	N4254	1216,3	+1442	S C	2468	10,2	80E	19S	14,6*	(II)	JN/JL	201
1967I	ANON	1526,0	+2849	S B		16,5	3W	6N	18,5*		JUN	202
1967J	ANON	0108,3	+3258	S B		15,7	9W	9N	17,5*		NOV	203
1967K	ANON	0126,9	+3335	S		15,6	13W	14S	15,0*		DEC	204
1968A	N1275	0316,5	+4120	E PEC	5160	13,0	7E	24S	15,5*	I	JAN	205
1968B	N4874	1257,2	+2814	SO	7171	13,7	12W	1N	17,4*		FEB	207
1968C	ANON	1057,5	+2659	IRR		18,0	1E	2S	17,8*		FE/MA	208
1968D	N6946	2033,8	+5959	S C	80	10,5	45E	20N	13,5*	II	FEB	209
1968E	N2713	0854,7	+0306	SBB		12,9	6E	14S	13,5*		MAR	211
1968F	N3834	1141,0	+1922	E		15,1	8W	11N	V16,2*		FEB	212
1968G	ANON	0813,0	+2037	S C P	12800	17,5	12E	2S	17,4*	I	APR	213
1968H	ANON	1255,9	+2724	S B	12050	16,6	9W	2N	16,6	I	MAY 7	214
1968I	N4981	1306,1	-0631	S C		12,3	2E	7N	13,5*	I	APR	215
1968J	ANON	1404,1	+5322	SO		15,4	2W	12S	16,6)		MAY	216
1968K	ANON	1600,8	+1719	S)		18	8E	000	17,5*		MAY	217
1968L	N5236	1334,3	-2937	S C	506	8,0	5W	000	V11,*)	II	JUL	218

1968M	ANON	2318,5	+1457	S B		18	5W	7S	15,4		JUL31	21
1968N	ANON	0022,6	+2946	SB	4773	15,5	34E	19S	19,0)		JUL25	22
1968O	ANON	0022,0	+2900	S	4749	14,8	7W	6N	15,8*		AUG 8	22
1968P	ANON	2335,1	+2632	S A		16,5	5E	3S	V16,8*		JUL	22
1968Q	ANON	0116,3	+0440	S		19	5W	5N	18,5*		AUG	22
1968R	ANON	0127,3	-0257	SO			9W	13S	15,5*		AUG	22
1968S	ANON	2142,3	+0251			16,5	2E	2S	16,5*		SEP)	22
1968T	ANON	1625,3	+4122	S C		15,2	2E	14S	V18,0*		OCT	22
1968U	N4183	1210,7	+4358			13,5	20W	95N	14,5*		OCT)	22
1968V	N2276	0710,5	+8551	S C	2391	12,3			V15,7*	(II)	JAN	22
1968W	N2276	0710,5	+8551	S C	2391	12,3			16,6*		MAR)	23
1968X	N4939	1301,7	-1005	S B		12,2	35W	8N	16,0*		NOV	23
1968Y	ANON	0129,7	+3203	S		15,7	1W	5N	18,5*		DEC	23
1968Z	N7768	2348,4	+2653	E		14,0	24W	61N	17,9*		SEP)	25
1969A	ANON	0334,9	-3625	S A		16,0			17,*)		JAN	23
1969B	N3556	1108,5	+5556	S C	680	10,7			V16,*)		JA/FE	23
1969C	N3811	1138,6	+4758	S	3120	13,0	9E	6N	13,7	I	FEB	23
1969D	ANON	0515,0	+0547	S		16,5	2W	5N	17,5*		FEB	23
1969E	N4526	1231,5	+0758	SO	447	10,6	11W	27S	16,0*		MAR	23
1969F	ANON,	1317,4	-1652	DBL E		16,0	4W	10S	16,0*		APR)	23
1969G	ANON,	1231,2	+0610	E		17	4E	7N	18,0*		APR	24
1969H	N4725	1248,0	+2546	S(B)B	1114	10,2	18E	10N	15,0*	I	JUN)	24
1969I	ANON,	1555,6	+1937	S		18	000	9N	17,0*		JUN	24
1969J	ANON,	0123,7	+3122	S B		15,0	5E	7N	17,0*		OCT)	25
1969K	ANON,	2338,2	+2633	SO		15,1	13E	19S	17,5*		NOV)	25
1969L	N1058	0240,3	+3708	S C	480	11,8	190E	110S	12,8	II	DEC 4	25
1969M	ANON,	0326,5	+3951	S		16,5	15E	1S	17,5*		DEC)	25
1969N	ANON,	0127,8	-0112	IR,DW		19	9W	4S	16,0*		DEC	26
1969O	ANON	0103,8	+0253			18,	6W	7S	18,2*		SEP	27
1970A	I3476	1230,2	+1420	S C		13,5	17W	26S	14,0		FEB	26
1970B	ANON	1050,4	+1421	S/IRR		16,5	1E	7N	15, *		MAR	26
1970C	ANON	1258,0	-0612	S C		14,5	9W	29N	16, *		MAR	26
1970D	ANON	1035,2	+1101	S(B)B		15,1	6E	12N	17, *		APR	26
1970E	ANON	1201,5	+5212	S		17,5	2E	5S	18,5*		MAY	26
1970F	ANON	1148,1	+5320	S		16,5	4E	5N	19,0*		MAY	26
1970G	N5457	1401,5	+5435	S C	266	8,7	97W	370S	11,0	II	JUL	27
1970H	ANON	2220,2	+3546	S C		16	13W	15N	17, *		AUG	27
1970I	ANON	2315,1	+0544	S		16,0	12W	5S	18,0*		SEP	27
1970J	N7619	2317,7	+0755	E 3	3757	12,7	27W	30S	14,5		SEP	27
1970K	ANON	0239,4	+3620	S A		16,7	<1E	8S	18,5*		OCT	27
1970L	N2968	0940,2	+3210	IRR/P	1608	13,1	120E	75N	15,5*		OCT	27
1970M	ANON	1045,6	+1419			16,5	13E	1N	16,5*		NOV	27
1970N	ANON	0102,0	-3523			.	000	17N	18,8*		AUG)	29
1970O	ANON	1223,8	+2802			.			13,7*		FEB)	35
1971A	ANON	1115,6	+2833	S	9808	14,7	18E	13S	16,5*		JAN	280
1971B	ANON	1149,9	+4924	S C		16,0	22W	20S	17,5*		JAN	281
1971C	N3904	1146,7	-2902	E 3		11,9	167W	60S	V15,3*		JAN)	285
1971D	N5861	1506,4	-1108	S C		12,4	33E	2N	V15,5*		FEB)	286
1971E	ANON	1523,1	+2633	S		16,5	9E	8S	17,5*		MAR	289
1971F	ANON	1410,9	-3222	S B		16,0	44W	6N	17,3*		MAR	293
1971G	N4165	1209,7	+1331	S C		14,7	3E	30S	14, *		APR	295
1971H	ANON	1118,2	+2834	S		18,	11E	30S	18, *		APR	298
1971I	N5055	1313,5	+4217	S B	+519	9,7	2W	147S	V11,8*	I	JUN	299
1971J	ANON	1203,5	+1406	IRR)		18,5	1E	1N	18, *		MAY	300
1971K	N3811	1138,6	+4758	S C	3120	13,0	30W	17S	16,1*		JUN)	302
1971L	N6384	1730,0	+0706	S B	1751	13,2	27E	20N	V13,0	I	JUN)	303
1971M	ANON	1658,6	+6616	S		17,	13E	6S	19,0*		AUG)	304
1971N	ANON	0009,7	+2822	S		16,5	5W	5N	17,0*		AUG)	305
1971O	ANON	2325,9	+1811	S		16,5	7E	2S	16,8		AUG)	306
1971P	N7319	2233,8	+3343	SBB	6657	14,8	27E	18S	16,8*		AUG)	307

971Q	ANON	0107,9	+3118	S		16,5	8W	3N	18,5*		AUG)	308
971R	I4798	1853,6	-6210			.	12SE		13, *		OCT)	309
971S	N 493	0119,6	+0041	S C		13,0	48E	38N	15,5*		NOV	310
971T	N1090	0244,0	-0027	SBC		12,8	17W	13S	V16,0		NOV)	311
971U	ANON,	1045,7	+2651	S C		14,5	10E	3S	V15,5*	I	DEC	312
972A	ANON	1512,6	+2750	S		16,5	1E	1N	17,0*		JAN	313
972B	ANON	1111,9	+5446	S		17,	6E	3N	V18,5*		JAN	314
972C	N3947	1150,8	+2102	SBC		14,2	28E	9N	16,0*		JAN	327
972D	ANON	1039,0	+1226			18,5	2E	1N	18,5*		APR)	332
972E	N5253	1337,1	-3124	IRR	403	10,8	38W	100S	B 8,5	I	MAY11	336
972F	ANON	1204,6	+5357	S 0		16,0	17E	15S	16, *		MAY	337
972G	ANON	1719,8	+6558	E		18,5	7W	6N	18,0*		MAY	338
972H	N3147	1012,7	+7338	S B		11,3	31E	37N	15,4*		AUG)	344
972I	ANON	2312,7	+1442	SBB		16,0	7W	9N	17,5*		JUL	345
972J	N7634	2319,1	+0836	S 0		13,7	5W	30S	B14,3*		AUG	346
972K	ANON	0106,2	+3157	S		16,5	1W	28N	16,0*		SEP)	347
972L	N 735	0153,7	+3356	S		13,9	28E	21S	15,0*		SEP)	348
972M	N7564	2313,1	+0701	S		15,6	3E	6S	18,0*		OCT)	349
972N	ANON	0025,7	+3032	S(B)C		14,3	41W	22N	V17,0*		OCT)	350
972O	ANON	0516,4	+0557			18,	1W	5S	V18,5*		OCT)	351
972P	ANON	0116,9	+0016	S		17,5	1E	6N	V18,0*		OCT)	352
972Q	N4254	1216,3	+1442	S C	2468	10,2	15E	97N	15,8*		DEC	353
972R	N2841	0918,5	+5112	S B	631	9,9	46W	70S	16,0*	IPEC	DEC)	354
973A	N3574	1109,5	+2753	E		15,7	2W	30N	V18,0*		JAN)	355
973B	ANON	1512,7	+0255			16,5	27W	16S	V15,0*	I	JAN	356
973C	N3656	1120,8	+5407	S PEC		13,4	11W	6S	V17,0*		JAN)	357
973D	N3570	1109,4	+2751	E		15,0	16E	10S	V17,0*		JAN)	359
973E	ANON	1254,5	-0445	IRR		16,4	000	4S	16,2*		APR)	372
973F	N4944	1301,5	+2828	S0		13,3	27E	3N	V17,0*		MAR	373
973G	ANON	1129,2	+5352	S0		16,5	42W	9S	V18,0*		APR)	376
973H	ANON	1202,9	-0402	SBA		15,7	3E	16S	18,5*		APR)	377
973I	ANON	1315,7	-0640	IRR		19,5	6E	14S	20,5*		APR)	378

piece of information which, potentially, has a very powerful bearing on this question is the distribution in integrated color of the parent galaxies. If supernovae result from the evolution of high mass stars then we would expect that blue galaxies of a given mass should have a higher supernova rate than red spirals. Now spiral galaxies exhibit a fairly large range in intrinsic *UBV* colors (de Vaucouleurs, 1961). This in turn may be understood in detail by supposing that all spirals have the same average mass distribution for the stars at birth and a range in the rate of fall-off of the rate of star formation with time (Searle *et al.*, 1973). According to this interpretation the bluest spirals have experienced a roughly constant rate of star formation for the past 10^{10} yr while in the reddest spirals the present rate of star formation is a few percent of the initial rate 10^{10} yr ago. We should therefore expect blue galaxies of a given luminosity to have a higher supernova rate than redder ones. Is this borne out by observation? Unfortunately the data on the colors of those galaxies in which supernovae have been observed are rather sparse. They are summarized in Table VII, in which we list in line 2 the number of galaxies for which de Vaucouleurs (1961) gives colors in each 0.1 mag. interval of $(B-V)_0$, the observed integrated color corrected for internal reddening. In line 3 a similar distribution is shown for those galaxies of known color in which supernovae have been observed. In line 4 the galaxies of a given color with supernovae are expressed as a percentage of all galaxies with that color. Corresponding data for type II supernovae alone are given in brackets in lines 3 and 4. Line

5 of Table VII shows how the supernova frequency should depend on the integrated color of the parent galaxy according to the calculations summarized in Table II of the paper by Searle, Sargent and Bagnuolo. (The quantity in line 5 is proportional to the quantity R_B in that paper.)

It will be seen from Table VII that the variation of supernova frequency with the color of the parent galaxy is in satisfactory agreement with the theoretical prediction for the type II supernovae. That is, the data in Table VII are consistent with the idea that fairly massive stars become supernovae of type II.

However, there is a discrepancy between observations and theory for supernovae of type I, which, are much too plentiful in red galaxies. The masses of the parent stars of type I supernovae remain a mystery. The fact that these objects are found in very red galaxies would imply most straightforwardly that the parent stars are of fairly low mass – say around one solar mass. However, all four supernovae of well established type which have occurred in Irr I galaxies – 1895A and 1972E in NGC 5253, 1937C in IC 4182, and 1954A in NGC 4314 have been type I objects. The predominance of type I supernovae in Irr I and E galaxies is a puzzle; perhaps they result from some rare kind of binary system which involves low mass stars.

In any case it is fairly certain that studies of supernova frequency should take into account the color of the parent galaxy as well as its luminosity. For this reason we badly need more data on the colors of galaxies and more determinations of the types of observed supernovae.

Acknowledgments

We took over the Palomar supernova search from F. Zwicky following his (formal) retirement in 1968. It is a pleasure to thank him for his advice and encouragement since then. We also thank the National Science Foundation for their continuing support.

References

Kowal, C. T. and Sargent, W. L. W.: 1971, *Astron. J.* **76**, 756.
Searle, L., Sargent, W. L. W., and Bagnuolo, W. G.: 1973, *Astrophys. J.* **179**, 427.
Tammann, G.: 1970, *Astron. Astrophys.* **7**, 289.
Vaucouleurs, G. de: 1961, *Astrophys. J. Suppl.* **5**, 233.

DISCUSSION

F. Zwicky: You say that the increase in your Table V of the number of galaxy years necessary for the discovery of a supernova is due to the fact that the observers missed most of the supernovae in $m_p = 14.5$–15.7 mag. galaxies. This is totally unrealistic. Your effect is actually mostly due to the fact that most of the apparently fainter galaxies are also intrinsically fainter and produced relatively few supernovae.

From your table the average old frequency of 1 supernova per 300 yr in galaxies of the absolute range $\sim -20 < M_p < -15$ remains confirmed.

L. Searle: I disagree. The data clearly show that supernovae are being lost in the faintest magnitude intervals. It is clearly difficult to extract an overall supernova rate until these selection effects are better understood but a preliminary attempt to do so supports Tammann's conclusions.

G. A. Tammann: When Zwicky claims completeness of his search at a level of $m_{SN} = 19^m$ he must refer to the expected observable number of SNe, essentially as calculated in 'Stars and Stellar Systems'. These numbers have little to do (at large distances) with the true cosmic number of SN events. Sargent's numbers show convincingly that the known sample of SNe falls far short of the number of SNe which actually must be out there. I agree with Sargent's conclusion that there is very considerable loss at $m_{Gal} \sim 15^m$.

M. Friedjung: Are the masses of the galaxies in each column of your last table comparable? If not, one could not compare the supernova rate for galaxies of different colours.

L. Searle: The redder galaxies tend to be more massive so that if we are able to make mass into account (which one is not, because of inadequate information) I think the corrected observations would show the color term in an even more pronounced way.

SUPERNOVA SURVEY AT THE KONKOLY OBSERVATORY

L. DETRE
Konkoly Observatory, Budapest, Hungary

Speaking immediately after the Palomar report I feel proper to be very short and modest.

The Konkoly Observatory is situated only 8 km from the centre of Budapest, but in 1962 we installed a mountain station 100 km to the north-east of the city in the Mátra-mountain on the so-called Piszkéstető at an altitude of 960 m above sea-level. The station is equipped with a 60/90/180 cm Schmidt-telescope made by VEB Zeiss Jena. It is provided with two objective-prisms of UBK7 glass having refractive angles 5° and 2°, respectively. The field is circular with a diameter of 5°.

About a third of observational time in the moonless nights is devoted to supernova search, that means 35–40 nights annually. Other programs with the Schmidt-telescope are: spectral classification and *U*, *B*, *V*, photometry in open clusters and associations and search for flare stars in the Pleiades and Praesepe.

Observations started at the end of 1963. For the supernova survey two plates are taken in succession for every field with an exposure time of 15 min. On Kodak 103a-O plates the limiting magnitude is about 19.

The first supernova, 1964e, was discovered by Lovas on 13 March, 1964. According to two unwidened spectra obtained by B. Balázs with the aid of the 5° objective prism the supernova was of type I, a few weeks after maximum light (IBVS Nos. 50 and 53, 1964).

Till now 15 supernovae have been found, 10 by M. Lovas, 3 by I. Jankovics, 1 by L. Balazs, and the last one two weeks ago by Miss Mnatsakanian (Byurakan Observatory) while working at our observatory. She asked us to show her the plates on which supernovae were found. In her opinion there is a psychological effect: when somebody discovers a supernova, easily overlooks an occasional second one on the same plate. Indeed, she discovered on a plate on which Lovas had found SN 1967i, a second supernova in NGC 4975 (IBVS No. 785, 1973).

One supernova was found by Lovas on Field No. 155 of the Palomar Sky-Survey prints by comparing it with Schmidt-plates of our observatory (IBVS No. 612). The supernovae 1966g, 1967c, 1969b and 1969e were independent discoveries with Reaves, Chuadze, Wild and Rosino, respectively. One supernova discovered by Jankovics is not in the master list as the telegram announcing the discovery and sent to 'Satellites New York' did not arrive at Cambridge, Mass.

Among the supernovae found at our mountain station two are especially interesting: 1968a and 1970g, both discovered by Lovas. 1968a appeared in the active galaxy NGC 1275 (Per A) and is the only supernova found in a Seyfert-galaxy. 1970g is the brightest supernova of type II so far found, it appeared about 7″ to the northwest of the H II region NGC 5455 in M 101.

C. B. Cosmovici (ed.), Supernovae and Supernova Remnants, 51–53.

The plates on which SN 1970g was found, were taken under the most unfavourable conditions between clouds and at a zenith distance over 70°. They are the two weakest plates in our collection. I show you a part of this plate to demonstrate how observations under very bad circumstances can sometimes be very useful in supernova search. Of the splendid galaxy only the nucleus and the brightest parts of the inner spiral-arms can be seen.

Finally, I should like to make two remarks on the international supernova program initiated by Professor Zwicky.

(1) A great number of fields are under observations, for instance 80 fields in our program. That is very effective for increasing the number of discoveries, but in this way only fragmentary data are obtained for most of the discovered supernovae. Beside the general program 5–6 areas should be selected and every observatory participating in the supernova program should observe these special fields as many times as possible. I recommend the Perseus and Coma clusters, and some parts of the Virgo and Ursa Maior clusters. We could arrive at an agreement at this conference, or Professor Zwicky could be so kind to give us a final list. In this way, for the supernovae appearing in the selected regions their complete history could be recorded, in some cases the important ascending branch too.

(2) For a sound statistics of the frequency of supernovae not only the discoveries

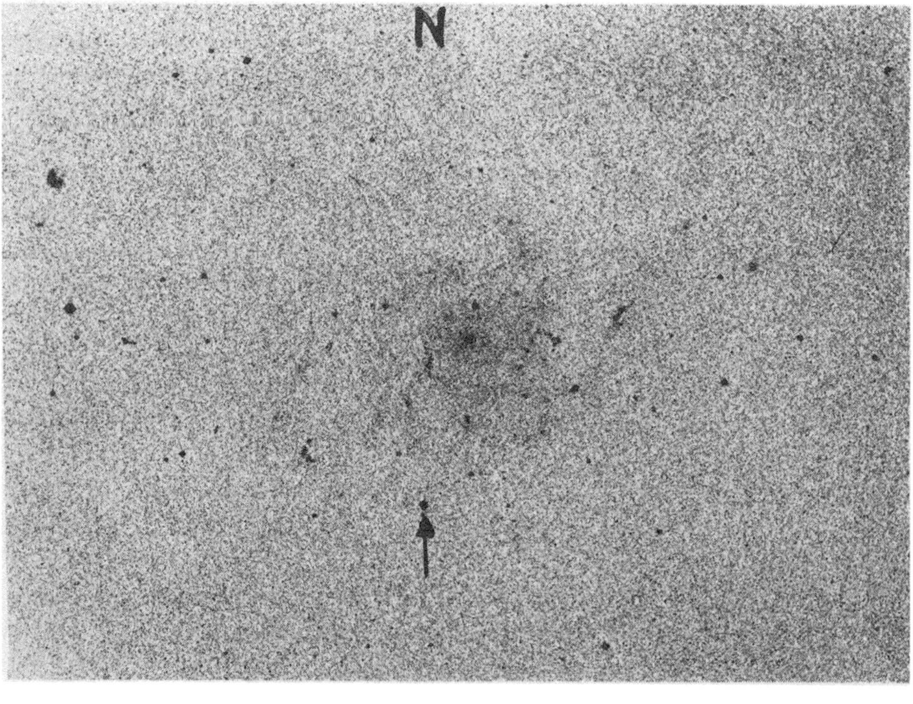

Fig. 1. The discovery photograph for SN 1970g in M 101 taken by Lovas on July 30, 1970. The supernova together with the H II region NGC 5455 is indicated by the arrow.

should be published, but also the number, position and dates of the plates on which no supernovae were found.

DISCUSSION

M. F. McCarthy: Do several of your astronomers examine independently the SN survey plates? I ask this to see if it is possible for one to miss an object on a plate (through fatigue etc.) which a colleague might recover in an independent examination of plate survey material.

L. Detre: At present only Lovas is working in the SN-program at our observatory.

W. M. Goss: Is this the extension on your photograph of 1970 due to the H II region NGC 5455 in which the SN is inbedded?

L. Detre: The extension is due to the SN itself. It was situated a little eccentrically in the association.

F. Zwicky: We should not forget to form a cooperative between the observatories engaged in the supernova search for buying emulsions from Eastman in large enough numbers, the cost for which are too great for a single observatory (best II a-0 or 103 a-F and newer emulsions). Father McCarthy agrees to this.

L. Detre: I suppose, all participants agree with this important suggestion.

A SEARCH FOR RADIO EMISSION FROM YOUNG EXTRAGALACTIC SUPERNOVA REMNANTS

A. G. DE BRUYN
Sterrewacht Leiden, Leiden, Holland

In the radio domain supernova remnants (SNR) are among the brightest discrete sources in our galaxy. Although their radio luminosity (10^{33-35} erg s^{-1}) is only small compared to the optical luminosity of the supernova (SN) event itself, the radio remnants are still of interest because the radio emission, caused by the synchrotron mechanism, points to the existence of a large energy reservoir in relativistic electrons and magnetic fields, typically of the order of 10^{49} ergs. No satisfactory theory thas yet been proposed for the origin of this energy reservoir and since some young SNR already exhibit many diverse properties, it is even doubtful whether one unique theory exists or should be looked for.

An observational way of attacking the problem of the injection of high-energy electrons and the origin of the magnetic field is by studying the radio evolution of very young SNR because of the expected relation between the radio and energetic evolution. On the basis of SNR-statistics a few SNR younger than Cas A (age about 250 years) certainly are to be expected in our Galaxy but we can not be sure of their existence nor do we know where to find them. Thus the only place to look for them is in external galaxies. The search for optical SN in external galaxies now has provided us with a list of more than 300 SN with precisely known positions and which have ages from 0 till 86 years. We have used the Westerbork Synthesis Radio Telescope to search for radio emission at the sites of in total about 35 historical SN. Some relevant data about the telescope and the observing period are given in Table I.

TABLE I
Some data about the WSRT

observing technique	earth rotation aperture synthesis
frequency, bandwidth	1415 MHz, 4 MHz
half-power beamwidth	$24'' \times 24''/\sin\delta$
rms-noise after 1×12^h	$\simeq 1 \times 10^{-29}$ W m^{-2} Hz^{-1}
observing period	sept. 1970–sept. 1971

Results

Of the optically detected SN none of the remnants has been detected above the three-sigma level. Typical upper limits to their flux density at 1415 MHz are $2–5 \times 10^{-29}$ W-m^{-2} Hz^{-1}. (The only positive detection (SN 1970g), made one year after the present survey ended, is reported on by Goss at this conference.) All the observed galaxies, their SN and the determined upper limits can be found in de Bruyn (1973). In a number

C. B. Cosmovici (ed.), Supernovae and Supernova Remnants, 55–59.

of the observed galaxies (M 51, NGC 4258, IC 342 and NGC 3184) weak radio sources of probably nonthermal character were detected. The two sources in M 51 and NGC 4258, for which the nonthermal character has been established already from unpublished 6 cm observations with the WSRT, are lying exactly on the well-defined spiral arms close to a giant H II region. This makes it likely that they are (type II) SNR and their radio luminosity then is of the order of 10 times Cas A. The fact that no optical SN event has been reported on these positions and the negative results for all the historical SN could indicate that these sources are SNR of age greater than about 50 years, viz. the corresponding SN events took place before the extragalactic SN search was started. We come back to the possible significance of these detections below.

What are the upper limits to the intrinsic radio luminosities of the historical SN? If we take the distances of the parent galaxies as given by de Vaucouleurs' (1970) group-classification and assume a distance of 3.4 kpc for Cas A, the upper limits of a dozen of the nearer ones are found to lie between 2 and 7 times that of Cas A. Integrated over the radio spectrum this means less than about 10^{36} erg s^{-1}. An exception is S And, the 1885 SN in M 31, for which an upper limit of 0.06 times Cas A could be established. If we compare these results with the empirical 'evolution-law' of (much older) galactic SNR and with the present radio evolution of Cas A, it appears that the upper limits fall significantly below extrapolations of these relations to smaller radii. This is shown in Figure 1 where the results of a dozen type I and type II SN are plotted. (the radii of these SNR are calculated from their age and an expansion velocity of 12500 and 7500 km s^{-1} respectively).

Since time is limited I shall not try to evaluate what these results mean for specific evolution models of young SNR proposed so far. I will, however, discuss some possible absorption effects in young SNR which also may be relevant to the detection of these socalled SNR-candidates in some nearby galaxies. Three well-known absorption mechanisms in a compact radio source embedded in or surrounded by a thermal plasma are: (i) synchrotron self-absorption, (ii) Razin-Tsytovich absorption and (iii) free-free absorption. It is not difficult to show that effects (i) and (ii) can only be important for SNR of at most a few years old and can not be the cause of the absence of radio emission of most of the observed SNR. Free-free absorption in the ionized SNR shell, on the other and, can be important for a much longer time. For example, for a type II SNR shell with the following parameters: mass $\sim 2\ M_0$, filling factor (due to filaments) $\sim 10^{-2}$, expansion velocity 7500 km s^{-1} and electron temperature 10^4 K, the optical depth for free-free absorption at 1415 MHz is greater than unity up to an age of 50 years. For a type I SNR with mass $\sim 0.5\ M_0$ and expansion velocity 12500 km s^{-1} this becomes 17 years. As most of the observed SNR had ages in the interval 10 to 50 years and since the various parameters have large uncertainties, we can not be sure (except perhaps for a few 50-year old type I SNR) that the observed upper limits are intrinsic to the radiation source, i.e. not influenced by absorption. In the presence of free-free absorption at 1415 MHz, the inverted radio spectrum offers us a better chance of detection of 25–50 year old SNR at higher radio frequencies.

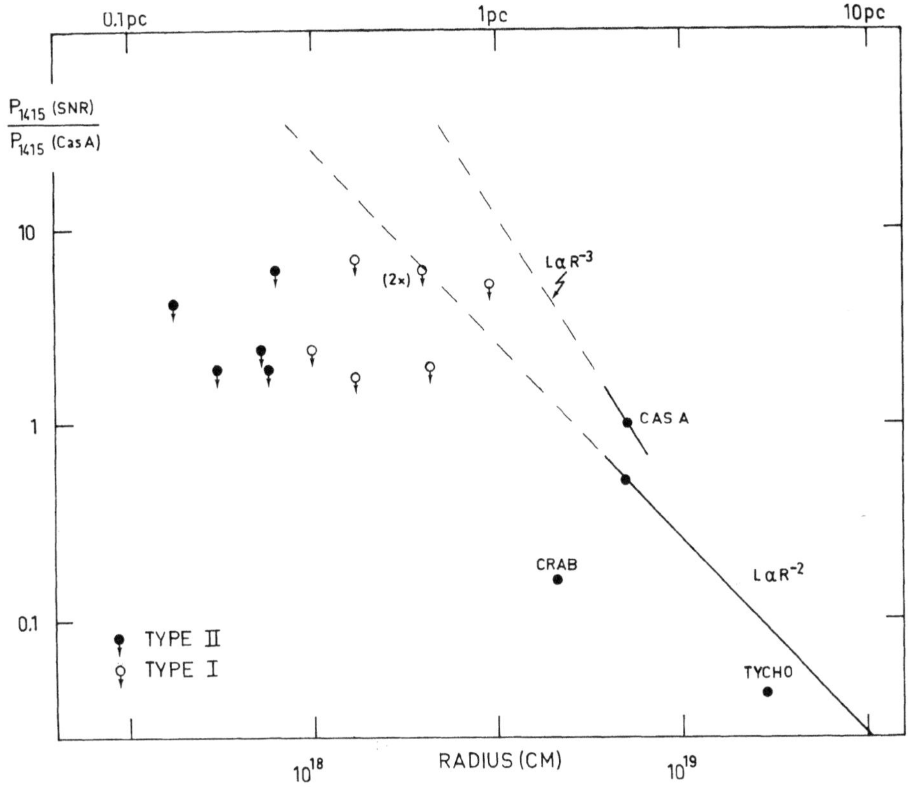

Fig. 1. Representation in the luminosity-radius plane of the empirical evolution law of old galactic SNR and Cas A. The plotted upper limits are the results for a dozen extragalactic SNR aged 20-75 years.

However, as the following argument shows, the chances of detection are still better than for older SNR at 1415 MHz or lower frequencies. Assume that in the optically thin part of the spectrum the following relations hold: $S \propto \nu^{-\alpha}$ and $S \propto t^{\beta}$ where S, ν and t stand for flux density, frequency and time respectively. It is easy to show then that, due to expansion of the SNR shell, the flux density of the peak in the radio spectrum varies like $S_{max} \propto t^{\beta+2.5\alpha}$. For young SNR where generation of relativistic electrons and magnetic fields is probably still taking place (leading to a positive or small negative number for β), the value of S_{max} is then increasing with time for a typical value of 0.75 for α. Note that this is opposite to the case of the well-known variable expanding radio source models. Figure 2, case a, shows a possible spectrum evolution of a young SNR in that case. The consequence of this is that, for a given detection limit in terms of milli-flux-units, we have the best chance of detecting SNR at low frequencies and when they are old. And as mentioned above, we have detected a number of these possibly 100-year old SNR. A prediction of this model is that these SNR-candidates also should show a low-frequency cutoff in the radio spectrum, maybe somewhere around 500 MHz. Observations at 600 and 5000 MHz of the galaxies in

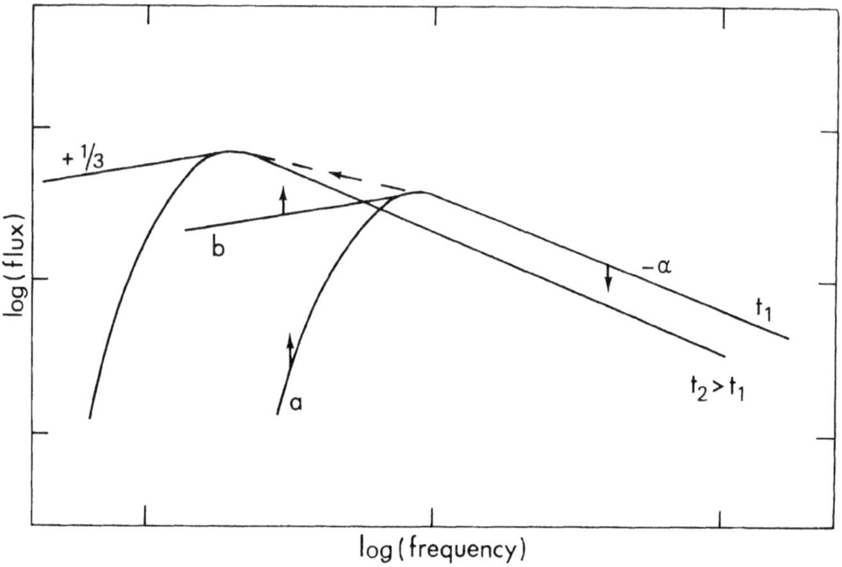

Fig. 2. Possible schematic spectrum evolution of a young SNR for the two cases described in the text: (a) free-free absorption and (b) deficiency of low energy electrons. For simplicity the dashed line has been given the same slope for the two cases. The line $L \propto R^{-2}$ has been taken from Ilovaisky and Lequeux (1972).

which these SNR-candidates were detected are planned or being done. In Figure 2 (case b) we have also drawn a possible spectrum evolution in the case a bend in the case a bend in the spectrum is caused by a low-energy cutoff in the electron spectrum. In this figure we have arbitrarily assumed that S_{max} has the same time-dependence as in the case of free-free absorption. In fact, S_{max} could have a much steeper dependence on time if the lower energy electrons in young SNR are injected at progressively later phases. Combined with a decreasing magnetic field in the SNR shell this then could lead to a very rapidly decreasing frequency at which the spectrum bends to a slope of $+\frac{1}{3}$. This model has been discussed by Kulsrud *et al.* (1972).

Acknowledgement

The Westerbork Radio Observatory is operated by the Netherlands Foundation for Radio Astronomy with the financial support of the Netherlands Organization for the Advancement of Pure Research (ZWO).

References

de Bruyn, A. G.: 1973, *Astron. Astrophys.*, in press.
Kulsrud, R. M., Ostriker, J. P., and Gunn, J. E.: 1972, *Phys. Rev. Letters* **28**, 636.
Ilovaisky, S. A. and Lequeux, J.: 1972, *Astron. Astrophys.* **18**, 169.
Vaucouleurs, G. de: 1970, in D. S. Evans (ed.), 'External Galaxies and Quasi-Stellar Objects', *IAU Symp.* **44**, 353.

DISCUSSION

J. R. Dickel: Could any of your detected pointsources in these galaxies be very old?

A. G. de Bruyn: No, they have intrinsic luminosities of about 10 times Cas A hence it is very unlikely they are older than a few hundred years.

H. Arp: Since your radio luminosity for detection of S And is about a factor of 100 less than the rest of the SN observed are you not surprised you did not observe it?

A. G. de Bruyn: The few data about the lightcurve and the spectrum (colours) of this SN indicated that it did not belong to one of the two main classes of SN. Because it is the only SNR for which I have established such a low upper limit I do not know whether the result is very significant.

H. Arp: Are you saying then that S And is peculiar in radio wavelengths as well?

A. G. de Bruyn: That may well be the case.

AN EXPERIMENT TO SEARCH FOR PROMPT EMISSIONS FROM SUPERNOVAE AT MICROWAVE FREQUENCIES

J. V. JELLEY*
Nuclear Physics Division, AERE, Harwell, Berkshire, U.K.

W. P. S. MEIKLE and R. W. P. DREVER,
Dept. of Natural Philosophy, Glasgow University, Glasgow, U.K.

G. G. C. PALUMBO
Laboratorio di TESRE, CNR, Bologna, Italy

F. BŌNOLI
Osservatorio Astronomico, Bologna, Italy

H. SMITH and T. DELANEY
Physics Department, University College, Dublin, Ireland

1. Introduction

A growing interest in recent years has developed in astronomical phenomena which occur on short time-scales, this interest being sparked off first by the discovery of pulsars, and enhanced by the work of Weber and others, on pulse sattributed to gravitational radiation. As Bondi (1970) has pointed out this new field of 'pulse astronomy', has been until recently somewhat neglected, though interest has at last been stimulated.

While many astronomical objects release vast quantities of energy, and some of them continuously, e.g. QSO's and Seyfert Galaxies, the supernova (SN) phenomenon is unique, in that the energy is released in a single catastrophic event, it arises from an object of only stellar dimensions, and the total duration of the phenomenon is extremely short, on astronomical time-scales. Energies $\sim 10^{48}$–10^{52} ergs are involved (Weekes, 1969) while the timescales may be set either by phenomena in the stellar envelope (minutes to hours), by the collapse time of the core regions ($\sim 1/\sqrt{G\varrho}$, i.e. $\sim$4 s to 4 ms, for $\varrho = 10^8$ and 10^{14} g cm^{-3} respectively), or by shock-wave phenomena ($\sim$tens of μs).

All discoveries of SN to date (Kowal and Sargent, 1971) have been made either by the photographic surveys initiated by Zwicky (1969) involving mainly Schmidt telescopes at a variety of observatories, of which the programme at Asiago here in Italy is one (Rosino and Di Tullio, 1974), or by the relatively new photoelectronic scanning technique (Dunlap *et al.*, 1972) in use at the Corralitos Observatory, U.S.A. As far as the author is aware, no SN has ever been discovered within less than a day of the outburst, and the optical emission which is then recorded lasts over periods of days to months.

We are here concerned with *prompt* emissions from SN, and we wish to describe a coincidence experiment which has been set up, with installations at four widely spaced stations, to search for short-lived bursts of microwave radiation which may occur

C. B. Cosmovici (ed.), Supernovae and Supernovae Remnants, 61–71.

when SN erupt in clusters of galaxies of high population. The four stations fully engaged in this experiment are at the Nuclear Physics Division of the AERE, Harwell, U.K.; the Department of Natural Philosophy at Glasgow University, Glasgow, U.K.; Medicina, the outstation of the Laboratorio de TESRE, CNR, Bologna, Italy; and at the Physics Department of University College, Dublin, Ireland.

Professor R. B. Partridge, Haverford College, Pennsylvania, has also joined us informally for those parts of the observing programme for which the considerable terrestrial longitude differences permit a reasonable overlap time in such a coincidence experiment.

There are essentially two objectives in this very exploratory work. In the first place, if prompt emissions from SN are discovered, in whatever region of the electromagnetic spectrum, some interesting physics will be generated. Secondly, if successful, the technique might be developed so that a new and conceivably more efficient method would be available for detecting SN. Moreover, again if successful, SN might be discovered soon enough for studies of the earlier phases of the light curves and spectra.

A few words on the background of the work. Some years ago Colgate (1968) predicted, on his shock-wave model of SN, that substantial X-ray and γ-ray emissions should accompany SN outbursts, and that the flux levels were such that from the 'general field' of galaxies (out to 75 Mpc) the event rate should be as high as one detectable SN per 8-h night of observation. The X-ray flash was expected to last some tens of microseconds and would be detectable with omnidirectional photodetectors on the ground, through upper-air fluorescence.

Armed with this prediction and realising the technical simplicity of the installations required, three groups set up installations to look for such pulses, namely NASA (Ögelman *et al.*, 1968) the Smithsonian Astrophysical Observatory (SAO) (Elliot, 1972) and AERE. (Charman and Jelley, 1972). All three groups indeed found occasional pulses of light, though for various reasons it was concluded that these were not of celestial origin. At this point the theory was improved (Colgate, 1970), resulting in a vast reduction in the predicted rate, and most of the experiments were abandoned.

Colgate (1971) then showed that SN outbursts should also be accompanied by bursts of coherent R. F. radiation (peaking at $\sim$300 MHz). An experiment was performed (Colgate *et al.*, 1972), with negative results, these being consistent with the modified theory.

In order to detect isolated and rare events of short duration it was soon appreciated that coincidence or correlation techniques, between widely spaced stations, were especially helpful in rejecting local spurious phenomena. The coincidence technique, so familiar in nuclear and elementary particle physics, has indeed been used in recent years by the groups (Meikle *et al.*, 1972; Partridge and Wrixon, 1972) involved in the search for EM pulses which might accompany the events recorded by Weber (1969), attributed to gravitational radiation.

It was realised at the outset that to perform a ground-based coincidence experiment to search for prompt emissions of EM radiation from SN, it would be essential to operate at radio frequencies, otherwise the fractional coincidence overlap time would be intolerably low. At optical frequencies, be this a direct light flash from the SN, or a flash in the ionosphere from X-ray fluorescence, the coincidence efficiency is rather low, even at desert sites, when allowance is made for the fractional period of total darkness, moonlight, and stray man-made lighting etc.

Even in the radio frequency band, coincidence observations at the lower frequencies (metre wavelengths) can be misleading in daytime hours, due to interference from the Sun (Charman *et al.*, 1970) (type III solar bursts). In addition to this, negative results at VHF (O'Mongain, 1973), as well as certain practical considerations to be discussed later, led to the decision to operate at microwave frequencies.

Points in favour of choosing microwave frequencies in this work are the following:

(1) Low absorption and dispersion, in the local plasma around the SN, in the interstellar media in the parent galaxy of the SN and in our own, and in the intergalactic medium. For a critical frequency of 10 GHz ($\lambda \sim 3$ cm) the plasma density would be $n_0 = 1.3 \times 10^{12}$ electrons cm^{-3}.

(2) No interference from the Sun; observations can run on a 24-h schedule.

(3) At microwave frequencies, sources of man-made, atmospheric, and ionospheric interference, are less serions than at VHF.

(4) Large bandwidths are available.

(5) Antennas having beam widths which match the angular diameters of the major clusters of galaxies ($\sim 3°–12°$) are relatively small, are therefore cheap to construct, and are easily adapted for automatic tracking.

Against these factors there are however certain disadvantages in the choice of microwaves; these are listed as follows:

(1) The effective collecting areas of the antennas are relatively small, in relation to the required beamwidths.

(2) If analogy with pulsars is in any way meaningful and if bursts of coherent radiation arise in SN, the yield at microwave frequencies would be expected to be much less than that at lower frequencies.

(3) The spectra of all radio sources, steady sources and pulsars included, all slope so that the fluxes are lower at the higher frequencies. Against this however one should mention that a SN outburst is not an equilibrium situation, so that little can be predicted.

An alternative approach to that of the coincidence technique is to make use of the interstellar dispersion to provide a signature, by searching for differences in arrival time of isolated pulses, at two or more different frequencies. This indeed has been used (O'Mongain, 1973) in recent experiments to look for radio pulses from the galactic centre. However, some evidence has already been obtained for such pulses (Hughes and Retallack, 1973) even without this method.

This technique has however a limitation in our experiments. The problem is that if the two or more chosen frequencies both lie in the microwave bands, the delay-differences may be exceedingly small compared with the duration of the event in the individual bands.

2. Theoretical Considerations

To date, Colgate's shock-wave model of SN (Colgate and White, 1966), the validity of which is in any case disputed by Arnett (1967) and others, is the only model in which any prompt EM radiation is clearly predicted, though bursts of prompt neutrinos have also been suggested as a means of detecting SN. The essential theoretical problem relevant to the present exploratory experiment hinges on whether or not any of the exceedingly energetic phenomena in the rapidly collapsing core regions of the star can be transmitted to or through the outer envelope on the time-scales involved.

In one specific model of stellar collapse, LeBlanc and Wilson (1970) show that for a rotating magnetised star, a kinetic energy dissipation in an axial jet will occur, at a rate $\sim 10^{51}$ ergs s^{-1} for a period ~ 0.3 s, with a mass ejection rate $\sim 10^{32}$ g s^{-1}, over the same period. One crucial question is whether such a jet can penetrate the outer envelope to be in any way effective in producing radiation from particles outside, in a density region such that microwave radiation, if produced, is not absorbed.

On this same model, magnetic field energy dissipation also occurs, at a rate $\sim 10^{50}$-ergs s^{-1} which may also contribute to generating accelerated particles and hence EM-radiation. Since the hope of observing an EM pulse on a short time-scale depends crucially on the outer envelope of the star, the situation may well be very different for the two main types of SN.

A. ENERGY REQUIREMENTS

If we take the LeBlanc-Wilson model of SN and assume a figure of $\sim 10^{51}$ ergs for the total prompt energy released in a SN outburst, in all forms, and assume this occurs in ~ 1 s, we find that with our available sensitivity of 4×10^{-11} ergs cm^{-2} $(\text{event})^{-1}$ (see below Section 4), for a SN going off in a galaxy at a distance of say 10 Mpc, the conversion efficiency must be $\geqslant 2 \times 10^{-9}$ to enable it to be detectable with our installations; this then is the minimum fraction of the total available energy which must appear in the form of microwave radiation within the 40 MHz bandwidth of our receivers, for the SN to be detected, assuming isotropic emission at the source.

B. CONSTRAINTS ON THE TIME SCALE

Cavallo and Ventura (1972) have recently shown from quite general considerations that for a given threshold sensitivity of a receiver, there are restructions on the duration of the pulse, so that the further away the source the longer the pulse. They consider four alternative radiation mechanisms, and conclude that for extragalactic sources it is unlikely that pulses will be detected of duration less than ~ 0.1 s. It should be emphasized however that the above theory does not include the case of relativistic sources, or effects of beaming of the radiation.

3. Supernova Frequencies and the Choice of Clusters

Some years ago the SN frequency was found to be approximately 1 per 380 yr, per galaxy (Zwicky, 1969), though more recent surveys (Katgert and Oort, 1967; Tammann, 1970; Kukarkin, 1965: Wamsteker *et al.*, 1973) have shown that for certain types of galaxy the rate is very much higher, ~1 per 15 yr to 1 per 30 yr. Zwicky's earlier figure is however also supported by Barbon (1963) averaging over all galaxies. Even if the highest of these rates is accepted, it is clearly unfeasible to run the experiment on isolated galaxies, in spite of somewhat enhanced rates among a few selected galaxies, (Zwicky, 1970). However, with the beamwidths available in our small radiometers, it is possible to track continuously on the more concentrated regions of the larger clusters, thus simultaneously encompassing large numbers of individual galaxies.

The following factors, some inter-related, have to be taken into account in the choice of the best regions to search, in order to optimise the probability of detecting an event:

(1) The SN rate as a function of the morphology of the galaxy.

(2) The SN rate related to the Luminosity Class of the galaxy.

(3) The populations, and magnitude distributions, of galaxies in the clusters.

(4) The SN rate as a function of the radial distance from the centre of the parent galaxy (i.e. corrections for selection effects; SN are possibly missed in the conventional surveys by over-exposure of the nuclear regions, and by resolution).

(5) The angular sizes of the clusters, and the radial distribution functions of the number of galaxies within the cluster.

(6) The distances of the clusters.

(7) The declination of the search-region in relation to the latitudes and longitudes of the observing stations, and

(8) The difference in the RA of the cluster and the RA of the Sun during the period of the observing programme on a given cluster.

Factors (1) to (4) largely influence the choice of clusters on purely astronomical grounds, to attain the maximum basic rate, while (5) and (6) determine the required beamwidth, hence the collecting area, and in turn the sensitivity. (7) determines the fraction of time the cluster is above any local horizon and therefore the overall observing efficiency, while (8) is important in relating any observed multi-station coincidences with the established supernova photographic surveys. A rapid follow-up confirmation would be extremely important and this could clearly not be executed if the sun was close to the area in question.

One of us (Bōnoli, 1973) has carried out a preliminary survey of the various factors listed above, using data available prior to this conference, in order to derive 'figures of merit' for the various larger clusters, so as to optimise the frequency of occurrence of SN within the antenna beam. The future observing programme will be determined largely by this and similar surveys, while also taking into account various practical and operational considerations.

The analysis presented here, Table I, is only of course a first attempt and should be revised following the very thorough survey presented by G. A. Tammann at this Meeting.

The clusters included in our analysis, Column A, were selected from the Zwicky

TABLE I

A	B		C	D	E	F	G	H	I				J			J		K	L		M
Cluster	α	δ +	Character	ϕ deg	Distance class	d Mpc	p	p^1	Coma-like (Galaxy Type				Virgo-like (Galaxy Type)			Integral SN. $(yr)^{-1}$ occurring		f %	Integral SN. $(yr)^{-1}$ detectable		Relative signal strength So
									Ir	E	SO	S	Ir	E	S	(Coma like)	(Virgo like)		(Coma type)	(Virgo type)	
Virgo	1224	13.20		15		25	3000	100					2	22	76		5.2	57		3	1
Coma	1257	28.06	C	11	Near	175	2150	1000	50	400	550	30	8			25.8		68	17		0.02
Perseus	0303	41.25	MC	30.	Near	125	5130	600	3	240	330	18	12	132	456	15.4	31.6	83	13	26	0.04
Zwicky Catalogue	1054	20.17	MC	4.3	MD	~ 500	1134	550	3	220	300	16	11	121	418	14	28.9	62	9	18	0.002
	1105	32.57	MC	4.2	MD	~ 500	885	450	2	180	247	13	9	99	342	11.5	23.7	72	8	17	0.002
	1142	21.26	MC	9	Near	~ 200	1895	800	4	320	440	24	16	176	608	20.5	42.1	62	13	26	0.02
	1211	18.30	MC	3.4	D	~ 875	1420	1300	6	520	715	39	26	286	988	33.3	68	61	20	41	0.009
	1422	17.32	Open	4.5	Near	~ 200	803	500	2	200	275	15	10	110	380	13	26	60	8	16	0.02
	1503	18.53	Open	4.4	Near	~ 200	1093	720	4	288	396	22	14	158	574	18	38	55	10	21	0.02

catalogue (1961–1967) and were restricted to those with more than 800 member galaxies, which at the same time had a northern declination, column B. The clusters were then grouped into classes of character, angular diameter φ, and distance, columns C, D and E respectively, with column F representing approximate absolute distances d (assuming the recent value for the Hubble constant, of $H = 40$ km s^{-1} (Mpc)$^{-1}$, (Branch and Patchett, 1973)).

The total populations p are listed in column G, and after allowing for the population distributions over the clusters; the number of galaxies p' which lay within our antenna beam (taken as constant, at 3° diam, to the −3db points) are then tabulated in column H. The treatment of the variation of SN-rate with galaxy type was carried out as follows: For the Coma and Virgo clusters as prototypes, the fractions of the total population of galaxies of different types were taken from the works of Rood and Baum (1968), and Brown (1968), respectively. For the other clusters the analysis was carried out on the assumptions that they were either Coma-like or Virgo-like, column I. For these prototype galaxies, from Rood and Baum (1968) and Brown (1968), the percentage figures used were:

Type	Coma	Virgo
Ir	0.5	2
E	40	22
SO	55	–
S	3	76

The SN frequencies, differentiated by galaxy-type, were taken from the data in Katgert and Oort (1967) and Tammann (1970) respectively, it being felt that the figures in Wamsteker *et al.* (1973) were over optimistic. We thus use the following frequencies:

ν(S-type galaxies) = 0.062 SN gal^{-1} yr^{-1}. Tammann (1970)

ν(E+SO) = 0.025 SN gal^{-1} yr^{-1}. Katgert and Oort (1967)

and

ν(Ir) = 0.0033 SN gal^{-1} yr^{-1}. Zwicky

The overall integral frequencies for SN going off in the clusters, column J in Table I, were estimated from the expression:

$$\nu(\text{SN-yr}^{-1}) = \{N_{\text{Ir}}\,\nu(I_r) + N_{\text{E+SO}}\,\nu(\text{E}+\text{SO}) + N_{\text{S}}\,\nu(\text{S})\} \tag{1}$$

Column K gives the fraction f of the time that any given cluster is above the local horizon, for the most southerly station, i.e. Bologna, and column L represents the overall SN-rate expected, with the above assumptions, for those SN which are above the Bologna horizon at the time of their occurrence. Finally, in column M, we tabulate the relative signal strength for a standard source strength $S_0 = 1$ placed in the Virgo cluster, simply to indicate the importance of distance.

4. The Microwave Installations

As it is intended to describe the Bologna radiometer in detail elsewhere (Jelley *et al.*, 1974) only the essential features of the instruments will be presented here. The radiometers at all four sites are essentially similar, so a description of one will suffice. The radiometer's main parabolic reflector antenna, which has a diameter of either 91 cm (beamwidth $\sim 3°$) or 30 cm (beamwidth $\sim 12°$), is mounted equatorially and driven continuously at the sidereal rate. The superheterodyne receiver is of the switched type, so that the signal in the main antenna is compared with that in a reference beam ($\sim 30°$ beamwidth) at a switching frequency of 270 Hz. The main antenna tracks the selected cluster, and the reference beam is directed roughly towards the North celestial pole. The receiver system temperature is ~ 3000 K, the bandwidth is 40 MHz (two bands each of 20 MHz width), and the post-detector integration time is currently set at $\tau = 0.3$ s. The output is displayed on standard chart paper moving at 15 cm h^{-1}, and time is determined from a local crystal periodically checked with time-signals from OMA, MSF or HBG, the Czech, English and Swiss time-signal transmitters respectively. Although the precision in the time signals is at least 0.6 s, resolution on the recording charts is reduced to about 20 s when comparing records at different stations. The recording circuit encompass about three decades of range in time, i.e. 0.1 s–100 s, while it is hoped to add additional channels down to 1 ms, i.e. an additional two decades.

The overall output sensitivity is such that with the larger antennas, with τ set to 1 s, three times peak-to-peak noise on the chart corresponds to ~ 15 K or 8×10^{-8} erg, or 4×10^{-11} ergs cm^{-2} $(event)^{-1}$.

5. Observations of the Coma Cluster

Prior to setting up the radiometers at Medicina, Bologna, and University College, Dublin, the main programme of cluster observations was begun by Harwell and Glasgow. Observations were carried out between April 19, 1972 and September 5, 1972, tracking on the Coma Cluster. In 857 hours of overlap coincidence time, no correlated events were detected. This integrated observing period excluded times (a) when the cluster was below either of the local horizons, (b) when the DC chart recorders were off-scale, and (c) when either of the two stations were experiencing severe interference, such as radar, heavy rain etc.

With a liberal resolving time of one minute, 10 coincident pulses were observed. Of the Glasgow pulses 8 were from the main antenna (highgain dish), one from the reference channel (horn antenna), and the remaining one was bipolar. These 10 events we will call observed coincidences. Adding and subtracting 5 min to the Harwell charts for each Glasgow event, to obtain chance coincidences, generated an average of 8.5 delayed events. Apart from these statistics however, none of the 10 events appeared sufficiently similar at the two stations to suggest that any of them were celestial in origin.

It may be noted that in this observing time, 0.098 yr, we would have expected statistically only two SN, if the figure of 17 SN $(yr)^{-1}$ shown in column L is used for the comparison.

Several months of observation on the Perseus cluster have now been carried out, but the chart analysis is so far incomplete.

In conclusion it may be mentioned that the present experiment has received con-

Fig. 1. The tracking radiometer designed and built at AERE, Harwell, for the CNR of Italy. The photograph shows the installation on site at Medicina, their field station outside Bologna.

siderable stimulus from the recent report (Klebesadel *et al.*, 1973) that isolated single bursts of gamma-rays have been detected in coincidence between Vela satellites, these bursts probably arising from supernovae in distant galaxies.

6. Acknowledgements

In a joint experiment involving four groups, we acknowledge assistance from many people. Both the Harwell and Bologna installations were largely constructed by Mr J. H. Fruin and Mr W. McKellar of AERE, Harwell, for whose work we are especially grateful. The Glasgow group wish to thank Mr G. W. Lessnoff for maintenance of the microwave system during part of the Coma Cluster observations, and Mr A. McKellar for constructuing mch of the electronics at this site. The Bologna group is indebted to Dr G. Cavallo for theoretical discussions, and to Dr G. Morigi for electronic work and maintenance of the installation at Medicina. The Dublin group would like to thank Dr J. White for designing and constructing the equatorial mount for their radiometer.

The Glasgow and Dublin groups are also grateful to assistance from Harwell, under extra-mural research contracts Nos. 2154 and 1561 respectively, on the instrumentation side. The Harwell and Bologna groups acknowledge with gratitude the partial support of this research by NATO, through Research Grant No. 653, and finally, one of us (W.P.S.M.) wishes to gratefully acknowledge financial support from the Royal Commission for the Exhibition of 1851, during part of the period of this work.

References

Arnett, D.: 1967, *Canad. J. Phys.* **45**, 1621.
Barbon, R.: 1963, *Astron. J.* **73** (10 Part 1), 1016.
Bondi, H.: 1970, *Quart. J. Roy. Astron. Soc.* **11**, 443.
Bōnoli, F.: 1973, Thesis, University of Bologna, Italy.
Branch, D. and Patchett, B.: 1973, *Monthly Notices Roy. Astron. Soc.* **161**, 71.
Brown, F. G.: 1968, *Monthly Notices Roy. Astron. Soc.* **138**, 527.
Cavallo, G. and Ventura, A.: 1972, *Astrophys. Space Sci.* **19**, 431.
Charman, W. N. *et al.*: 1970, *Nature* **228**, 346.
Colgate, S. A. and White, R. H.: 1966, *Astrophys. J.* **143**, 626.
Colgate, S. A.: 1968, *Canad. J. Phys.* **46**, 476.
Colgate, S. A.: 1970, *Astrophys. Space Sci.* **8**, 457.
Colgate, S. A. and Noerdlinger, P. D.: 1971, *Astrophys. J.* **165**, 509.
Colgate, S. A., McKee, C. R., and Blevins, B.: 1972, *Astrophys. J. Letters* **173**, L87.
Dunlap, J. R., Hynek, J. A., and Powers, W. T.: 1972, *Advances in Electronics and Electron Physics* (ed. by J. D. McGee, D. McMullan and E. Kahan), Vol. 33B, p. 789.
Elliott, J. L.: 1972, Smithsonian Astrophysical Observatory, Report 341.
Hughes, V. A. and Retallack, D. S.: 1973, *Nature* **242**, 105.
Jelley, J. V. and Fruin, J. H.: 1974, Harwell report, to be published.
Katgert, P. and Oort, J. H.: 1967, *Bull. Astron. Inst. Neth.* **19**, 239.
Klebesadel, R. W., Strong, I. B., and Olson, R. A.: 1973, *Astrophys. J. Letters* **182**, L89.
Kowal, C. T. and Sargent, W. L. W.: 1971, *Astron. J.* **76**, 756.
Kukarkin, B. V.: 1965, *Astrofizika* **1**, 465.
Le Blanc, J. M. and Wilson, J. R.: 1970, *Astrophys. J.* **161**, 541.

Meikle, W. P. S., Drever, R. W. P., Haynes, R. F., Shakeshaft, J. R., Charman, W. N., and Jelley, J. V.: 1972, *Monthly Notices Roy. Astron. Soc.*, **160**, 5; see also: 1971, *Nature*, **232**, 177.
Ögelman, H., Bertsch, D., and Fichtel, C.: 1968, *Bull. Am. Phys. Soc.* **13**, 1478.
O'Mongain, E.: 1973, *Nature (Phys. Sci)*, **242**, 136.
Partridge, R. B. and Wrixon, G. T.: 1972, *Astrophys. J. Letters* **173**, L75; see also: 1971, *Phys. Rev. Letters* **26**, 912.
Rood, H. J. and Baum, W. A.: 1968, *Astron. J.* **73**, 442.
Rosino, L. and Di Tullio, G.: 1974, this volume, p. 19.
Tammann, G. A.: 1970, *Astron. Astrophys.* **8**, 458.
Wamsteker, W., Wisniewski, W. Z., Lee, T. A., and Wdowiak, T. J.: 1973, *Nature (Phys. Sci.)* **241**, 7.
Weber, J.: 1969, *Phys. Rev. Letters* **22**, 1320.
Weekes, T. C.: 1969, *High-Energy Astrophysics*, Chapman and Hal, London, p. 32.
Zwicky, F.: 1969, *Supernovae and Their Remnants* (ed. by P. J. Brancazio and A. G. W. Cameron), Gordon and Breach, New York, Ch. 1, p. 1.
Zwicky, F.: 1970, *Trans. IAU* **XIVA**, *Reports on Astronomy* (ed. by C. de Jager), D. Reidel Publ. Co., Dordrecht, Holland, p. 316.
Zwicky, F., Herzog, E., and Wild, P.: 1961–1967, *Catalogue of Galaxies and Clusters of Galaxies*, Vol. I–VI, Cal. Inst. Tech. Pasadena.

DISCUSSION

A. S. Wilson: What is your sensitivity in terms of J m^{-2} Hz^{-1} $(pulse)^{-1}$?

J. V. Jelley: In general we use somewhat different units and you could convert them quite easily. When we are using a 90 cm dish, with a full bandwidth of 40 MHz and $\tau = 1$ s, the sensitivity (for a signal/noise of $3 \times$ peak to peak, i.e. ~ 150) is 4×10^{-11} erg cm^{-2} $(event)^{-1}$ or $\sim 10^{-21}$ J m^{-2} $(event)^{-1}$.

THE DISTANCE OF NGC 7319 IN STÉPHAN'S QUINTET AND ITS SUPERNOVA

J. HEIDMANN
Observatoire de Meudon, France

Stéphan's quintet is a compact group of five galaxies which have widely different redshifts (see Figure 2 in Arp's Invited Paper): the galaxy on the South-East, NGC 7320, has a redshift cz$\sim$800 km s^{-1}, while the four other galaxies have redshifts cz$\sim$6000 km s^{-1}.

This year (Balkowski, Bottinelli, Chamaraux, Gouguenheim and Heidmann, *Astron. Astrophys.* **25**, 319) we made 21-cm neutral hydrogen line observations of the two spirals NGC 7320 and 7319. From the fluxes and widths of the lines, combined with optical data on magnitudes and diameters, we estimated the distances of these two galaxies using various distance criteria worked out in previous studies. There are five criteria, among which three are independent. We recall that these criteria are independent of redshifts and of the Hubble constant.

The results are shown on Figure 1:

(i) for NGC 7320 the mean value of the distance is 12 Mpc, in agreement with the distance D_z deduced from redshift and expansion of the universe;

(ii) for NGC 7319 we obtain a mean distance (22^{+15}_{-9}me) Mpc while D_z should be (84^{+28}_{-17}me) Mpc. The probability for the distance of NGC 7319 to be equal to or larger than D_z is only 1.3% and it is then probable that this galaxy has an anomalous redshift.

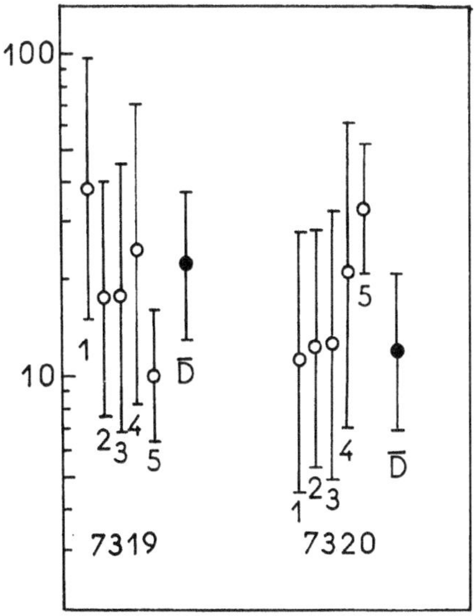

Fig. 1. Distance determinations according to five distance criteria and mean value of the first three ones for NGC 7319 and 7320. Vertical scale in Mpc; bars are mean errors.

C. B. Cosmovici (ed.), Supernovae and Supernova Remnants, 73–74.

This result is in agreement with the recent study of Arp on H II region diameters and interactions in Stéphan's quintet (see his paper).

But it is contradicted by the supernova discovered in NGC 7319 by Rosino in 1971. According to him this supernova is at normal redshift distance.

In this short communication I just wished to point out the probable discrepency between our distance determinations and the supernova properties for NGC 7319.

HISTORICAL OBSERVATIONS OF SUPERNOVAE

F. R. STEPHENSON

Dept. of Geophysics & Planetary Physics, School of Physics, University of Newcastle upon Tyne, England

1. Introduction

As is well known, although several supernovae have been observed in a number of individual external galaxies (e.g. NGC 3184, 4321, 5236 and 6946) during the present century, none has been seen in our own Galaxy since AD 1604 (excluding η Car. of 1843 which is similar to Zwicky's type V). We thus possess no telescopic observations of galactic supernovae. The early observations which we do have access to are very crude by present day standards, but it is desirable to make the best use that we can of such data. Apart from the fact that the identification of a SNR with a historically observed supernova draws special attention to the remnant, leading to its more detailed investigation, historical observations of supernovae are important for a number of reasons.

(i) If we can say definitely that a non-thermal galactic radio source is the remnant of a supernova which occurred in a particular year, then we know precisely how long the remnant has been developing. Much more would be known about the evolution of SNR's if the ages a of few further remnants could be established. Perhaps this is not a vain hope.

(ii) From the recorded variation in light or the duration of visibility it may prove possible to distinguish between types I and II (regarding type II as including types III and IV). This has been demonstrated very successfully by Baade (1943 and 1945) for the supernovae of 1572 and 1604 (see below), but observations of other galactic supernovae are much more difficult to interpret.

(iii) A number of records allow an estimate of the apparent brightness at maximum. If we can make a rough estimate of the interstellar absorption, e.g. from the recorded colour of the star, it may be possible to derive a useful value for the parallax. Minkowski (1968) assuming a type I absolute magnitude of -19 obtained distances of 6.7 and 1.3 kpc for the supernovae of 1604 and 1006.

(iv) Ilovaisky and Lequeux (1972b) showed that it is possible to obtain some helpful statistics on the frequency of occurrence of supernovae in the galaxy. Using their previously derived (1972a) radio luminosity function at 1 GHz for galactic SNR's together with estimates of the diameter of 6 remnants of known or assumed age they obtained a weighted mean interval between supernova outbursts in the galaxy (τ) of 50 ± 25 yr. This result is in agreement with other lines of evidence, cf. the work of Milne (1970), Caswell (1970), Tammann (1970) and Downes (1971) which leads to values of τ between about 25 and 75 yr.

A more general approach is also helpful. We possess fairly detailed records of 'guest stars' from the beginning of the Han dynasty in China, 203 B.C. The author

C. B. Cosmovici (ed.), Supernovae and Supernova Remnants, 75–85.

(Stephenson, 1971) compiled a list of new stars of long observed duration – at least 3 months – recorded in Far Eastern history (see Table I). Most of these stars were in very low galactic latitude, contrary to what we would expect for novae which were close enough to attract the attention of the unaided eye. It thus seems that as many as 12 supernovae may have been observed in the last 2200 years (the star 1592A was possibly Mira Ceti). An estimate of the actual number of supernovae occurring in the entire galaxy during this time is difficult to make owing to varying interstellar absorption, incomplete records etc. but an average frequency of 1 in 50 years seems much more acceptable than, say, 1 in 200 years.

TABLE I

Early guest stars observed for several months

Year (A.D.)	Constellation	Duration	b^{II}
185	Centaurus	8 months	$+3°$
369	Draconis, etc.	5 months	—
386	Sagittarius	4 months	-9
393	Scorpius	7 months	0
902	Cassiopeia	1 year	$+10$
1006	Lupus	10 years (?)	$+5$
1054	Taurus	22 months	-5
1181	Cassiopeia	6 months	$+2$
1572	Cassiopeia	18 months	$+1$
1592 A	Cetus	15 months	-70
1592 B	Cassiopeia	4 months	-3
1592 C	Cassiopeia	3 months	-3
1604	Ophiuchus	1 year	$+5$

As already mentioned, no galactic supernova has been seen in the last 370 years, whereas if a frequency of 1 in 50 years is supposed, 7 or 8 would be expected to have taken place. Assuming that supernovae occur randomly in time, this apparent lacuna need not trouble us too much. On the other hand, it is possible that several have occurred since 1604 and passed unnoticed, owing to heavy interstellar absorption. The outburst which produced Cas A (3C 461) would appear to come into this category. From the rate of expansion of the filaments, van den Bergh and Dodd (1970) estimate the age of the remnant as 306 ± 8 years. Gull (1973) prefers an age of 200 ± 40 years. A catalogue of non-thermal galactic radio sources such as that of Downes (1971) or Ilaiovsky and Lequeux (1972a) shows that a number of sources have comparable radio surface brightness (at present the only general indication of age since few distances are known) to 3C 10 and 3C 358 – the remnants of SN 1572 and SN 1604 – and could be somewhat younger.

2. Supernovae for which a Remnant Has Been Proposed

Only two SNR's (3C 10 and 3C 358) are known with certainty to be the remnants of historically observed supernovae. Before discussing other proposed associations, I feel that it is important to emphasise Minkowski's (1968) argument that both the supernovae of 1572 and 1604 were of type I. This is clear from the light curves deduced by Baade (1943 and 1945) from the observations of Tycho Brahe and Kepler (and others in 1604). Baade treated the various estimates of brightness (with reference to selected comparison stars and planets) as purely experimental data and compared the derived light curves with that of a type I supernova which occurred in the galaxy IC 4182 in 194 (see Figure 1, which is reproduced from Baade's 1945 paper). The rem-

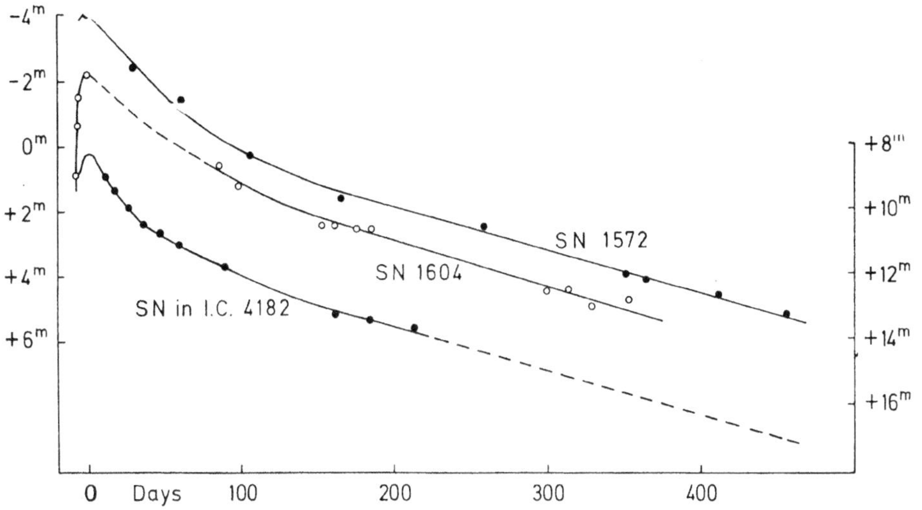

Fig. 1. Light curves of the Supernovae of 1572 and 1604 deduced by Baade (1945).

nants 3C 10 and 3C 358 ought to be regarded as type I prototypes, for in no other example can we be sure of the classification. Minkowski (1959) has pointed out that it is necessary to assume that Tycho Brahe's measurement of R.A. was 5^s in error. However, this discrepancy, although surprising, is trivial in comparison with the angular extent of 3C 10.

Other identifications which have been proposed are:

SN 185 = 1439–62
SN 1006 = MSH 14–415
SN 1054 = 3C 144 (Crab Nebula)
SN 1181 = 3C 58

Section 3 is devoted to an analysis of the various arguments for and against the acceptance of the Crab Nebula as the remnant of the new star of 1054. Brief comments on SN 1006, SN 185 and SN 1181 and their suggested remnants are given in Section 4.

3. The Crab Nebula as the Remnant of SN 1054

Until recently, 3C 144 was unanimously regarded as the remnant of SN 1054. However, Ho *et al.* (1970) have criticised this identification. As so much importance is attached to the Crab Nebula, I have thought it worthwhile to examine their arguments in some detail here. The following is a summary of a much fuller discussion to be published elsewhere. Translations of the various oriental descriptions of the star are given by Ho *et al.* (*op. cit.*) and these are not reproduced here except where considered necessary.

The principal argument against the acceptance of the Crab Nebula is that the only two fundamental sources (both Chinese) which describe the position of the star in detail both place it to the south-east of *T'ien-kuan* (ζ Tau), whereas the Crab Nebula is 1.1 deg. north-west of it. Additionally, although conceding that the lengthy visibility (some 22 months) is 'certainly a strong argument against our object being a comet', Ho *et al.* conclude that a comet seems to be indicated. Let us consider these two objections in turn.

Firstly there can be no question that *T'ien-kuan* is ζ Tau (a relatively isolated star of the 3rd magnitude), as Ho *et al.* have convincingly demonstrated, and we may note that the proper motions of both the Crab Nebula and ζ Tau are negligible. The relevant descriptions of the position of the new star in relation to *T'ien-kuan* are contained in the *Sung Shih* (compiled in 1345 but based on contemporary material) and the *Hzü tzu-chih t'ung-chien ch'ang-pien* (written before 1184). Both sources give the same date of appearance of the guest star which corresponds to 1054, July 4 (there is a slight confusion involving two very similar characters). More significantly the two descriptions are identical, proving that they share a common origin – in all probability the records of the imperial astronomers. Hence we may conclude that the evidence that the guest star was to the south-east of *T'ien-kuan* rests on a single statement and consequently a textual error in the relative positions of the two stars cannot be ruled out.

In either text the description may be translated as follows: "A guest star appeared possibly several inches (*ts'un*) south-east of *T'ien-kuan.*" Ho *et al.* have made the valid remark that the above translation is not the only possible one. An acceptable alternative is that the object was 'possibly several inches in length'. On the assumption that the Chinese astronomers were referring to the scale on their armillary sphere, the approximate dimensions of which are known, Ho *et al.* concluded that the 'several inches' of the text were equivalent to some 10 deg. An object of such apparent size could not fail to escape detection in the West so that Ho *et al.* inferred that the star was some 10 deg south-east of ζ Tau, a considerable distance from the Crab Nebula.

An immediate objection to such a location is the lack of any reference to asterisms other than *T'ien-kuan* both in the Chinese and Japanese records. The suggested position would place the star in or near the asterims *Shui-fu* and closer to several other asterisms than *T'ien-kuan* (cf. the oriental star maps drawn by Ho, 1962). Additionally, both the author (1971) and Kiang (1972) have shown from an analysis of various types of astronomical observation at arbitrarily selected periods in Chinese history that inches were roughly synonomous with tenths of a degree. This result is loosely confirmed by the frequent references to cometary tails several *ch'ang* (of 100 *ts'un*) in length. Kiang analysed 22 7th century observations of the position of the ecliptic in relation to the lunar mansions and deduced that 10 *ts'un* were equivalent to 1.50 ± 0.24-deg. If we accept this result, and combine it with the definition of *shu* ('several') given by Ho *et al.* as between 3 and 9 then the guest star was anything up to 1.5 deg from ζ Tau. This is in excellent accord with the position of the Crab Nebula apart from the direction being reversed.

Coming now to the nature of the star, it is significant that both the Chinese and Japanese records avoid the use of the term *hui-hsing* ('broom star'), the normal expression for a tailed comet. The object was described as a 'guest star' (*k'o-hsing*) in China and as a 'rayed star' (*po-hsing*) in Japan, which could refer to either a nova (or supernova) or a comet without a perceptible tail. Further, the various accounts remain silent on any motion of the star during the 22 months of visibility. That the star was still in the vicinity of ζ Tau on August 27 – more than 4 months after its first appearance – is proved by the account in the *Sung Hui-yao* (a contemporary chronicle) in which the 'Astronomer Royal' discusses the astrological implications of the observation that the star did not 'invade' the lunar mansion *Pi*. This asterism, which contains *a* Tau, is less than 15 deg from ζ Tau.

Both the *Sung Shih* and the *Sung Hui-yao* describe the guest star as 'guarding' (*shou*) *T'ien-kuan*. Ho (1966) gives several definitions of this term, e.g. 'to attach and stay by the side of (a constellation or another celestial body)' or 'to stay (in a particular place) without leaving'. Ho *et al.* attach some importance to an alternative definition of *shou* – 'moving backwards and forwards without going away'. Such an apparent motion could result from the relative motion of the Earth, which completed nearly two full orbits while the star was visible, but the description would only be appropriate if the object remained at a considerable distance – several AU – from the Earth. Since oriental records of comets of long duration usually give a detailed account of any motion, we may conclude that the guest star either showed no apparent motion or at most a slight parallactic displacement.

If the new star of 1054 was a comet the lengthy period of visibility, absence of a tail and lack of any significant motion cannot be reconciled with the *Sung Hui-yao* observation that it was seen in daylight for 23 days. Daylight comets are by no means infrequent (cf. Chambers, 1909), but they are characterised by a lengthy tail (several tens of degrees), rapid apparent motion at brightest (some 10 deg daily), short period of visibility in daylight (usually only 2 or 3 days at most) and short period of visibility to the unaided eye (never more than about 6 months). Accordingly we must infer that the star of 1054 was either a nova or supernova. The 'pointed rays' described in the *Sung-Shih* as emanating from the object may then be explained as the result of distortion within the eye. This phenomenon is commonly observed with Venus, especially when near maximum brightness.

Circumstantial evidence allows us to conclude that the star was not a fast nova. The star was very well placed for Northern observers so that the apparent lack of any reference to it in European or Arabic history suggests that it was never very much brighter than Venus. Twenty three days after discovery the star had just ceased to be visible in daylight and yet after a further four months it could still be described (in the

Sung Hui-yao) as 'bright and lustrous'. The typical magnitude range for a slow nova is given by Payne-Gaposchkin (1957, p. 28) as 9.8 – 1 mag. less than for a fast nova. Mayall and Oort (1942) could find no post-nova brighter than +8 near ζ Tau, which corresponds to a magnitude range of 13. Payne-Gaposchkin (*op. cit.*, p. 27) lists a few novae with ranges somewhat greater than this but all were extremely fast.

Bearing in mind the very great period of visibility (longer than that of the supernovae of 1572 and 1604) there can be little doubt that the star of 1054 was a supernova. However there does not seem to be enough information to discriminate between types I and II. Minkowski (1968), assuming the identity of the guest star with the Crab Nebula, derived an absolute visual magnitude at maximum of −18.2 using a value for the interstellar absorption of 1.08 mag. kpc^{-1} and a distance of 1.9 kpc. If we adopt an absolute magnitude at maximum of −19.5 (a fair estimate for type I) and reverse the calculation, we can deduce an upper limit for the distance of the guest star of about 2.5 kpc.

A young SNR at this distance should be an intense radio source. Even an intrinsically feeble SNR like 3C 10, which is at least twice as far away as the guest star, ranks as one of the most powerful sources in the sky (flux density 50 fu at 1 GHz). However, there is no suitable source of high intensity and high surface brightness (indicative of a young remnant) within 20 deg of ζ Tau except 3C 144. Extrapolation of modern results for the rate of expansion of the optical filaments of the Crab Nebula give a date of explosion within a century or so of 1054 (cf. Scargle, 1967 and Trimble, 1971). In view of possible acceleration or deceleration of the expansion this agreement is probably as good as could be expected. Apart from the extremely low probability (0.0001) of two supernovae appearing so close to one another and to the Earth within about a century, there is no oriental record of any event which could correspond with the Crab Nebula except that noted in 1054. Regardless of type, the supernova which produced the Crab Nebula should have been comparable in brightness with Venus. Even if it appeared while close to the Sun it would still be a prominent object a few days after heliacal rising. The Crab Nebula lies extremely close to the ecliptic. As a constant watch was maintained in the Far East on the motion of the Moon and planets oriental records of new stars appearing near the ecliptic should be particularly complete. However in Ho's (1962) catalogue of comets and novae recorded in Far Eastern history, which is based on an extremely thorough search of the available literature, no star other than of an obviously cometary nature is described as appearing near *T'ien-kuan* in a period of more than 2000 years except that in 1054.

In the author's opinion, the various independent lines of evidence discussed above fully justify the reversal of the recorded direction of the guest star of 1054 with respect to ζ Tau and permit the reinstatement of the Crab Nebula as one of the few select SNR's of precisely known age.

4. Remarks on the Supernovae of 1006, 185 and 1181

Both occidental and oriental records of the supernovae of 1006 have been thoroughly discussed by Goldstein (1965) and Goldstein and Ho (1965). The only comment which I propose to make here concerns the S. Gallen record, one of the very few European accounts of the star. The description in the *Annales Sangallenses Maiores* (Pertz, 1826) may be translated as follows: "1006. A new star of unaccustomed size appeared, glittering in aspect and dazzling the eyes, causing alarm. In a wonderful manner this was sometimes contracted, sometimes diffused and, moreover, sometimes extinguished. It was seen for three months in the inmost extremities of the south (*in intimis finibus austri*)...."

Oriental records indicate that the star was in a large southerly declination. Although the S. Gallen record does not give a precise description of position, its importance lies in the fact that the monastery (latitude 47°25′) lies considerably further north than any other place where the star was reported, thus setting a limit on the declination. I have made a detailed study of the *Annales Sangallenses Maiores* and it is clear that the *Pars Altera* containing the record of the star, which covers the period from 919 to 1044, is fundamentally a chronicle of S. Gallen, with occasional notices of important events occurring elsewhere in Europe. There is nothing to suggest that the frequent astronomical and meteorological observations recorded in the annals were made elsewhere than in the monastery itself. From the examination of a large scale relief map of the area, I have calculated that a star of declination less than -37.0 ± 0.5 would not rise above the apparent horizon of S. Gallen. Correcting for precession, the equivalent declination at 1950 is $-41\overset{\circ}{.}0 \pm 0\overset{\circ}{.}5$.

Gardner and Milne (1965) proposed that the radio source MSH 14-4*15* is the remnant of the new star. More recently Milne (1971) has discussed this source in detail and concludes that there are no serious objections to it being identified with SN 1006. A star in the position of this source (declination $-41°45'$ at 1950) might just rise above the visible horizon at S. Gallen, which would accord well with the description *in intimis finibus austri*, but a detailed survey of the southern horizon as seen from S. Gallen is needed to confirm this.

In marked contrast to the supernova of 1006, there is only a single (Chinese) record of SN 185. This is to be found in the *Hou Han Shu* ('History of the Later Han Dynasty' – written A.D. 450) and describes the star as appearing 'at' (*yü*) *Nan-mên* for 8 months. From Ho Peng Yoke (1962), the probable constituents of *Nan-mên* are *a*, *β* and ε Cen. Hill (1967) showed that the positions of two non-thermal galactic radio sources agree well with the recorded location of the new star. These sources are 1439–62 and 1356A. Hill preferred the former because it shows optical filamentary structure, a conclusion supported by Westerlund (1969), but a definite decision has yet to be made. Additionally, Downes (1971) and Ilovaisky and Lequeux (1972a) catalogue several other probable SNR's in the vicinity of 1439–62 and 1356A. There are few bright stars near *a*, *β* and ε Cen. so that the description 'at *Nan-mên*' could refer to a fairly extensive area of sky. Apart from 1439–62 and 1356A, the sources KES 20B, KES 20A, MSH 14–57, MSH 15–52 and KES 24, which are common to both catalogues, could all fit the Chinese description satisfactorily. Identification of the stars forming the asterisms adjacent to *Nan-mên* might help to clarify the situation.

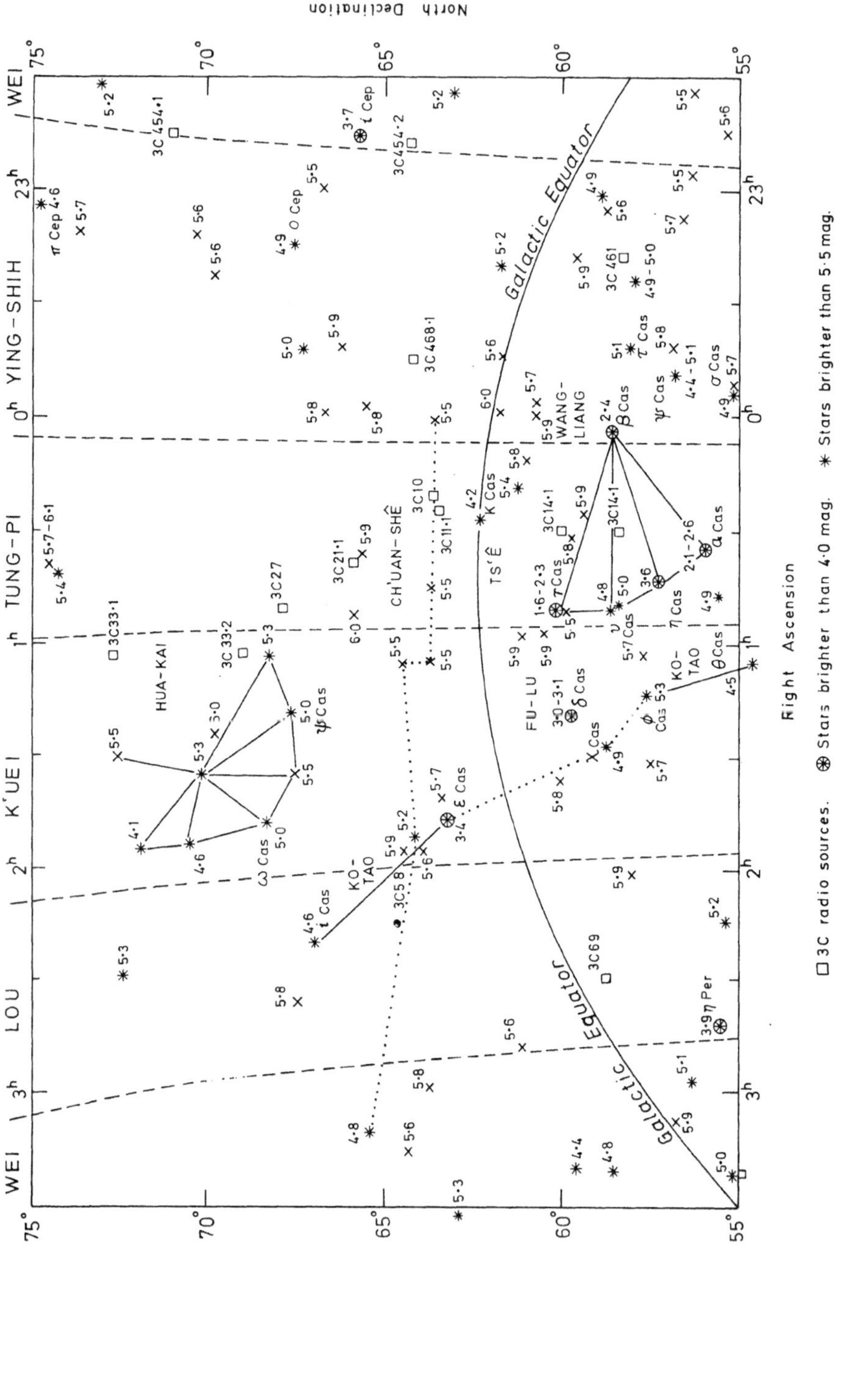

Fig. 2. Chart of the Cassiopeia region.

The author (Stephenson, 1971a) has expressed special interest in SN 1181 since apart from the supernovae of 1006, 1054, 1572 and 1604 it is the only historical supernova for which there are several independent observations. At least eight individual records of this star are known (from N. and S. China and Japan), and from these we can glean the following information about the star:

(i) it was visible for about 6 month (185 days in S. China, 156 days in N. China),

(ii) it seems to have been about as bright as Saturn,

(iii) it was 'reddish-blue' in colour. If this is taken to mean a very deep red, heavy interstellar absorption would be indicated,

(iv) there is no hint of motion of the star in any of the records

(v) as many as four asterisms are named in specifying the position.

The star was described as in *K'uei* lunar mansion (which roughly indicates the R.A.), near *Wang-liang*, at *Hua-Kai* and (in two dependent records) 'guarding' and 'invading' *Ch'uan-shê*. These last three asterisms are neighbouring groups of stars in Cassiopeia.

Apart from the use of the term 'guarding', discussed in Section 4 above, the lengthy period of visibility and lack of any reference to motion rule out the possibility of the object being a comet. We may confidently presume that the star was a supernova rather than a nova since it remained visible for so long and the galactic latitude was very low, certainly less than 5 deg. Additionally, it would appear that as the astronomers of S. China were able to follow the star for a month longer than their counterparts in N. China, the brightness diminished very slowly. This is characteristic of type I behaviour but type II cannot be ruled out.

In order to deduce the position of the star as accurately as possible, the author found it necessary to collate a number of early oriental star maps with the aid of a modern star catalogue. Figure 2 is taken direct from the author's 1971 paper and shows thc most probable formation of the asterisms mentioned in the guest star records together with the discrete radio sources listed in Bennett's (1962) survey which fall within the boundaries of the chart. It would, of course, be very desirable if the radio source could be restricted to those listed in the published catalogues of SNR's but these catalogues appear to be very incomplete as yet. On the other hand the number of 4C sources is excessive and this catalogue (Gower *et al.*, 1967) probably contains very few galactic sources which are not listed by Bennett.

It is clear that the radio source 3C 58 fits the observed descriptions extremely well. This occupies the correct lunar mansion and lies extremely close (0.7 deg) to one of the probable constituents of *Ch'uan-shê*, in accord with the statement that the star 'invaded' (*fan*) the asterism. This expression is a technical term implying an approach to within about 1 deg of another celestial body (Stephenson, 1971). 3C 58 is a well established SNR (cf Weiler and Seielstad, 1971) but to make the proposed identification certain, a detailed examination of other radio sources in the vicinity is required.

5. Conclusion

The author would welcome renewed interest among radio astronomers in the proposed remnants of SN 185, SN 1006 and SN 1181. Of the list of probable supernovae given in the table of Section 1, six (observed in A.D. 369, 386, 393, 902 and 1592 – two stars) have yet be to identified satisfactorily with a remnant, and it is felt that detailed examination of the oriental records would pay dividends.

References

Baade, W.: 1943, *Astrophys. J.* **97**, 119–129.
Baade, W.: 1945, *Astrophys. J.* **102**, 309–317.
Bennett, A. S.: 1962, *Mem. Roy. Astron. Soc.* **68**, 163–172.
Bergh, S. van den and Dodd, W. W.: 1970, *Astrophys. J.* **162**, 485–493.
Caswell, J. L.: 1970, *Astron. Astrophys.* **7**, 59–64.
Chambers, G. F.: 1909, *The Story of the Comets*, Clarendon Press, Oxford, p. 8.
Downes, D.: 1971, *Astron. J.* **76**, 305–316.
Gardner, F. F. and Milne, D. K.: 1965, *Astron. J.* **70**, 754.
Goldstein, B. R.: 1965, *Astron. J.* **70**, 105–114.
Goldstein, B. R. and Ho Peng Yoke: 1965, *Astron. J.* **70**, 748–753.
Gower, J. F. R., Scott, P. F., and Wills, F.: 1967, *Mem. Roy. Astron. Soc.* **71**, 49–142.
Hill, E. R.: 1967, *Austr. J. Phys.* **30**, 297–307.
Ho Peng Yoke: 1962, in A. Beer (ed.), *Vistas in Astronomy*, vol. 5, Pergamon Press Ltd., Oxford, pp. 127–225.
Ho Peng Yoke: 1966, *The Astronomical Chapters of the Chin Shu*, Mouton, Paris, p. 36.
Ho Peng Yoke, Paar, T. H., and Parsons, P. W.: 1970, in A. Beer (ed.), *Vistas in Astronomy*, vol. 13, Pergamon Press Ltd., Oxford, pp. 1–13.
Ilovaisky, S. A. and Lequeux, J.: 1972a, *Astron. Astrophys.* **18**, 169–185.
Ilovaisky, S. A. and Lequeux, J.: 1972b, *Astron. Astrophys.* **20**, 347–356.
Kiang, T.: 1972, *Mem. Roy. Astron. Soc.* **76**, 27–66.
Mayall, N. U. and Oort, J. H.: 1942. *Publ. Astron. Soc. Pac.* **54**, 95–104.
Milne, D. K.: 1970, *Austr. J. Phys.* **23**, 425–444.
Minkowski, R.: 1959, in R. N. Bracewell (ed.), 'Paris Symposium on Radio Astronomy', *IAU Symp.* **9**, Stanford University Press, pp. 315–322.
Minkowski, R.: 1968, in B. M. Middlehurst and L. H. Aller (eds.), *Nebulae and Interstellar Matter*, Chapter 11, University of Chicago Press.
Payne-Gaposchkin, C.: 1957, *The Galactic Novae*, North Holland Publishing Co., Amsterdam.
Pertz, G. H. (ed.): 1826, *Monumenta Germaniae Historica, Scriptores*, vol. 1, Hanover. Reprinted Kraus Reprint Corporation, Stuttgart, 1963.
Scargle, J. F.: 1967, *Astron. Soc. Pac.*, Leaflet No. 457.
Stephenson, F. R.: 1971, *Quart. J. Roy. Astron. Soc.* **12**, 10–38.
Tammann, G. A.: 1970, *Astron. Astrophys.* **8**, 458–475.
Trimble, V.: 1971, in R. D. Davies and F. G. Smith (eds.), 'The Crab Nebula', *IAU Symp.* **46**, 3–11.
Weiler, K. W. and Seielstad, G. A.: 1971, *Astrophys. J.* **163**, 455–478.
Westerlund, B. E.: 1969, *Astron. J.* **74**, 879–881.

DISCUSSION

M. Friedjung: The nova light curves shown were for fast novae. Slow novae have been observed. Nova Delphini was bright for many months, so one cannot say all long lived objects were supernovae.

F. R. Stephenson: I feel that the smoothness of the light curve of a SN encourages extended observation however, in any case the distribution in galactic latitude of guest stars of long (several months) duration is very much narrower than we should expect for novae of similar apparent brightness, in view of their nearness to us. I think that there is a good case for regarding practically all of the stars listed as SN.

F. Zwicky: Polynesians and others record about zero magnitude stars popping up in the magellanic clouds. Has anybody checked on stars disappeared from the BD and Cordoba Durchmusterung plastered on images of galaxies?

F. R. Stephenson: I regret I am unable to comment.

W. M. Goss: H I absorption measurements indicate that 3C 58 is either at the edge of the galaxy or is extragalactic. Recent near IR work by Scargle and Spinrad has failed to detect any optical evidence for a supernova remnant near 3C 58.

F. R. Stephenson: I believe the distance is estimated at 7 or 8 kpc. Both Tycho's SN and Kepler's SN have very feeble optical remnants and yet these stars were about 4 and 2 mag. brighter than the star in 1181. On this basis alone there is no reason why heavy interstellar absorption should not prevent detection of 3C 58.

A. S. Wilson: Could you tell me the positional agreement between the Chinese guest star of A.D. 1181 and the radio source 3C 58?

F. R. Stephenson: There are 7 stars in *Ch'uan-shê*, but the agreement in position between 3C 58 and the brightest star of that asterism is excellent – about 0.6 deg. This separation is in accord with one of the Chinese descriptions.

PART II

PHOTOMETRIC STUDIES OF SUPERNOVAE

PHOTOMETRY OF SUPERNOVAE

HALTON ARP

Hale Observatories, Carnegie Institution of Washington,
California Institute of Technology, Pasadena, Calif., U.S.A.

Supernova light curves began to become available in the 1930's. Figure 1 of this paper shows a collection of five of these early, best observed type I light curves. (The pioneering observations of Baade, Zwicky and Minkowski are reviewed, for example, in Zwicky (1965) or Pskovskii (1969)). The major result is that the well observed type I light curves that have been available to this date have encouraged the conclusion that all type I supernovae are extremely similar. It is well known, for example, that after maximum a type I declines rapidly for $2\frac{1}{2}$ to 3 mag. Abruptly then a steady decline

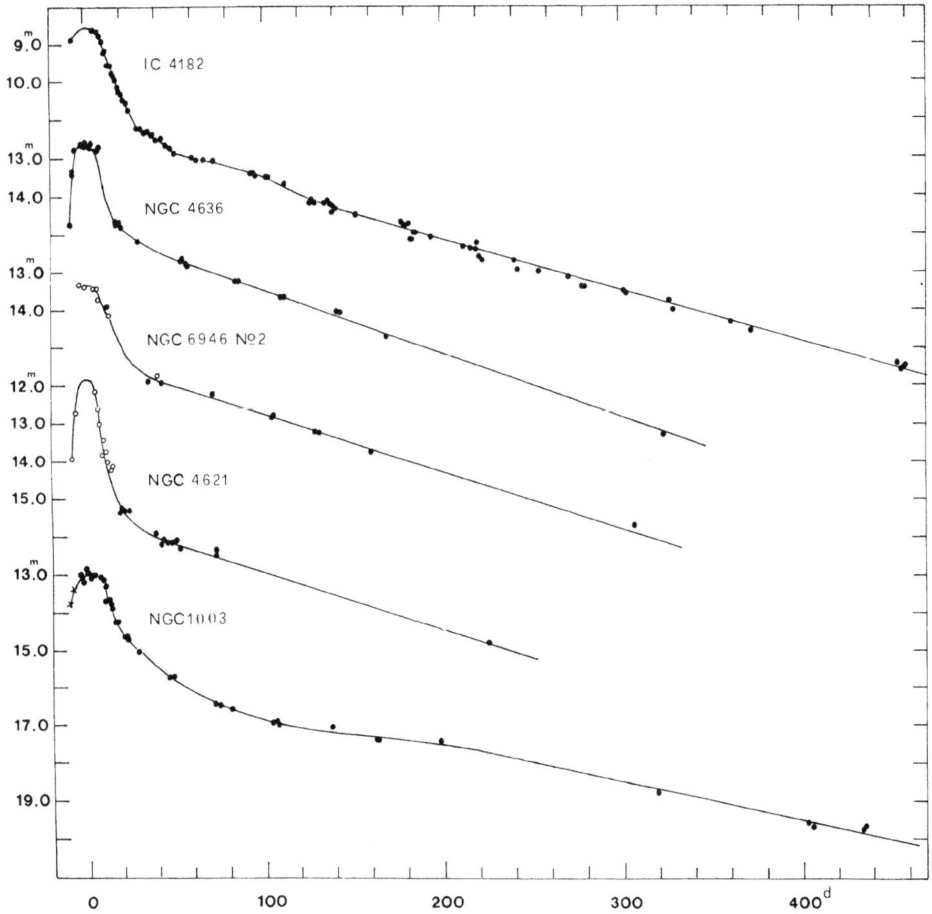

Fig. 1. Well observed type I supernova light curves. Material gathered from early work by Zwicky, Baade and Minkowski.

C. B. Cosmovici (ed.), Supernovae and Supernova Remnants, 89–97.

sets in which is at a rate of one magnitude per 50 to 70 days (Zwicky, 1965). The work of Minkowski (1939) showed that for several different supernovae the complex bands and features of the type I spectrum changed in a characteristic way with time and therefore could be related to light curve phase. A composite, standard type I light curve is shown in this conference by Rosino, Barbon and Ciatti.

The overall conclusion that has manifested itself in later years therefore, is that only one spectrum and a few apparent magnitudes need be obtained for any type I supernova in order to reconstruct the apparent magnitude at maximum light. Since maximum and pre maximum observations of supernovae are difficult to obtain, this belief makes it simpler to observe more galaxies on an extended schedule and obtain more data for statistics on the assumption that all light curves behave as the standard type I curve.

I will show evidence later in this paper that indicates that all type I are not identical and that the most important cases are perhaps nonstandard and require special observations. But before passing to the details of that supernova in NGC 7319 I would like to briefly comment on the situation for type II supernovae.

The light curves for type II supernovae were always believed to be quite diverse. Figure 2 of the present paper shows a selection of five of the best type II light curves

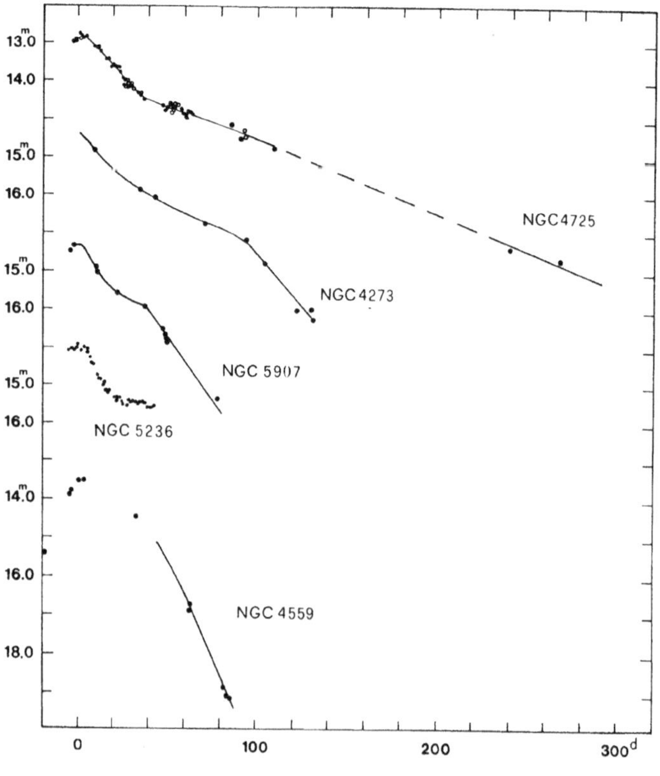

Fig. 2. Light curves of type II supernovae. Material from same sources as in figure 1 but light curve coverage is far sparser.

that were known in the early days. The radically different decline rates encouraged the belief that there was no standard type II light curve. The data on type II was much sparser, however, than on type I and therefore a special effort had to be made to observe a type II supernova with extreme thoroughness.

That effort was successful when Humason, in 1959, discovered a type II supernova

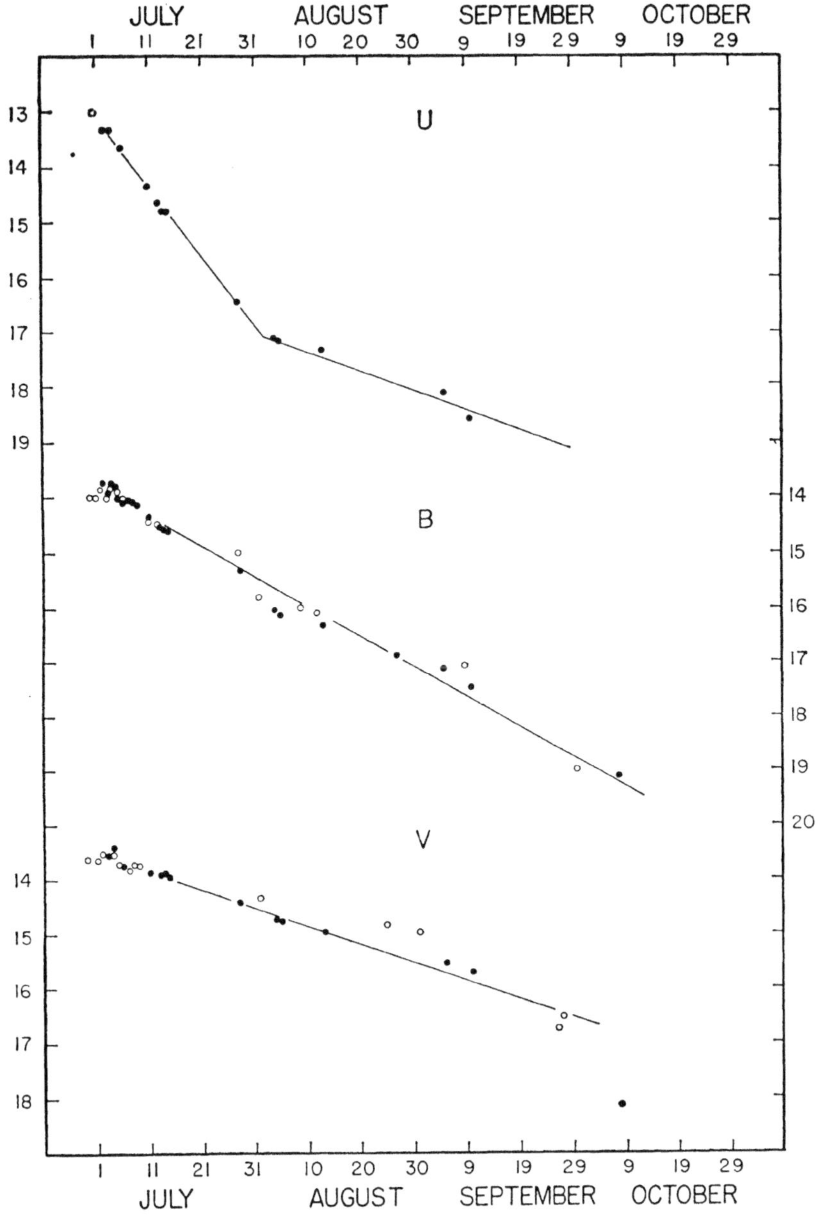

Fig. 3. Light curves in *U*, *B* and *V* wavelengths from best observed type II supernova in NGC 7331 (from Arp, 1961).

in NGC 7331. The resulting series of observations is a model example of what can be accomplished when a large number of observers cooperate together for certain observational result. In the span of 3 months that the supernova was early visible, 53 separate photometric observations were made by eleven different observers on four different telescopes in southern California (Arp, 1961). A number of spectroscopic observations were also made (Greenstein and Minkowski, 1973). As organizer of that observational effort I can report that it was not difficult since the observers were all extremely cooperative, it only required one person to initiate requests for observations and collate data. It seems strongly desirable that further intensive observational programs be undertaken on additional, bright supernovae.

The photometric light curve that resulted from the project just described is shown in Figure 3. Unlike many type II supernovae there is no break or feature in the light curve, it seems to descend monotonically toward fainter magnitudes. The absolute magnitude also seems rather bright for the average type II supernova so it may be worthwhile to consider whether it, and perhaps the recent type II in M101, form a subclass of type II, in contradistinction to other classes of type II. Further comments will be made on this subject in the Conference Summary Paper. But it

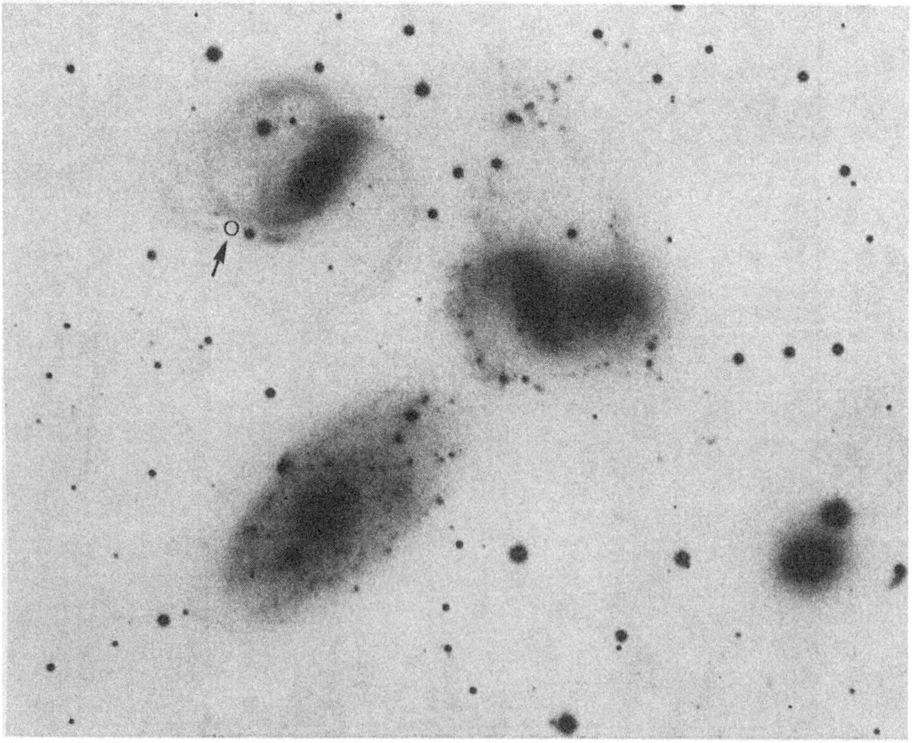

Fig. 4. Photograph of Stephan's Quintet with 200-in. Hale telescope Arrow points to where supernova occurred. Plate is 098-04, with RG2 filter exposed for 120 min on 13 August 1971.

would be important to obtain light curves comparable to the one in NGC 7331 for a number of type II supernovae in order to investigate the homogeneity or possible subclassification of this important type of supernova.

Now I would like to report on a specific supernova which was discovered in August 1971 by Rosino in the peculiar galaxy NGC 7319. A recent summary of numerous observations in Stephan's Quintet including that of supernova is given by Arp (1973.) As is well known, NGC 7319 is a violently interacting member of Stephan's Quintet. Figure 4 below shows a photograph of Stephan's Quintet and the arrow points to the place where the supernova appeared in NGC 7319. Unfortunately the presence of the supernovae was not known until about a month after the discovery plate was taken. Therefore there are not many observations near maximum light. Later more observations were obtained both at Asiago and at Palomar. What observations are known to be available are plotted in Figure 5 where it is seen that the light curve roughly resembles that of a type I supernova. A spectrum taken by Arp shows features typical of a type I supernova.

The light curve, however, has an almost horizontal still stand and then a sharp break away to fainter magnitudes after almost 40 days of nearly constant light. Even though the supernova magnitudes are visual estimates from the plates, and the sequence is extrapolated from a bright sequence around NGC 7331, it does not seem possible that more accurate measures would change the essential features of the light curve. There seems to be a conspicuous, and for a type I supernova, an unprecedented still stand after the initial decline.

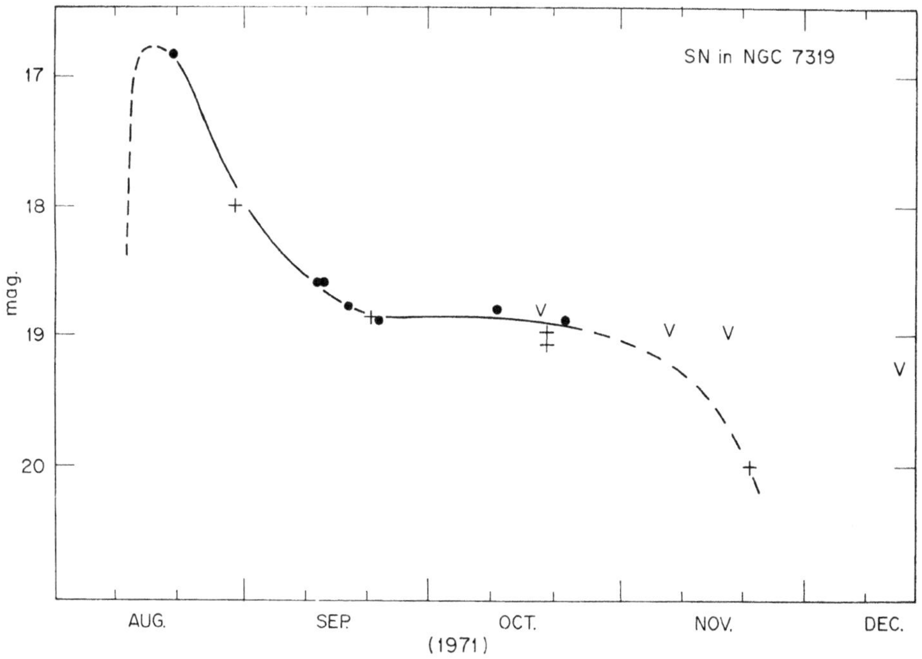

Fig. 5. Observations of supernova in NGC 7319 with best fit light curve suggested.

For the sake of discussion, we can examine the fit with a standard type I light curve. Figure 6 shows the same measured magnitudes with the composite type I curve of Rosino, Barbon and Ciatti passed through them. The last observed point falls about $\frac{1}{2}$ mag. fainter than the composite curve and the next to last four points fall about $\frac{1}{4}$ mag. too bright. But the most serious discordance is encountered in the time of maximum light. Figure 6 predicts that the maximum occurred 47 days prior to 21 September 1971, the date on which the Arp spectrum of the supernova was taken.

But the spectrum clearly indicates a phase of only about 24 days after maximum. The phase of the spectrum can be established in two ways: First, Leonard Searle kindly consented to classify the spectrum in the time sequence of forms outlined by Minkowski (1939) for type I supernovae. Searle reports the spectrum matches best a phase 20–24 days after maximum and is clearly different from phases as early as 18 days or as late as 44 days. Arp has checked this determination by actually measuring the position of the major emission band in the spectrum, the 4600 Å band. He finds, after correcting for 6700 km s^{-1} redshift, that this band occurs at a wavelength of $\lambda = 4610$ Å ± 5 Å. By using the relationship presented by Pskovskii (1969) between the shift of this feature and the phase of the supernova, Arp derived a phase of 25 days $\pm$ ± 15 days. Although the shift of the 4600 Å feature is connected with the appearance of the spectrum which Searle used to classify the phase, the two measures are semi-independent, agree very well, and set limits for the phase which are clearly exceeded by the fit of the standard type I light curve.

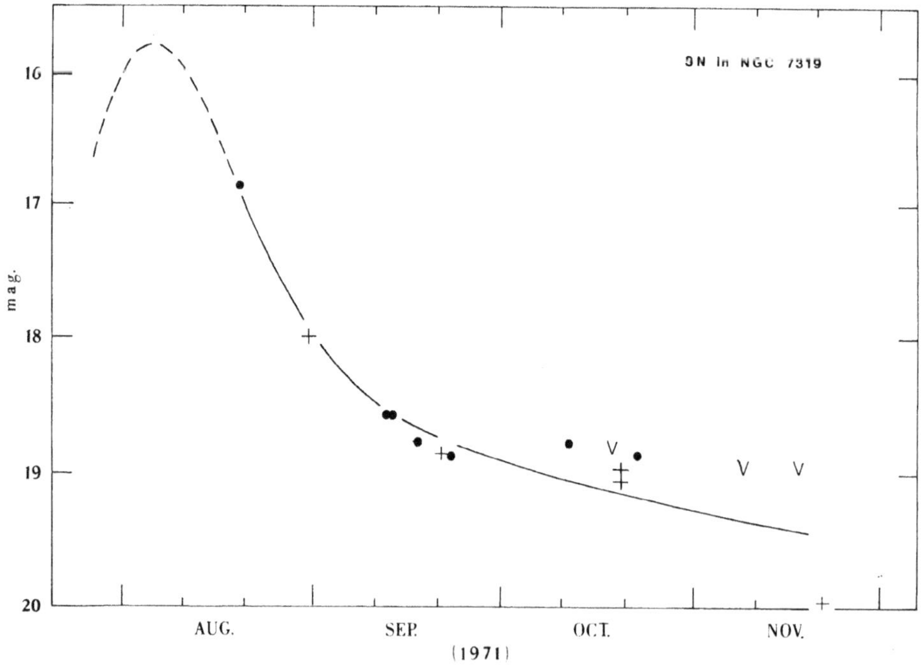

Fig. 6. Some observations of supernova as in Figure 5 with best fit standard type I light curve. Standard type I curve is from a composite by Rosino, Barbon and Ciatti.

Therefore there seems to be quantitative evidence that this particular supernova is peculiar. The importance of the possible peculiarity will now be discussed in relation to its host galaxy.

The critical point about NGC 7319 is that it is one of the high red shift members of Stephan's Quintet. Its redshift corrected for galactocentric rotation is $V_0 = 6935$ km s^{-1}. Using a Hubble constant of $H_0 = 53$ km s^{-1} Mpc^{-1} (Sandage and Tammann, 1972) its distance, on the conventional interpretation of the red shift, should be of the order of 122 Mpc. But there are a whole series of indicators which give a distance for NGC 7319 which is an order of magnitude closer. Briefly they are:

(1) The sizes of H II regions in NGC 7318 A & B, a high red shift galaxy which NGC 7319 is interacting with, give a distance of about 10 Mpc to the system.

(2) NGC 7320 is a low redshift member of Stephan's Quintet which everyone agrees is about 10 Mpc in distance. But the size and distribution of H II regions in both the high and low redshift members of the Quintet indicate they are interacting and at the same distance.

(3) A long interacting tail from NGC 7320 indicates the low redshift galaxy is interacting and at the same distance as the high red shift members.

(4) Proximity, radio sources, and faint luminous filaments all indicate that the peculiar system of Stephan's Quintet is associated with the nearby (10 Mpc), large spiral NGC 7331.

Now, very recently, and described in this conference, J. Heidemann has measured the hydrogen content of NGC 7319 with the Nançay radio telescope tuned to 6700 km s^{-1}. This independent distance estimate gives 22 Mpc for the distance of NGC 7319. Essentially Heidmann has found with this important observation, yet another quantitative piece of evidence that the high red shift NGC 7319 is at the close distance of NGC 7331 and NGC 7320.

This affects our discussion of the supernova in NGC 7319 in the following way: There is a relation between apparent magnitude reached at maximum and the red shift of the galaxy in which the supernova appeared. (The most recent presentation of the data for this relationship was given by Kowal and Sargent, 1971). This relation is conventionally interpreted, as the Hubble relation is, as a redshift-distance effect. According to this relation, the supernova in NGC 7319 should have reached an apparent magnitude of about mpg = 15.2 mag. With the fit of the light curve made by Arp in figure 5 the apparent magnitude at maximum high was probably around mpg = $= 16.6 \pm 0.3$ mag. (No correction for obscuration by our own galaxy has been made in this estimated maximum magnitude.) Therefore we see that the conventional interpretation of the apparent magnitude of the supernova at maximum would place it at an even greater distance than the conventional redshift distance of NGC 7319. On the other hand, if NGC 7319 were really as close as the low redshift members of the NGC 7331 – Stephan's Quintet system – then the supernova would be expected to reach an apparent magnitude of between mpg = 10 and 11 mag., on conventional assumptions.

Even though there were no observations of the supernova for many months prior to its inferred maximum, it is probable that it never reached as high an apparent magnitude as 10 or 11. Therefore we are faced with a difficult choice of conclusions if we accept the evidence previously discussed which showed NGC 7319 to be at the close distance of about 10 Mpc. If we accept the close distance for NGC 7319 then we must accept one of two alternatives:

(1) That in galaxies which have appreciable components of nonvelocity redshift, that the type I supernova phenomenon appears the same as in normal galaxies, but only the luminosity is scaled down,

(2) That in galaxies which have appreciable components of non-velocity red shift, that the type I supernovae are different from normal type supernovae, are much lower in luminosity and peculiar in other respects.

The latter alternative, seems much more plausible. We would expect other differences to be manifest along with reduced luminosities of these particular supernovae. Now it is apparent why the peculiarities we have discussed for the supernova observed in NGC 7319 are so important. As far as the observations go, the supernova does seem to have bona fide peculiarities. The observations are not decisive, however, and therefore it places an extreme order of importance on more and better observations of other supernovae in peculiar and suspected anomalous galaxies. Any high redshift companions of large nearby galaxies would be extremely important in which to discover and follow thoroughly any new supernovae. For example, observing a supernova in NGC 7318 A or B would be almost decisive. There we would expect the supernova to be intrinsically more normal in luminosity like the H II regions are intrinsically more normal in size and luminosity, despite their large component of non-velocity red shift.

So we have now come back to our original point discussed in the introduction to this paper. It is true that a great deal has been learned about supernovae since the early 1930's. Enough has been learned to set up types and categories of behavior. But this only opens up the possibility now of using these models of observed behavior to test, in critical cases, the fundamental nature of the supernova phenomenon, to test crucial propositions about the nature of galaxies. It is of the utmost importance to discover early in their outburst a large number of supernovae, and in those cases where there is special interest to follow through with a vigorous, well coordinated observing program.

References

Arp, H.: 1961, *Astrophys. J.* **133**, 883.
Arp, H.: 1973, *Astrophys. J.* **183**, 411.
Greenstein, J. L. and Minkowski, R.: 1973, *Astrophys. J.* **182**, 225.
Kowal, C. I. and Sargent, W. L. W.: 1971, *Astron. J.* **76**, 756.
Minkowski, R.: 1939, *Astrophys. J.* **89**, 143.
Pskovskii, Y. P.: 1969, *Sov. Astron. J.* **14**, 798.
Sandage, A. R. and Tammann, G. A.: 1972, *Astrophys. J.*, in press.
Zwicky, F. 1965, in Aller and McLayghlin (eds.), *Stars and Stellar Systems*, Vol. VIII, Chapter 7, The University of Chicago Press.

DISCUSSION

A. G. de Bruyn: Is there any galactic obscuration in the direction of this SN and galaxy?

H. Arp: The absorption in this field is variously estimated as from $A_B = 0.3$ to 0.7 mag. I have not applied this to the apparent magnitude at maximum because of the large uncertainty in that maximum magnitude.

P. Wild: Isn't the SN of last December in NGC 2841 similar to the NGC 7319 SN, in showing spectral peculiarities and a rather long standstill on the light-decline?

H. Arp: It would be interesting to compare the two SN, but the early phases of the light-curve of the NGC 2841 SN may not be well enough known.

L. Sartori: The last point on the light curve indicates a sudden decline, certainly not characteristic of SN I. Do you consider that point to be reliable?

H. Arp: That point was from a single plate taken on the 48-inch telescope. The supernova was faint but certainly above the plate limit. The estimated error of the final point is $\pm\frac{1}{2}$ mag. so that the light curve is still indicated as shelving away faster than a normal type I, even allowing for the probable error of the final magnitude determination.

G. A. Tammann: The occurrence of the SN in NGC 7319 is a first, still inconclusive – but independent and predictable indication that the galaxy is a (background) giant spiral. If more SN could be found in Stephan's high-redshift quartet it would be a strong argument for the giant nature of its members and therefore for its large distance.

H. Arp: I am not a statistics fan but statistics of one leave me even more unenthusiastic than usual. There is simply not enough known about the statistics of Supernovae in Stephan's Quintet to make any comment.

F. Zwicky: Stephan's Quintet has more members and also belongs, as one agglomeration, to a larger cluster.

A sixth member ($V_s = 6400$ km s^{-1}) I announced some time ago. About stagnations of light curves we must be careful because they are very likely due to collisions of expanding shells with preexisting interstellar clouds.

H. Arp: There is one group of 5 galaxies tightly clustered around NGC 7320 in Stephan's Quintet which have high (5700–6700 km s^{-1}) red shifts. There is another group of three of the same redshift tightly grouped around NGC 7331. There are two more about 7000–8000 km s^{-1} grouped around NGC 7331. There are one or two other galaxies scattered within 30′ of NGC 7331 which have the low redshift of NGC 7331. But the high redshift galaxies in the area cluster, in my opinion just around NGC 7331 and NGC 7320 and do not form an extended cluster. To your second comment about still stands and collision of gas clouds: I would not be surprised to see such effects in type II supernovae, but the SN in NGC 7319 was apparently either type I or type I peculiar and would not generally, therefore, be expected to explode in a region of dust and gas clouds.

L. Rosino: About the precision of the date of maximum as derived by spectra of type I SN, I should like to anticipate what Dr Ciatti will say in a later communication that the SN in NGC 2841 recently found by Wild, also show a disagreement between the epoch of maximum derived by the light curve and the epoch derived by the spectrum.

So I would rather think that the date of maximum derived by an analysis of small dispersion spectra may be somewhat uncertain.

H. Arp: What was the nature of the spectral peculiarity of the recent SN in NGC 2841?

L. Searle: I know nothing about the light curve of supernova in NGC 2841 but we did get scans of it at Palomar. It is a type I but its band strengths are a little unusual and it differs in detail from typical SNI's like that 1972 e in NGC 5253.

F. Bertola: I would like to ask Dr. Searle whether the supernova in the Stephan's Quintet could be considered of type I pec (like SN NGC 1073) in which both the absorption at 6150 Å, and the progressive redshift of the blue bands are missing.

L. Searle: I have no spectra of my own of the supernova that enable me to answer your question. Dr Arp asked me to classify a spectrum that covered the region in the blue and I fitted it, as best I could, to Minkowski's maps of the spectrum of SN 1937 in IC 4182.

H. Arp: My spectrum did not reach to longer wavelengths that about 5500 Å. I have the spectrum with me, however, and would like you to examine it say whether you think it is peculiar in the blue spectral region.

ON THE LIGHT CURVE OF TYPE I SUPERNOVAE

R. BARBON, F. CIATTI and L. ROSINO
Asiago Astrophysical Observatory, University of Padova, Italy

Many authors have spoken about an 'average light curve' for type I supernovae, giving also parameters to define it, but no one has shown so far how this curve can be built up using the data available in the literature. With this average curve at hand, one has the possibility to see whether SN I form a homogeneous class and find their general properties from the photometric point of view, above all the intrinsic dispersion at maximum which is of prime importance if these objects are to be used as distance indicators. To this aim we collected the blue light curves of 38 supernovae classified as type I, with sufficient photometric observations. Then, by shifting the individual curves in both coordinates and aiming to the least dispersion, we got the result shown in Figure 1. Here the shift on the time axis has been obviously limited to the uncertainty on the epoch of maximum as reported in the sources. The scatter seen in the figure should be largely due to the intrinsic dispersion of the 'shape' of type I light curves. Of course, especially in the linear decline, also errors coming from the photometry are present.

The importance of this average curve lies in the fact that one can determine from it epoch and apparent magnitude at maximum of supernovae which have not been well observed, by simply fitting their observed blue magnitudes to the average curve.

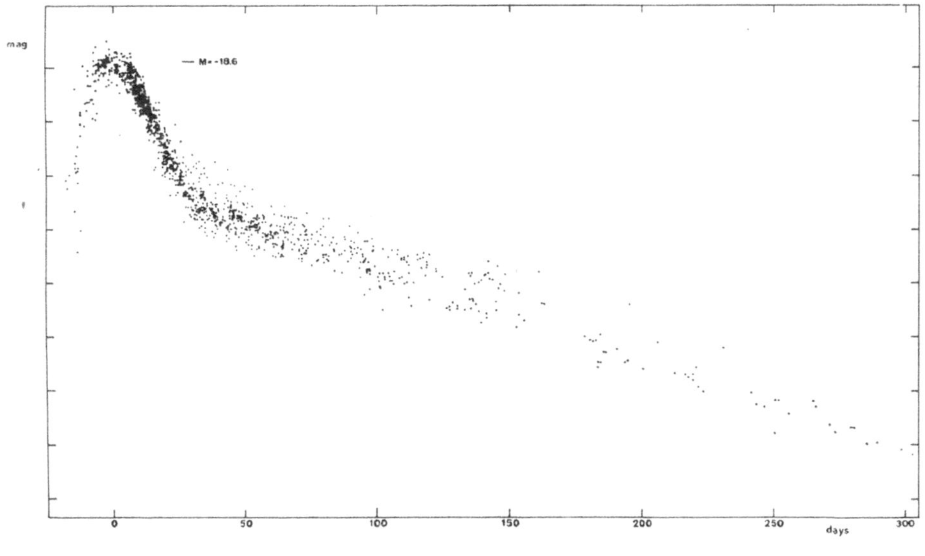

Fig. 1. Composite blue light curve obtained by the fitting of the observations of 38 type I supernovae. One magnitude intervals are marked on the ordinates.

C. B. Cosmovici (ed.), Supernovae and Supernova Remnants, 99–102.

Moreover, if we assume that SN I always behave alike, i.e. we are faced by a phenomenon occurring always on the same scale, one can calibrate the average curve in terms of absolute magnitude and use it as a powerful relative distance indicator. According to Kowal (1968) we assumed as absolute photographic magnitude of SN I at maximum the value −18.6. Then, if our basic assumption is correct, the scatter at maximum of the average curve (mean deviation 0.11) has to be attributed to the intrinsic dispersion of the SN I maxima.

In building up the average curve some difficulties in obtaining a good fit for all the available data were found. Some of the light curves had sharp maxima with large amplitude, while others showed less steep and broader maxima. This led to the possibility of a further division of SN I in two subclasses, according to the following three parameters: (a) width of the light curve two magnitudes below maximum, (b) amplitude from maximum to the inflection point, (c) rate of the final decline. The mean values of these parameters deduced by the light curves of well observed supernovae, caught in the early rise before maximum and thoroughly followed up to the final decline, are as follows:

ΔT	(width of the light curve two magnitudes below maximum)	32^d	38^d
Δm	(amplitude from maximum to the inflection point)	$3^m\!.25$	$2^m\!.50$
R	(rate of decline from 30 to 150 days past maximum in mag day^{-1})	0.016	0.012

Figures 2 and 3 show the average curves of the two subclasses, hereafter called 'fast' and 'slow' SN I. Data of eleven and fifteen supernovae respectively were included in building up these curves. Although the criterion *c* appears weak in discriminating between the subclasses, the differences shown in Figures 2 and 3 are beyond photometric errors and we do believe that the discrimination is real and has a physical meaning. To support this opinion, it is to note that all the well observed supernovae could be assigned to either class with no case in between. If the discrimination were due to photometric errors, one should expect to find a continuous sequence covering the range from fast to slow SN I. Using the supernovae which appeared in galaxies of known distance, no definite conclusion could be reached on whether the two subtypes reach the same absolute magnitude at maximum. It is to be noted that a much larger scatter in M_{max}, by a factor of 4, is present for the fast SN I.

Assuming the reality of the distinction, it is interesting to see how the objects belonging to the two subclasses are distributed according to the morphological types of their parent galaxies. We found the following figures:

	E+SO	S+SB	Irr
fast	0	8	3
slow	3	12	0

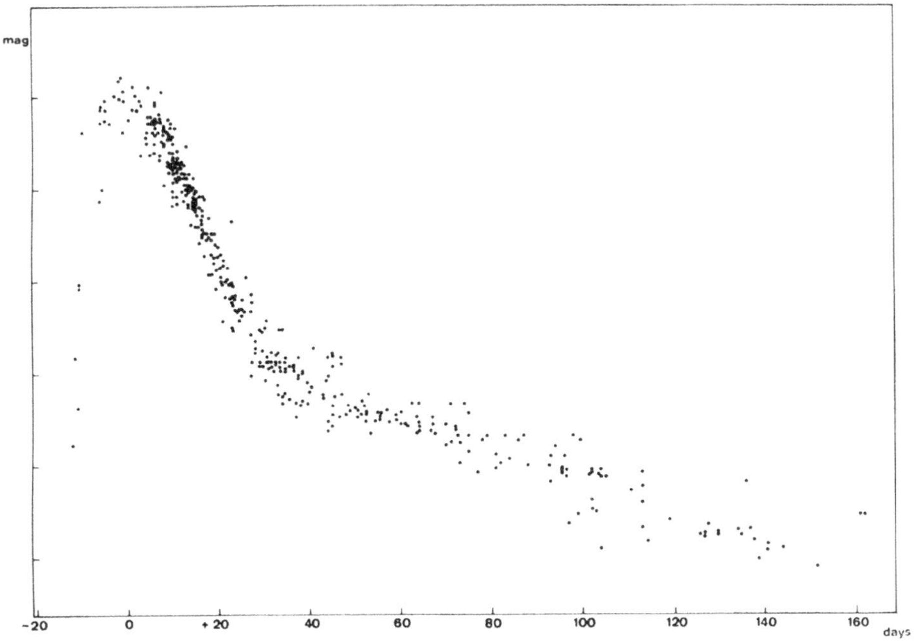

Fig. 2. Composite light curve of 11 'fast' SN-I.

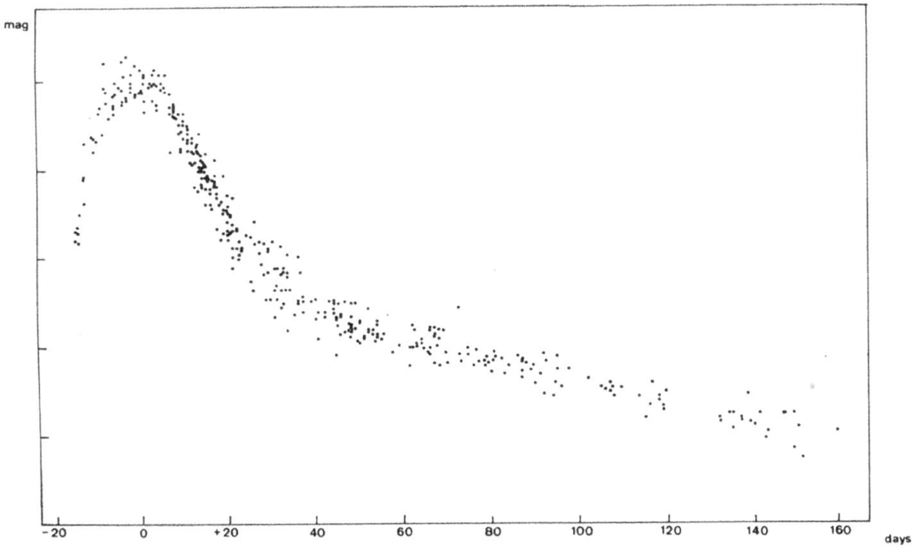

Fig. 3. Composite light curve of 15 'slow' SN-I.

Although the numbers are small, we can argue that the fast and slow subclasses may be related to different types of stellar population in galaxies. This should be also in agreement with an earlier suggestion by Minkowski (1964), Bertola and Sussi (1965) and more recently by Dallaporta (1974) of the existence of two kinds of SN I related to stellar population I and II respectively. To definitely ascertain this point we need more well studied supernovae.

A detailed investigation on this subject will appear in Astronomy and Astrophysics.

References

Bertola, F. and Sussi, M. G.: 1964, *Contr. Asiago*, No 176.
Dallaporta, N.: 1974, this volume, p. 187.
Minkowski, R.: 1964, *Ann. Rev. Astron. Astrophys.* **2**, 247.
Kowal, C. T.: 1968, *Astron. J.* **73**, 1021.

DISCUSSION

R. M. Duin: Is there any difference in spectra between the fast and the slow type I SN?

R. Barbon: We have made mainly a photometric analysis. Since type I SN show a common spectral evolution we don't expect a large spectroscopic discrimination.

A. Ardeberg: With the peculiar energy distribution of SN, the interrelation between different B systems must be time-dependent. Then after your shifts of the different sets of data, considerable residuals may remain. Did you investigate these residuals?

R. Barbon: No we did'nt, but the majority of the data are from Asiago observations in which this effect doesn't apply.

PHOTOMETRY AND SPECTROSCOPY OF THE 1972 SUPERNOVA IN NGC 5253

A. ARDEBERG and M. DE GROOT
ESO (European Southern Observatory) Santiago, Chile

1. Introduction

Apparently very bright supernovae are especially interesting from two points of view. Firstly, they provide the possibility for detailed and exact study during maximum luminosity. Secondly, they give us a favourable opportunity to obtain a good coverage of supernova evolution.

Our best chance in later years has been the 1972 supernova in NGC 5253, announced by Kowal (1972) on May 15, 1972. It is the fourth apparently brightest extragalactic supernova ever recorded (Kowal and Sargent, 1971). It is quite interesting to note that also the second apparently brightest extragalactic supernova, SN 1895B, belonged to NGC 5253.

The announcement of the 1972 supernova in NGC 5253 was received at the European Southern Observatory in Chile on May 17, 1972. The same night we started photometric as well as spectroscopic observations of the object. The photometric observations are still going on, whereas the spectroscopic observations were continued until the middle of January, 1973. The first night of observation we devoted to repeated high-resolution observations. We then tried to follow these observations up by moderately frequent re-observations.

Figure 1 is reproduced from a Palomar E chart. It shows NGC 5253. The position of the supernova and our local standard star for photometry are indicated. It can be seen that the supernova is situated far enough from the main body of the galaxy to be photometrically fairly undisturbed.

2. Photometric Observations

All the photometric observations, in the UBV system, have been made photoelectrically with the ESO 1 m telescope (Hooghoudt, 1966), equipped with the ESO photometer (de Vries, 1966). This ensures that all the measurements obtained belong to the same magnitude and colour system.

The accuracy of the photometry has been good, except for those measurements that had to be done at very high air-masses and against bright sky. Those observations, made at the end of the first season and at the beginning of the second, were included to obtain as good a coverage of the evolution as possible. The 'normal' accuracy was during the first observing season about $0\overset{m}{.}01$ in all three colours. The measurements made in the second season are accurate to within $0\overset{m}{.}02$–$0\overset{m}{.}03$ in V and B and to within $0\overset{m}{.}05$ in U.

C. B. Cosmovici (ed.), Supernovae and Supernova Remnants, 103–113.

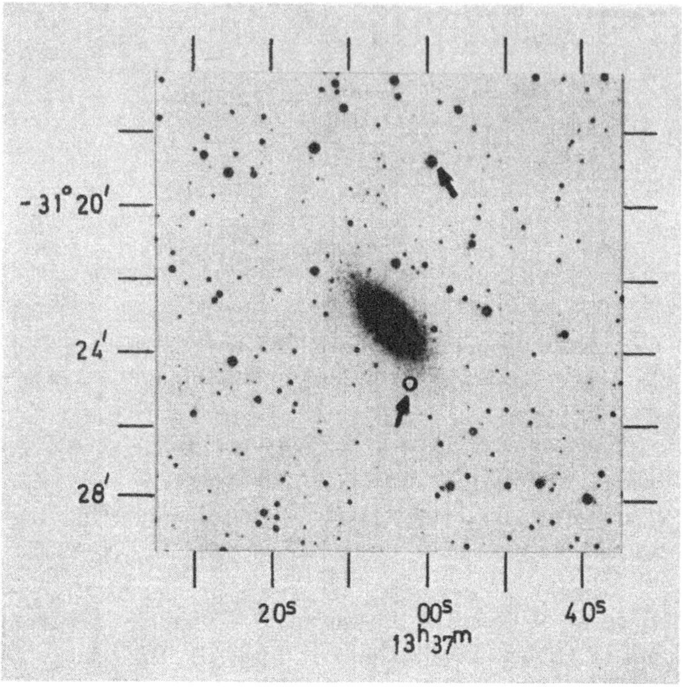

Fig. 1. Finding chart for the 1972 supernova in NGC 5253. The print is an enlargement from the Palomar E chart of the field. The coordinates given are for 1950. Arrows indicate the positions of the supernova and the standard star for photometry, CD − 30° 10790.

Figure 2 shows the results of comparisons between our results and data published by Wisniewski *et al.* (1972), Osmer *et al.* (1972), Walker *et al.* (1972), Przybylski (1972) and Cousins (1972). The small differences displayed are easily explained by the peculiar spectral energy distribution of supernovae. After the preparation of Figure 2, new results have been published by Lee *et al.* (1972). They are also in good agreement with our values.

3. Light Curves

The light curve in V is shown in Figure 3. The filled circles denote our own measurements. The four crosses stand for data taken from Cousins and Przybylski, included to fill the gap in our observations. To these data we have applied small corrections to our system as obtained from the comparison in Figure 2.

Figure 4 displays the results in B with the same notations as in Figure 3. The two squares indicate prediscovery photographic magnitudes obtained by Austin (1972). From prediscovery plates we also get the upper limit to the luminosity indicated by an arrow in Figure 4.

Finally, Figure 5 gives the U light curve with notations as in Figure 3.

From the light curves it is quite clear that the supernova is of type I. Furthermore,

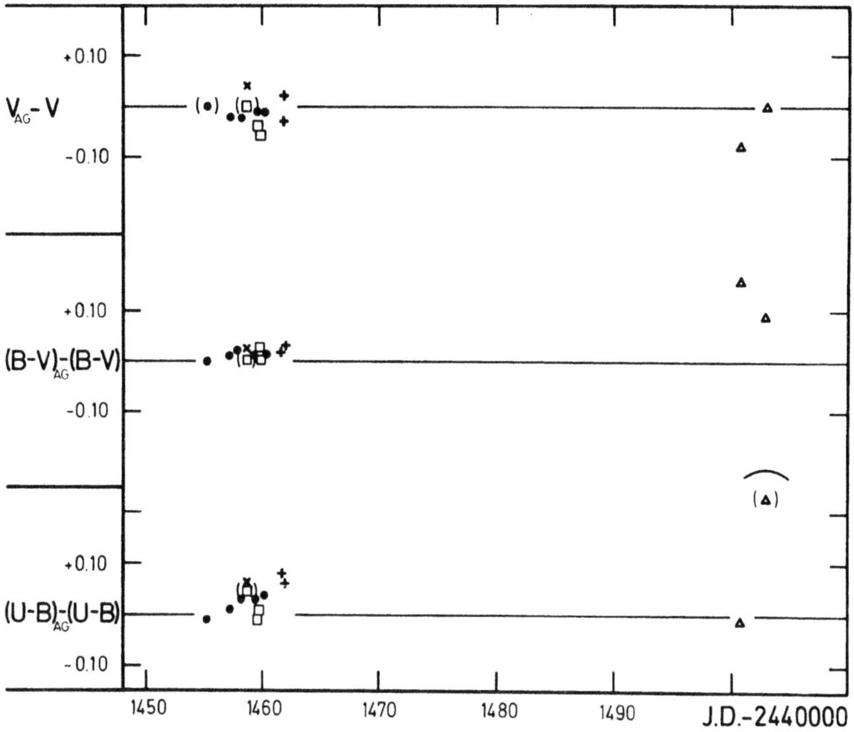

Fig. 2. Comparison between published *UBV* data for the supernova and corresponding data interpolated from our measurements. Filled circles denote measurements by Cousins, squares measurements by Osmer *et al.*, horizontal-vertical crosses stand for data from Walker *et al.*, diagonal crosses for data from Wisniewski *et al.* and triangles for measurements by Przybylski.

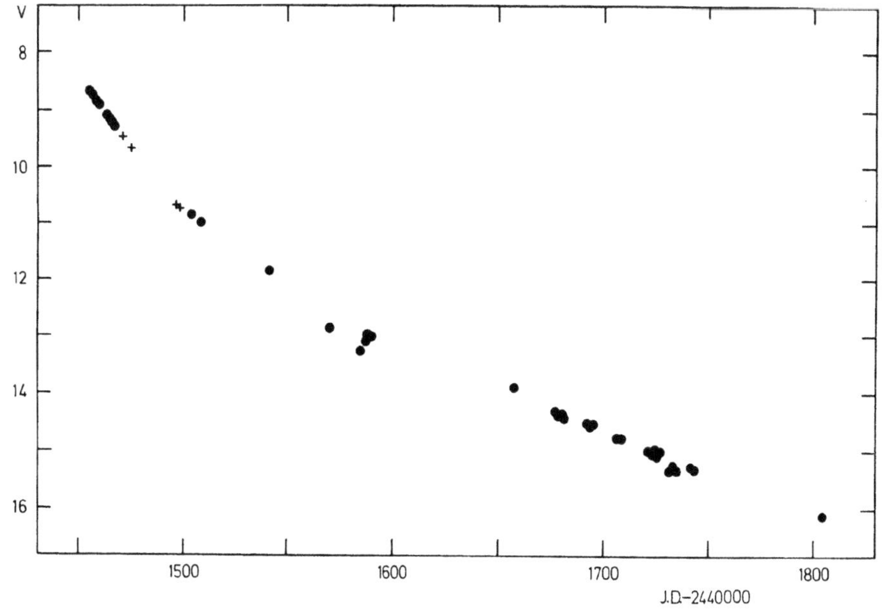

Fig. 3. *V* magnitudes plotted versus Julian Dates. Filled circles denote our own measurements and crosses magnitudes taken from Cousins and Przybylski and corrected to our system.

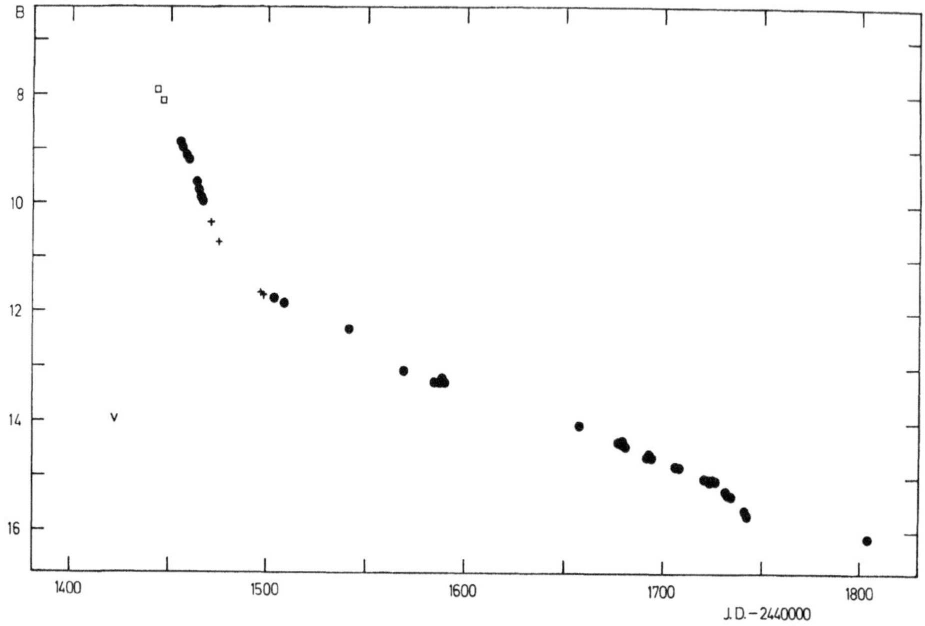

Fig. 4. *B* magnitudes plotted versus Julian Dates. Filled circles and crosses have the same meaning as in Figure 3. The squares indicate photographic prediscovery magnitudes and the arrow an upper limit to the luminosity.

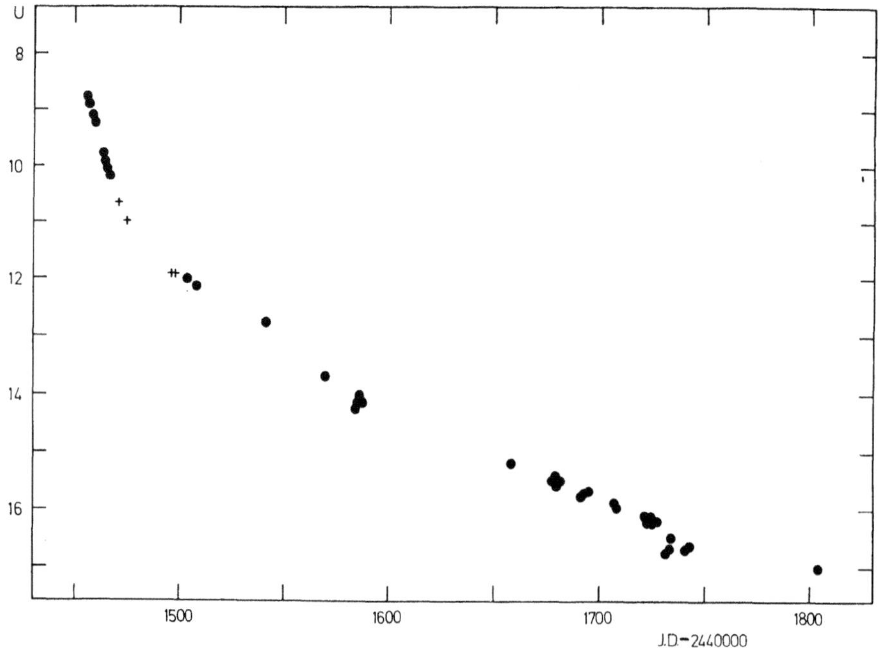

Fig. 5. *U* mag. plotted versus Julian Dates. Filled circles and crosses have the same meaning as in Figure 3.

for the B light curve the maximum peak can be reconstructed fairly easily. We get a maximum luminosity corresponding to $B=7.75 \pm 0.10$. The time of the luminosity maximum obtained is JD = 244 1438.0 ± 2.0.

From distance determinations by van den Bergh (1960) and Sérsic (1960, 1968) we conclude an uncorrected distance modulus of 28.0. This means an absolute photographic magnitude for the luminosity maximum of $B_0 = -20.25$. If we compare this with existing compilations of absolute magnitudes at luminosity maximum (van den Bergh, 1960; Minkowski, 1964) we find the 1972 supernova in NGC 5253 to be on the bright side of those values. However, if we apply more recent results for the galactic foreground absorption (Shane and Wirtanen, 1967) to these compilations a better agreement is obtained. Still, however, the supernova must be regarded as a luminous one.

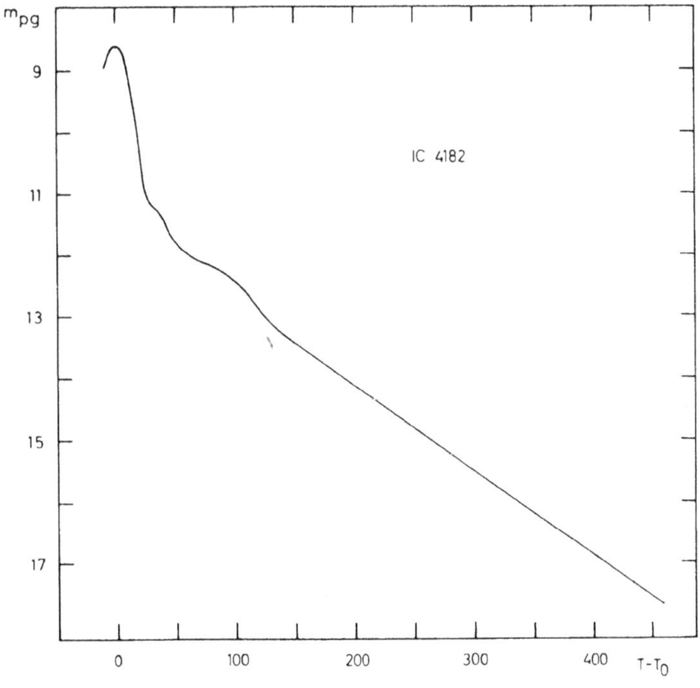

Fig. 6. The photographic light curve of the 1937 supernova in IC 4182 as obtained by Baade (Minkowski, 1964).

The B light curve as a whole resembles that of the 1937 supernova in IC 4182, given in Figure 6 (Minkowski, 1964). Especially the non-linear behaviour after the maximum peak should be pointed out.

For the other Zwicky parameters (Zwicky, 1965) we find for the B light curve a peak width of about 50 days, a peak height of about 3.0 mag., and a decline gradient after the maximum peak of around 60 days mag^{-1}. These values are all in very good agreement with those normally found for supernovae (Zwicky, 1965; Pskovskii, 1970).

4. Colour Curves

The $(B-V)$ curve is shown in Figure 7. The notations are the same as in Figure 3. The curve shows a marked maximum about 42 days after the B luminosity maximum. The maximum apparent colour is about $+1.07$, corresponding to an intrinsic colour of $+0.82$. Assuming a linear $(B-V)$ relation for the time around luminosity maximum we obtain an intrinsic $(B-V)$ colour of about -0.6 at the time of maximum luminosity. The last part of the curve shows quite a stabilized colour.

Our $(B-V)$ curve may be compared with existing compilations of $(B-V)$ values for type I supernovae by Mihalas (1963) and by Pskovskii (1970). The results of those compilations are displayed in Figures 8 and 9. The first part of our curve is in good

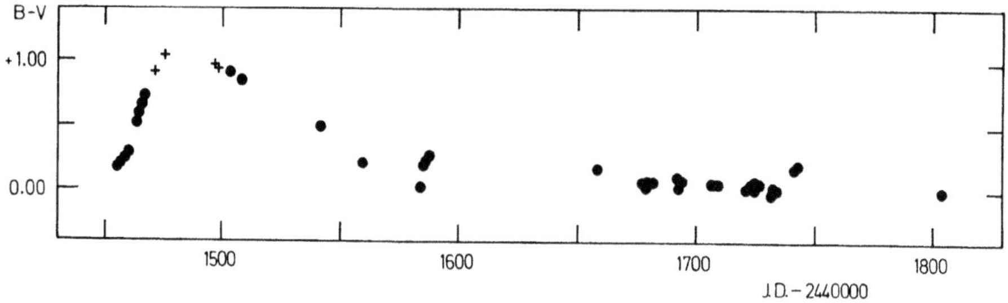

Fig. 7. $(B-V)$ colour indices plotted versus Julian Dates. Same notations as in Figure 3.

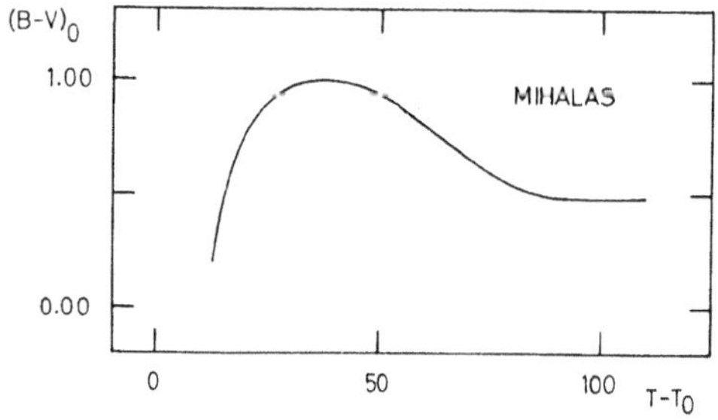

Fig. 8. Mihalas' mean $(B-V)$ curve for supernovae of type I (Mihalas, 1963).

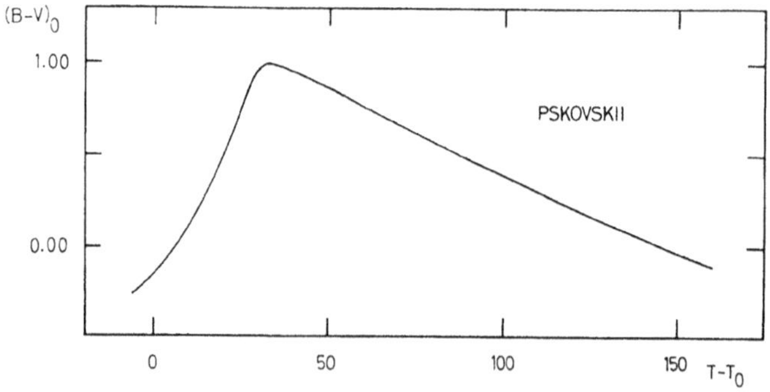

Fig. 9. Pskovskii's mean $(B-V)$ curve for supernovae of type I (Pskovskii, 1970).

agreement with the Mihalas curve. The agreement between our curve and that of Pskovskii is excellent.

Our $(U-B)$ curve is given in Figure 10. Two plateau parts are clearly visible. With the same assumption of linearity around maximum luminosity as for the $(B-V)$ curve we get an intrinsic $(U-B)$ colour of -0.8 for the time of luminosity maximum.

Our $(U-B)$ curve can be compared with Pskovskii's compilation (1970), Figure 11. The two curves are entirely different.

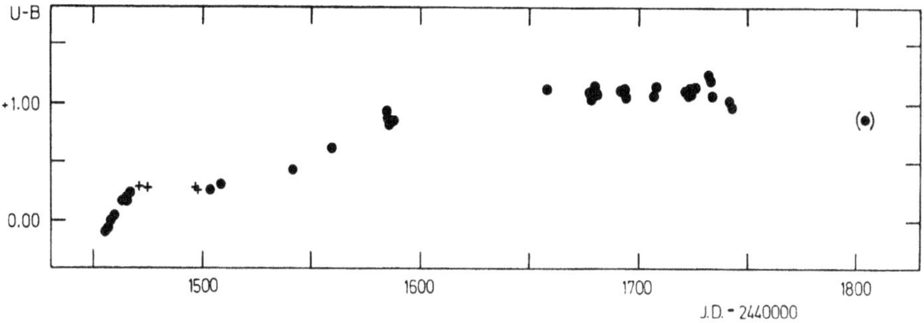

Fig. 10. $(U-B)$ colour indices plotted versus Julian Dates. Same notations as in Figure 3.

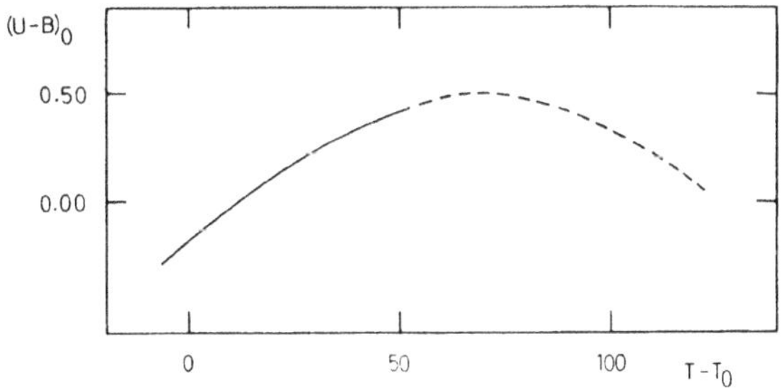

Fig. 11. Pskovskii's mean $(U-B)$ curve for supernovae of type I (Pskovskii, 1970).

5. The First Night of Photometry

A combination of standard *UBV* photometry and high-speed monitoring was acccomplished the first night of observation. Every set of *UBV* measurements was divided into several integrations with a length of one-three seconds. First a series of *B* measurements were made, then a set of *V* measurements and finally a set of *U* measurements. The first hour of observation the length of the individual integration was kept at one second. It was then increased to three seconds for another 1.5 h of observations. All the time the filters were changed approximately every 15 s. In this way we got accurate and fairly frequent UBV data at the same time as the possibility to investigate each of the three colours for rapid variations.

6. The Repeated *UBV* Measurements During the First Night of Observing

The magnitudes and colours from the repeated measurements during the first night are plotted versus fraction of Julian Date in Figure 12. Also the first-order least-squares relations are indicated.

The results displayed in Figure 12 may be used for a test of the small-scale constancy of the early decline. Table I gives the results of such a test. For the magnitudes and

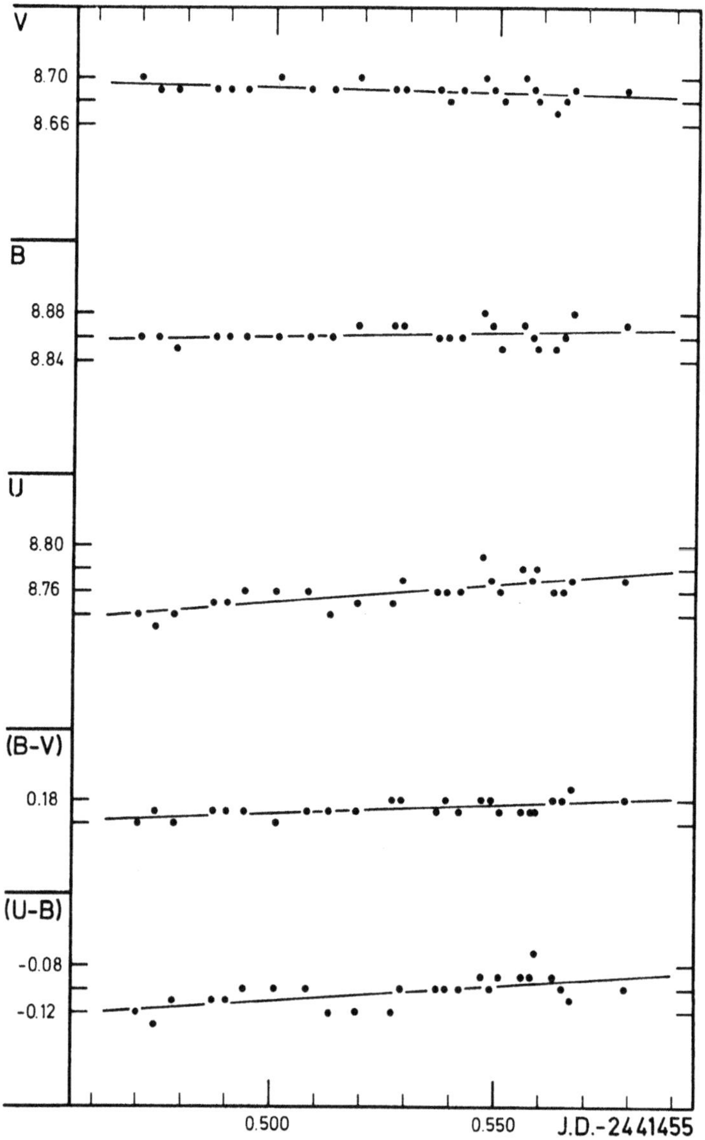

Fig. 12. Magnitudes and colour indices plotted versus Julian Dates from our repeated measurements on JD = 2441455. First-order least-squares relations are indicated.

TABLE I

Decay gradients in mag. per day as obtained from mean values from first and fourth nights and from first and second nights and from repeated measurements during first observing night. Standard deviations are given of single measurements with respect to the fitted first-order relations for the first night.

Time base	dV/dT	dB/dT	dU/dT	$d(B-V)/dT$	$d(U-B)/dT$
First 4 nights	+0.053	+0.080	+0.114	+0.027	+0.034
First 2 nights	+0.057	+0.082	+0.115	+0.025	+0.033
0.109 days	−0.086	+0.065	+0.318	+0.151	+0.253
σ(0.109 days)	0.007	0.008	0.009	0.005	0.009

colours decay gradients are given as defined by the night mean values of the first and fourth observing nights, the first and second observing nights and as defined by the fitted linear relations from the first night. Finally, standard deviations of a single measurement are given with respect to these fitted relations.

From Table I it is quite obvious that the large-scale decay during the first four nights of observations is quite constant. It is equally obvious that this constancy does not hold on a small time scale. It should be pointed out that frequent measurements of the local standard star excludes the possibility of spurious gradients during the first night. Consequently the gradient values together with the small spread around the relations clearly point to the existence of a second-order decay fluctuation with a period between 0.1 and 1.0 days.

7. The High-Speed Monitoring

For every series of high-speed integrations the mean value was calculated together with the standard deviation of a single integration. In Table II we give for the 1*s* and 3*s* integrations the mean values of these standard deviations, σ. Also the standard deviations of these standard deviations, σ_σ, are given.

All the series of 1*s* and 3*s* integrations have been searched for rapid fluctuations. Table III summarizes the results. With the assumption of a normal error distribution we have calculated the number of series with standard deviations for a single integration, σ_s, in the ranges given in the table.

Table III shows that the probability is low for a short-time activity. This result is in good agreement with that of Osmer *et al.* (1972). From high-speed monitoring carried out on May 22 they conclude that no activity is present on a short time scale.

We can then conclude that the only second-order decay fluctuation found has a period of 0.1–1.0 days. The cause of this fluctuation could be a density wave of short duration.

8. Spectroscopic Observations

All the spectrogrammes of the supernova have been taken with the ESO 1.5 m telescope (Fehrenbach, 1968). A total of 19 spectrogrammes have been taken with the

TABLE II

Standard deviations in magnitudes for 1*s* and 3*s* integrations in *V*, *B* and *U* for the 1972 supernova in NGC 5253. Measurements made on JD = 2441455.

	σ_V	σ_{σ_V}	σ_B	σ_{σ_B}	σ_U	σ_{σ_U}
1*s* integr. time	0.007	0.002	0.007	0.002	0.013	0.010
3*s* integr. time	0.004	0.002	0.003	0.001	0.005	0.003

TABLE III

Results of high-speed monitoring of the 1972 supernova in NGC 5253 made on JD = 2441455. σ_s denotes standard deviation of a single measurement series. σ and σ_σ have the same meaning as in Table II.

	V, 1*s*	*V*, 3*s*	*B*, 1*s*	*B*, 3*s*	*U*, 1*s*	*U*, 3*s*
$\|\sigma_s - \sigma\| \leqslant \sigma_\sigma$	5	13	5	10	7	10
$\sigma_\sigma < \|\sigma_s - \sigma\| \leqslant 2\sigma_\sigma$	3	3	3	5		6
$\|\sigma_s - \sigma\| > 2\sigma_\sigma$		1		2	1	1

standard Cassegrain spectrograph (Baranne *et al.*, 1969). The dispersion of these spectrogrammes is 74 Å mm^{-1} and the range is λλ3250–5000 Å. Three spectrogrammes have been taken in Coudé focus with a dispersion of 20 Å mm^{-1} and a range of λλ3400–4900 Å. Some additional spectrogrammes have been taken in Cassegrain focus.

The first night of observing 12 spectrogrammes were obtained in Cassegrain focus. Up till the middle of January, 1973, spectrogrammes were taken regularly enough to secure an adequate tracing of the spectral development.

As the light curves the spectrogrammes show the supernova to be of type I. The features visible at a visual inspection are quite similar to those normally observed in type I supernovae (Zwicky, 1965). Five intensity maxima are found between $\lambda = 4600$ Å and $\lambda = 3600$ Å. No small-scale feature can be readily observed.

Only the two redmost intensity maxima can be followed through all the period covered by the spectrogrammes. For these maxima visual estimates give displacements towards higher wave-lengths of the order of 100 Å during the time elapsed from our first to our last spectrogramme. As far as can be seen the other intensity maxima seem to follow this trend. This would be in good agreement with the form of the colour curves obtained.

The measurements of the spectrogrammes is in progress. Positions and strengths of all visible features are measured. Special attention is given to a search for small-scale features.

References

Austin, R. R. D.: 1972, *IAU Circular*, No. 2421.
Baranne, A., Maurice, E., and Prévot, L.: 1969, *Eur. South. Obs. Bull.* **7**, 11.
van den Bergh, S.: 1960, *Z. Astrophys.* **49**, 201.
Cousins, A. W. J.: 1972, IAU Inform. Bull. on Variable Stars, No. 700.
Fehrenbach, C.: 1968, *Eur. South. Obs. Bull.* **3**, 31.

Hooghoudt, B. G.: 1966, *Eur. South. Obs. Bull.* **1**, 28.
Kowal, C. T.: 1972, *IAU Circular*, No. 2405.
Kowal, C. T. and Sargent, W. L. W.: 1971, *Astron. J.* **76**, 756.
Lee, T. A., Wamsteker, W., Wisniewski, W. Z., and Wdowiak, T. J.: 1972, *Astrophys. J.* **177**, L59.
Mihalas, D.: 1963, *Publ. Astron. Soc. Pacific* **75**, 256.
Minkowski, R.: 1964, *Supernovae and Supernova Remnants in Annual Review of Astronomy and Astrophysics*, Vol. 2, p. 247.
Osmer, P. S., Hesser, J. E., Kunkel, W. E., Lasker, B. M., McCarthy, M. F., and Landolt, A. U.: 1972, *Nature Phys. Sci.* **238**, 21.
Przybylski, A.: 1972, *IAU Circular*, No. 2434.
Pskovskii, Yu. P.: 1970, *Astron Zh.* **47**, No. 5 (*Sov. Astron.-A. J.* **14**, 798).
Sérsic, J. L.: 1960, *Z. Astrophys.* **51**, 64.
Sérsic, J. L.: 1968, *Atlas de Galaxias Australes*, Observatorio Astronómico, Córdoba, Argentina.
Shane, C. D. and Wirtanen, C. A.: 1967, *Publ. Lick Obs.*, Vol. XXII, Part I.
de Vries, M.: 1966, *Eur. South. Obs. Bull.* **1**, 35.
Walker, Christie, Feasey and Freeth, 1972, *IAU Circular*, No. 2413.
Wisniewski, W., Lee, T., Wdowiak, T., and Michlovic, J.: 1972, *IAU Circular*, No. 2407.
Zwicky, F.: 1965, in L. H. Aller and D. B. McLaughlin (eds.), *Supernovae in Stellar Structure*, University of Chicago Press, Chicago and London, p. 367.

DISCUSSION

M. F. McCarthy: Would you explain the filling in of the absorption like depressions in spectrum obtained on 18 May.

A. Ardeberg: It is a photographic exposure effect in the production of the slide.

A. Treves: Has anybody looked for a pulsar in the supernova?

A. Ardeberg: To my knowledge only Osmer *et al.*, as I mentioned before. They did not find anything.

M. F. McCarthy: The work of the CTIO group searched for possible fluctuations and concluded that during the first days of observation of 1972e no fluctuations greater than 0.0008 mag. were noted. We estimated that a pulsar could be detected with present CTIO equipment if its absolute magnitude exceeded that of NP 0532 by ~ 5.

T. L. Cline: Could you give the exact date of the maximum?

A. Ardeberg: The B luminosity maximum was on JD $= 2441438.0 \pm 2.0$. assuming linear relations of the $(B-V)$ and $(U-B)$ curves around the time for luminosity maximum, we find the V luminosity maximum to have been reached 2–3 days after the B luminosity maximum and the U luminosity maximum 3–4 days before that of B.

J. Heidmann: In 1971 we made a study of NGC 5253 with the 21-cm line (*Astron. Astrophys.*). From it, it may be possible to obtain a distance evaluation. Do you think this could be useful for determination of the absolute magnitude at maximum of the SN, which appears to be quite on the bright side?

A. Ardeberg: Yes, absolutely. It would be very interesting, not only because the SN appears to be quite bright but also because existing distance determinations for NGC 5253 do not settle the question in a very definite way.

RECENT OBSERVATIONS OF SUPERNOVAE AT ASIAGO

R. BARBON, F. CIATTI and L. ROSINO
Asiago Astrophysical Observatory, University of Padova, Italy

More than 40 supernovae have been observed at Asiago from the spectroscopic and/or photometric point of view (for a summary on the Asiago supernovae program, see the communication of Rosino, this conference). We have thus collected a fairly large and homogeneous set of data, from which we intend to deduce the main properties of both type I and II, and see whether these objects can be used as reliable distance indicators. In this communication the most recent results are presented in advance of publication (Barbon *et al.*, 1973a, c). The most relevant data for nine supernovae are shown in Table I.

For the large distance from the nucleus of its parent galaxy the SN in NGC 2968 is to be remarked. It was discovered by Wild 'in the luminous bridge' connecting this irregular non-magellanic galaxy to NGC 2970 (in this connection see also Tammann, this conference). The first five supernovae very likely all belong to type I. For this classification we have spectra only of two objects (SN in NGC 4165 and in NGC 5055), but we trust on the light and color curves, very similar to each other and to the average curve for SN-I (Barbon *et al.*, 1973 b). A good fit is obtained by superimposing these light curves, all included indeed in building up the average curve.

The photometric observations of the SN in NGC 3147 seem again to indicate a type I SN, but a peculiarity is given by the sharpness of maximum, according to the pre-discovery observations by Goranskij (1973). The SN 1972-r in NGC 2841 (the third SN discovered in this galaxy), showing a constant rate of decline, is probably a SN-I during its final decline. From the fit to the average curve, we infer the maximum on November 5 or earlier, while Sargent and Searle (1973) deduced from the spectrum that this is a SN-I which attained its maximum on December 10.

The observations of the SN in NGC 5457 = M 101 and in NGC 4254 = M 99, both belonging to type II, are of interest since we have up to now only a few well studied SN-II with *UBV* photometry and spectra. Our spectrograms of SN in M 101, secured near maximum, show the characteristic broad emission bands with blueshifted absorption components ($v_{\text{abs}} = 7800 \text{ km s}^{-1}$) due to Balmer lines, probably He I, the blend at 4640 Å. The spectrum of SN-II does not differ from that of a nova, when much larger expansion velocity and mass of the ejected gas are involved. The light curves of these SN-II present the same shape, with a characteristic 'plateau' followed by a steep decline; in the last period SN-II seem to follow a linear decline with a smaller rate than SN-I.

It is always said that SN-II show a wide and disappointing variety in their photometric properties: even if this is true in several cases, we want to remark the close similarity among the SN-II here presented and the SN-II 1969-l in NGC 1058 (Ciatti *et al.*, 1971). This suggests that at least an important group of SN-II are very alike,

C. B. Cosmovici (ed.), Supernovae and Supernova Remnants, 115–118.

TABLE I

Supernova	Discoverer	Parent galaxy		Type of SN	Spectrum	Maximum			Distance (kpc)
		Name	Type			m_B	Date	M_B	
1970-j	Rosino	NGC 7619	E3	I	no	14.5	Oct 8	-18.8	7.7
1970-l	Wild	NGC 2968	I0	(I)	no	(12.5)	(Oct 5)	(-18.5)	10.7
1971-g	Rosino	NGC 4165	Sc	I	yes	13.5	Apr 13	-17.3	1.8
1971-i	Jolly	NGC 5055	Sb	I	yes	11.6	May 26	-17.6	3.3
1971-l	Logan	NGC 6384	Sb	I	no	13.6	Jun 29	-18.5	3.1
1970-g	Lovas	NGC 5457	Sc	II	yes	11.5	Jul 25	-16.8	8.8
1972	Goranskij	NGC 3147	Sbc	(I)	no	14.9	Jul 27	-17.8	6.7
1972-r	Wild	NGC 2841	Sab	I	yes	< 12.5	< Nov 5	< -16.8	2.4
1972	Rosino	NGC 4254	Sc	II	no	15.8	Dec 10	-17.7	5.3

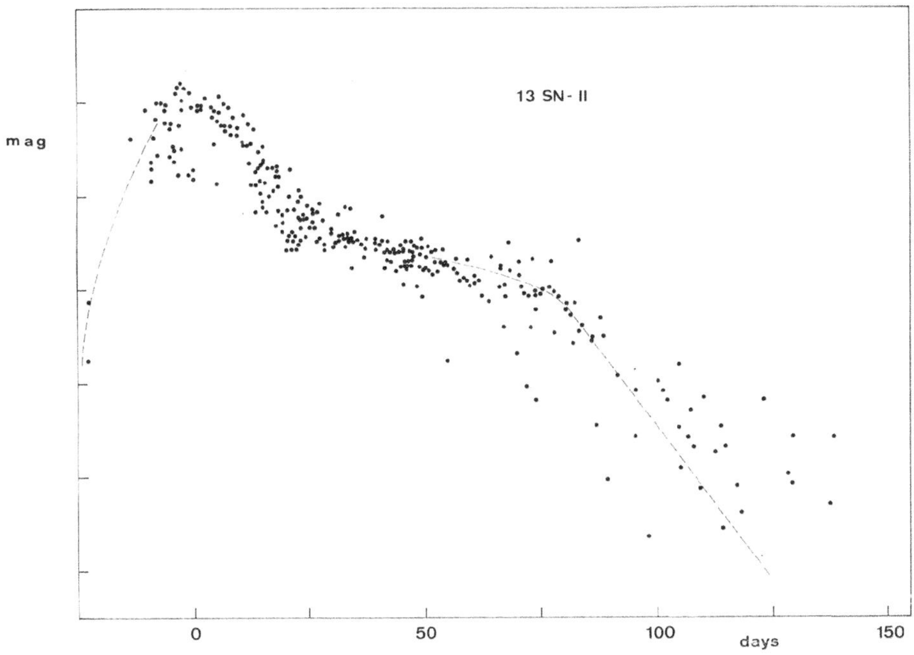

Fig. 1. Composite light curve obtained from 13 type II supernovae.

besides the spectral evolution, also in light and color curves and absolute magnitudes. Then, for this group, an 'average' light curve might be obtained, as we did for SN-I. Figure 1 shows a preliminary fit of 13 SN-II. Although the scatter is larger than for SN-I, the common trend allows the determination of relevant photometric data for supernovae lacking extensive observations.

References

Barbon, R., Ciatti, F., and Rosino, L.: 1973a, *Mem. Soc. Astron. Ital.* **44**, 65.
Barbon, R., Ciatti, F., and Rosino, L.: 1973b, *Astron. Astrophys.* **25**, 241.
Barbon, R., Ciatti, F., and Rosino, L.: 1973c, *Astron. Astrophys.*, in press.
Ciatti, F., Rosino, L., and Bertola, F.: 1971, *Mem. Soc. Astron. Ital.* **42**, 163.
Goranskij, V. P.: 1973, *Astron. Zirculiar*, No. 723.
Sargent, W. L. W. and Searle, L.: 1973, *IAU Circular*, No. 2498.

DISCUSSION

S. M. Hassan: Is it possible to find some type II supernovae showing constant rate of decline after max.?

F. Ciatti: There is the case of the SN in NGC 7331 studied by Arp.

A. Ardeberg: Do you suggest that your last diagram is mainly made up of two families of light curves? Or is there a more smooth distribution of different light curves?

F. Ciatti: According to Pskovski, a relation can be found between the rate of decline in the first period after maximum and the time of appearance of the 'shoulder'. The limiting cases are SN 1969 in NGC 1058 and SN 1970 in M 101.

W. P. S. Meikle: Can you say when the supernova in M 101 reached maximum?

F. Ciatti: Maximum was attained on July 25 $\pm$ 3 with a B magnitude 11.5, V=11.3.

G. A. Tammann: Bahcall, Kowal, and I have a photographic light curve of SN 1970-g. It shall be very interesting to compair our data.

OBSERVATIONS OF THE 1968 TYPE II SUPERNOVA IN NGC 5236*

ROGER WOOD
Royal Greenwich Observatory, Herstmonceux Castle, Hailsham, Sussex, England

1. Introduction

NGC 5236 is an Sc galaxy (Sandage, 1961) viewed almost face-on and is known to have produced four supernovae (e.g., Tammann, 1970). Two of these were of unknown type while that of 1923 shows a light curve (Lampland, 1936) suggestive of type II but apparently underluminous; the light curve is discussed by de Vaucouleurs elsewhere in these proceedings. The 1968 supernova was observed photometrically with the Radcliffe 1.88 m reflector following its visual discovery on July 16 by Pretoria amateur J. C. Bennett in the course of his search for comets. It is found to be typical of type II, although its site, superimposed on the nucleus of the galaxy, is unusual and in marked contrast to the three previous events which occurred far out in the spiral arm structure (Karpowicz and Rudnicki, 1968).

2. Photometry

The Radcliffe photometry by P. J. Andrews, is based on very short exposure Newtonian photography referred to a photoelectric sequence. On July 24 and August 24 several 10 to 30 s exposures of the field of NGC 5236 were obtained with the standard U, B and V plate and filter combinations, and with the telescope aperture stopped to 92 cm. Under these conditions no trace of the nucleus was present and the supernova and reference stars were measured with a Sartorius iris photometer.

Wamsteker's (1972) measurements of the supernova are about a magnitude brighter than the corresponding Radcliffe values, but systematic differences are ruled out by the excellent agreement for a common reference star. From a detailed description of his observational technique (private communication, 1972) it is clear that Wamsteker had underestimated the background contribution. However, he used an accurate off-setting procedure and any nuclear contamination would have been constant from night to night. By subtraction of this additional background from his measures, it was possible to reconcile the two sets of data quite satisfactorily. It should perhaps be mentioned that this background agreed very closely in $(B-V)$ with a photoelectric determination of the nucleus made in 1969. The resulting light curves show V to be more or less constant at about $10\overset{m}{.}0$ from discovery to mid-September while B drops steadily from $12\overset{m}{.}0$ at discovery to a plateau at $13\overset{m}{.}0$ which begins in about mid-August.

* The work summarised here was carried out jointly by P. J. Andrews of the Radcliffe Observatory, Pretoria, and the author. A full discussion of the observations, with diagrams, is presented elsewhere (Wood and Andrews, 1974).

C. B. Cosmovici (ed.), Supernovae and Supernova Remnants, 119–122.

Consequently $(B-V)$ is near zero at discovery, changes steadily to roughly 1.0 in mid-August and remains near that value. The similarity of these curves to those for the 1969 type II supernova in NGC 1058 (Ciatti *et al.*, 1971) is very striking and adds further weight to this morning's suggestions by Ciatti that these may define a sub-class of type II supernovae.

An upper limit for the absolute magnitude at maximum was derived by extrapolating the B light curve to the date of maximum indicated by the appearance of the spectra, July 12, and assuming an intrinsic $(B-V)$ of $-0\overset{m}{.}3$ at maximum. The total reddening (Galactic and in NGC 5236) is then $0\overset{m}{.}3$ giving $A_B = 1\overset{m}{.}2$ and on this basis $M_{B\,\mathrm{max}} = -17\overset{m}{.}3$ using de Vaucouleurs's (1966) distance modulus. Less reddening in the parent galaxy or a slightly redder $(B-V)_{\mathrm{max}}$ would make this figure fainter, but it is unlikely to be by more than $0\overset{m}{.}5$.

3. Spectroscopy

The fifteen spectrograms cover a period of 57 days after discovery, after which time the supernova faded into the nuclear background. During the first fortnight the spectrograms were obtained on Kodak IIaO emulsion and show the spectrum to evolve from almost pure continuum at discovery (indicating that maximum was only a few days before) to a well developed and typical type II structure a week later. Substantial depressions at 4250 Å and 4750 Å have generally been indentified with very broad, blue-shifted Balmer lines, Hγ and Hβ respectively, while other features are believed to be due to metals, although there is, as yet, no consensus as to precise identifications. On the earliest spectrograms the continuum fades smoothly to the clear plate level in the blue (because of increasing absorption in the spectrograph optics) but a week later there appears to be an absorption 'edge' at about 3900 Å.

Spectrograms taken two to eight weeks after discovery were on Kodak 103a F emulsion and show the structure out to Hα. In the blue the structure maintained essentially the same form as seen earlier, but becomes more and more 'washed out' as the supernova faded. The banded structure in the red is dominated by a very broad peak centred on about 6000 Å.

On July 19 the spectrum of the supernova and that a star of known energy distribution (Willstrop, 1965) were photographed under closely similar conditions. A comparison led to a determination of the shape of the energy distribution of the supernova light just above the earth's atmosphere; this distribution was found to resemble that of a black body at about 8000 K. However, taking into account that there may have been up to $0\overset{m}{.}3$ reddening in $(B-V)$, the intrinsic energy distribution a week after maximum was probably more like that of an 11 000 K black body (Arp, 1961). These figures tie in fairly well with those from scanner observations of type II supernovae (Kirshner *et al.*, 1973) where continuum distributions resemble black bodies of about 10 000 K about a month after maximum.

Wavelengths of intensity maxima and minima were measured on each spectrogram; some features were found to drift to the red as the supernova faded. In terms of

Patchett and Branch's (1972) identifications the absorptions due to Hβ and 'Fe II 4550' drift by about 60 Å, the shape in a wavelength-time diagram being very similar to that for the maxima in type I spectra (Minkowski, 1939). The steepness near discovery again suggests maximum having occurred only a few days previously. However, other minima do not show the redward drift, the examples for which the data are most complete being Hγ and 'Fe II 4233'. The maxima at 4600 Å and 4350 Å drift to the red, while that at 4800 Å is essentially constant. It is particularly interesting to note that the drift of the 4600 Å feature mimics its behaviour in type I supernovae.

4. Discussion

It seems that this supernova may fall into a sub-class of type II supernovae represented by the 1969 supernova in NGC 1058. Assuming a distance modulus of 28.0 (de Vaucouleurs 1966) and about $0\overset{m}{.}2$ of reddening in NGC 5236 the absolute B magnitude of maximum was $-17\overset{m}{.}3$, although this may be reduced by up to half a magnitude or so if the reddening was smaller or the intrinsic $(B-V)$ at maximum was redder than $-0\overset{m}{.}3$. The spectrogram of July 19 indicates that the energy distribution between 4000 Å and 5000 Å resembles that of an 11 000 K black body; and it is noted that, as proposed by Kirshner *et al.* (1973), the bulk of the energy flux appears to be carried by continuum radiation. Wavelength shifts of features in type II spectra now seem to be well established (Bertola, 1963; Ciatti *et al.*, 1971) and in this case the maximum at 4600 Å behaves in a similar fashion to the well documented drift seen in type I supernovae. The suggestion (Zwicky, 1965) that supernovae in NGC 5236 are under-luminous now seems to be dispelled: it would rather appear that those in the irregular galaxy NGC 5253 (generally supposed to be physically related to NGC 5236) are over-luminous.

References

Arp. H. C.: 1961, *Astrophys. J.* **133**, 874.
Bertola, F.: 1963, *Mem. Soc. Astron. Ital.* **34**, 33 (= Asiago Contrib. No. 135).
Ciatti, F., Rosino, L., and Bertola, F.: 1971, *Mem. Soc. Astron. Ital.* **42**, 163 (= Asiago Contrib. No. 255).
Karpowicz, M. and Rudnicki, K.: 1968, *Publ. Astron. Obs. Univ. Warsaw* **15**, 1.
Kirshner, R. P., Oke, J. B., Searle, L., and Penston, M. V.: 1973, *Astrophys. J.*, in press.
Lampland, C. O.: 1936, *Publ. Astron. Soc. Pacific* **48**, 320.
Minkowski, R.: 1939, *Astrophys. J.* **89**, 156.
Patchett, B. and Branch, D.: 1972, *Monthly Notices Roy. Astron. Soc.* **158**, 375.
Sandage, A.: 1961, *Hubble Atlas of Galaxies*, p. 28 (Carnegie Institute of Washington Publication, No. 618).
Tammann, G. A.: 1970, *Astron. Astrophys.* **8**, 458.
de Vaucouleurs, G.: 1966, *Stars and Stellar Systems*, Vol. 9, draft chapter (= McDonald Observatory Preprint, No. 9, University of Texas, 1971).
Willstrop, R. V.: 1965, *Mem. Roy. Astron. Soc.* **69**, 83.
Wamsteker, W.: 1972, *Astron. Astrophys.* **19**, 99.
Wood, R. and Andrews, P. J.: 1974, *Monthly Notices Roy. Astron. Soc.*, in press.
Zwicky, F.: 1965, *Stars and Stellar Systems*, Vol. 8, *Stellar Structure*, (ed by L. H. Aller and D. B. McLaughlin), p. 367.

DISCUSSION

M. Friedjung: Were the wavelength shifts constant with respect to radial velocity?

R. Wood: We did look into this very closely. It seemed to be so, though we did not have enough data. Using Branch and Patchett's identifications the radial velocity decreased from about 7000 to 4000 km s^{-1}.

J. V. Jelley: You said there was a total of four known SN in this one galaxy. As a point of interest, were the other SN of the same type?

R. Wood: Before 1968 the SNe in this galaxy were thought to be anomalous since they were apparently much fainter than in other galaxies. Only the 1923 SN had a light curve and this is reminiscent of type II. It has a well determined maximum at photographic mag. 14, making it 2 mag. fainter than normal, but there has been a suggestion that the 'maximum' is in fact the post-maximum plateau. The SNe of 1950 and 1957 were faint and neither spectra nor light curves were obtained.

H. Arp: Your H-alpha picture of NGC 5236 looked resolved and irregular right into the very center. Is the galaxy what Sersic calls a hot spot nucleus galaxy?

R. Wood: I'm not familiar with the term hot spot nucleus but there is certainly considerable, small scale blobby structure on both Hα and blue plates.

PART III

SPECTRA OF SUPERNOVAE AND THEIR INTERPRETATION

THE INTERPRETATION OF THE SPECTRA OF SUPERNOVAE

LEONARD SEARLE
Hale Observatories, Carnegie Institution of Washington, California Institute of Technology, Pasadena, Calif., U.S.A.

1. During the past decade substantial progress has been made with the difficult problem of interpreting the spectra of supernovae. The basic character of the spectrum is now apparent and there seems to be general agreement that some few spectral features have been successfully identified in the spectra of supernovae of type I as well as in those of type II. This recent progress stands in striking contrast to the relative stagnation that the subject experienced in the quarter of a century that followed the first detailed description of the spectrum of a supernova by Minkowski (1939). It results in part from the development of new methods of observation and in part from the abandonment of earlier preconceptions regarding the character of supernovae spectra.

Minkowski's (1939) description of his great series of observations of SN 1937c (IC 4182) has been, until recently, the starting point for all attempts to interpret the spectra of supernovae. It is still a fundamental source and has recently been made yet more valuable by the work of Greenstein and Minkowski (1973) which provides an intensity calibration of the original spectrograms. Most supernovae are found to have spectra that very closely resemble SN 1937c (IC 4182) when observed at comparable epochs but Minkowski (1941) made the important discovery that others do not. He called the first group type I's and the latter group type II's.

Since it subsequently turned out that the spectra of supernovae that are not type I's are not themselves all alike, it is clearly desirable to refine this classification. Zwicky (1965) has described the spectra of prototypes of groups that he names types III, IV and V. It certainly does seem useful to use the term type II in a more restricted sense than as the equivalent of 'not type I'. Most supernovae that are not type I's differ from type I's in having spectra with the Balmer series clearly present during the first three weeks following maximum light. I shall here use the term type II to refer to these supernovae, whatever other differences their spectra reveal*. Beyond the fact that some few do occur, little is known concerning the small group of supernovae that are neither type I nor type II.

The distinction between types I and II appears to be quite sharp and almost certainly has physical significance but I think that it is important to emphasize that recent observations have shown that many spectral features are common to super-

* This is purely a matter of convenience and is not meant to deny that there are significant differences among the type II's so defined. It was stressed at the conference that subdivisions of type I may also eventually become necessary. My type II includes Zwicky's types II, III and V but not his type IV.

C. B. Cosmovici (ed.), Supernovae and Supernovae Remnants, 125–130.

novae of both major types and that the basic character of the spectrum appears to be the same in supernovae of all types.

2. For a statement of the 'classical' view concerning the nature of the spectra of supernovae it is best to cite the words of Minkowski. In his paper of 1939, discussing the spectrum of SN 1937c (IC 4182) he wrote "... after the star has reached its maximum brightness, the appearance of the spectrum leaves little doubt that the spectrum is composed essentially of wide and partially overlapping emission bands." Concerning the complete lack of satisfactory identifications he wrote: "The assumption of low ionization, which would permit the identification of the band at about $\lambda 5900$ with the NaD lines, is contradicted by the complete absence of other lines which would be expected to accompany the D lines such as Ca I and Ca II. ... We are thus led to the conception that the bands ... are emitted by highly ionized atoms ... the failure to obtain identifications may be due merely to the incompleteness of our knowledge of the spectra of highly ionized atoms." Later when he described the spectra of type II supernovae Minkowski wrote: "Supernovae of type II differ from those of type I in the presence of a continuous spectrum at maximum and in its subsequent transformation to an emission spectrum whose main constituents can be readily identified."

Two studies of untypical supernovae – ones that had spectra with especially narrow and well-defined features – appear, in retrospect, to have been particularly important in suggesting new ideas and new directions for research. One of these concerned the spectrum of SN 1954a (NGC 4214) which bore some resemblance to the spectra of standard type I's. McLaughlin (1963) noted that it showed "an unusually smooth distribution of energy except for a few deep absorption like minima." The other concerned the spectrum of SN 1961v (NGC 1058) which, as I have defined the class, would be called a type II but which, in Zwicky's classification, is a type V. In any case its spectrum bore some resemblances to the more usual variety of type II's. Bertola (1963) reported that it also rather closely resembled the spectra of common novae shortly after maximum light. He identified spectral features with H, He I and Fe II and reported that these features had the character of emission lines flanked by violet shifted absorptions. A more detailed study by Branch and Greenstein (1971) confirmed Bertola's conclusions.

As for the nature of the overall energy distribution in supernovae spectra, Arp (1961) reached the important conclusion that the light curve and colors of the type II SN 1959d (NGC 7331) implied the existence of an opaque, cooling and expanding envelope that resembled in many ways the atmosphere of a supergiant star. Essentially the same conclusion was reached by Mustel (1971a, b) for type I supernovae using the spectrophotometry of Bloch (1964).

The idea that the spectra of normal type I's might consist of broad and overlapping absorption lines cutting into a continuum was first suggested by Pskovskii (1968). This hypothesis has been modified and developed by Mustel (1971a, b, 1972) who suggests that the absorption lines that occur are those expected in a tenuous envelope irradiated by a dilute radiation field with a color temperature of $\sim 6000°$. Lines arising from metastable levels of once-ionized abundant metals are prominent in his list of suggested identifications.

A related line of investigation developed from the study by Branch and Greenstein (1971) of SN 1961 v (NGC 1058). The conclusion that the identifications proposed in this case can also account for the broad spectral features of normal type II's has been reached by Patchett and Branch (1972). In a recent paper Branch and Patchett (1973) conclude that the main features in normal type I's can also be explained in the same way.

These investigations have had some qualitative success in accounting for the general location of the troughs and peaks in the spectra of supernovae. Now that extensive observations are being produced by photoelectric scanning of supernovae spectra it may become possible to make quantitative syntheses on the basis of these ideas. Until that is done the proposed identifications must remain in doubt. However, the important thing is that a view has emerged regarding the character of the supernova spectrum that is quite different from the classical one. This new view, which has now been decisively confirmed by photoelectric scanning observations, is that the spectrum of both type I's and type II's consists of a continuum modified by the superposition of broad emission and absorption lines produced in a low temperature envelope.

3. The studies of supernovae spectra which I have here briefly reviewed were almost entirely based on photographic spectra. The features in supernovae spectra are so broad compared to the wavelength range of the spectrograms and the intensity contrasts are so much greater than the latitude of the photographic emulsion that it is very difficult indeed to form a correct view of the nature of the spectra by the usual methods of photographic spectroscopy. A technique is required which records the spectrum over a larger wavelength interval and with a much larger dynamic range. One instrument well suited to the task is the multi-channel spectrophotometer (Oke, 1969) in use at the Cassegrain focus of the 200-in. Telescope. Since 1968 this instrument has been used to observe supernovae spectra. The occurrence of a number of relatively bright supernovae in recent years and in particular of SN 1972e (NGC 5253) has enabled observations to be gathered fairly rapidly and a preliminary atlas of spectral scans and a brief discussion of their interpretation is given in a paper* to be published in the October 1, 1973 issue of the *Astrophysical Journal* by Kirshner, Oke, Penston and Searle.

These new observations make the basic character of supernovae spectra very clear. There is a continuum in both type I's and type II's at all phases that have so far been observed. The observations obtained include phases as late as one year after maximum light for supernovae of both types. The continuum changes slowly and uniformly with time during the month following maximum light and thereafter remains unchanging. Soon after maximum the color temperature is about 10000° and this drops to about 5000° in the first month and then remains constant. The similarity between

* It is unnecessary to repeat the contents of that paper here although at the conference the observations were illustrated and described in some detail.

type I's and type II's in this behavior is very striking and the uniform and systematic change in the overall shape of the energy distribution of type I's shows quite clearly that the spectrum does not consist of overlapping broad emission lines. On the contrary the continuum carries the bulk of the flux radiated by supernovae of both types.

There are lines in the spectra of both type I's and type II's whose shape resembles those of lines seen in the spectra of P Cygni and of novae shortly after maximum light. These lines consist of broad emissions flanked on their violet edges by broad absorptions. The line shape suggests that the line-forming mechanism is resonance scattering in an expanding envelope and, if so, the lines mainly redistribute in frequency the continuum radiation of the underlying photosphere. In type II's, which have relatively few and weak lines, this line shape is clearly seen to be the general shape of most lines, although the profiles of different lines differ in detail. In type I's where the lines are strong and crowded this characteristic P Cygni profile is clearly discernible in a few particularly strong and isolated lines but it is probably also the basic shape of the components of the blended features in more crowded regions of the spectrum.

Although the spectra, and in particular those of type I supernovae, change with time, there are some lines that persist right through the evolution of the spectra and which are common to type I's and type II's. These lines all consist of an emission and an associated violet-displaced absorption. The strongest persistent lines which are common to supernovae of all types are the H and K resonance lines and the infrared triplet lines of Ca II. The emission peaks are at the rest wave-length (in the rest frame of the parent galaxy) while the minima of the associated absorptions are blue shifted by about 7000 km s^{-1} in type II's and by about 12000 km s^{-1} in type I's. The expansion velocities of the outermost layers of the supernovae envelopes are much greater than this. These velocities, estimated from the violet edges of the violet shifted absorptions, are 15000 km s^{-1} for type II's and 20000 km s^{-1} for type I's.

The identification of the Ca II lines seems quite secure and we have suggested identifications for a few other features. However most spectral features have not yet been successfully identified. No doubt many of the features are blended and the usual method of identification by wavelength coincidence is hardly applicable. A method of spectrum synthesis like that pioneered by Branch and Greenstein (1971) seems the most likely way to make progress with this difficult problem.

4. The most important aspect of the new interpretation of the spectra of supernovae is the possibility it opens up of determining the distance to supernovae by purely astrophysical means. It clearly would be most desirable to have a fundamental check on the involved and uncertain chain of observations and argument that goes into the present determinations of the extragalactic distance scale. To apply to supernovae Baade's (1926) method (i.e. his method of distance determination for variable stars) it is first necessary to be able to determine the effective temperature of the supernovae from the observed spectral energy distribution. A second requirement is to determine the expansion velocity of the photosphere (which is not, of course, a material layer)

from the line profiles of identified spectral features. Unfortunately, at the moment neither of these things can be done with respectable accuracy. The first requires a knowledge of the opacity sources in the envelope. The second requires a kinematic model of the line-forming region. The required understanding of supernovae envelopes is probably attainable but is certainly still some way off.

However the job can be done approximately by fitting the observed energy distributions to black body curves and by roughly estimating the expansion velocities. Such procedures have been applied by Branch and Patchett (1973) to type I supernovae and by Kirshner (1974) (using rather more realistic assumptions) to type II's. These authors arrive at a distance scale in essential agreement with that of Sandage and Tammann (1970). The importance of these results, in my view, is that they show that the interpretation of the supernova spectrum that has emerged from the investigations that I have surveyed in this talk, is almost certainly basically correct. That being so, of course, it really becomes very important to attempt to construct detailed models for supernovae envelopes and it is very much to be hoped that makers of model stellar atmospheres will interest themselves in this problem. Spectral scans of distant supernovae can readily be obtained and they surely contain all the information needed for a fundamental determination of the Hubble constant.

References

Arp, H.: 1961, *Astrophys. J.* **133**, 883.
Baade, W.: 1926, *A.N.* **228**, 359.
Bertola, F.: 1963, *Asiago Contr.*, No. **142**.
Bloch, M., Chalonge, D., and Dufay, J.: 1964, *Ann. Astrophys.* **27**, 315.
Branch, D. and Greenstein, J. L.: 1971, *Astrophys. J.* **167**, 89.
Branch, D. and Patchett, B.: 1973, *Monthly Notices Roy. Astron. Soc.* **161**, 71.
Greenstein, J. L. and Minkowski, R.: 1973, *Astrophys. J.* **182**, 225.
Kirshner, R. P.: 1974, private communication.
McLaughlin, D. B.: 1963, *Pub. Astron. Soc. Pacific* **75**, 133.
Minkowski, R.: 1939, *Astrophys. J.*, **89**, 143.
Minkowski, R.: 1941, *Pub. Astron Soc.. Pacific.* **53**, 224.
Mustel, E. R.: 1971a, *Astron. Zh.* **48**, 3.
Mustel, E. R.: 1971b, *Astron. Zh.* **48**, 665.
Mustel, E. R.: 1972, *Astron. Zh.* **49**, 15.
Oke, J. B.: 1969, *Pub. Astron. Soc. Pacific.* **81**, 11.
Patchett, B. and Branch, D.: 1972, *Monthly Notices Roy. Astron. Soc.* **158**, 375.
Pskovskii, Yu. P.: 1968, *Astron. Zh.* **45**, 945.
Sandage, A. R. and Tammann, G.: 1970, *Ann. Rep. Dir. Hale Obs.*
Zwicky, F. 1965, *Stellar Structure*, Ch. 7. (ed. by L. H. Aller and D. B. McLaughlin), University of Chicago Press, Chicago.

DISCUSSION

D. Branch: It seems well established that the *B-V* index for type I supernovae reaches a maximum about 40 days after maximum light, and then decreases. Do you know how to reconcile this with your statement that the continuum carries most of the flux and reddens monotonically?

L. Searle: No, I'm not sure why that happens, but presumably emission lines begin to play a role. The reddening of the continuum is most noticeable during the first month after maximum light. After that it ceases to redden but a continuum is clearly present for another two months or so. I have not tried to derive *UBV* colors from the detailed energy distribution but of course it could be done by anyone interested in interpreting broad band color data.

F. Zwicky: Your calculation of the absolute luminosity of SN by assuming sphericity of the expanding shells is incorrect by about a factor 10 because we know:

(a) from observation that all shells disrupt into a set of clouds and

(b) the shells, from theory cannot possibly expand as uniform spheres because of the precipitous instabilities associated with the generation of the large scale magnetic fields. The factor 10 above accounts then for a factor 3 driving the Lundmark-Hubble constant from 50 km s^{-1} 10^{-6} pc^{-1} to 150 km s^{-1} 10^{-6} pc^{-1}.

L. Searle: The luminosity and temperature determinations at two epochs give the ratio of the emitting surfaces independent of the details of geometry. Integrating the velocity curve gives the linear expansion of the surface. I don't see how you can pull a factor of ten out of that whatever the geometry may be.

D. Branch: The beautiful homogenity of type I SN suggest to me that the spherical approximation may not be too bad, at least during the optical phases.

C. Gordon: If you assume a black body distribution for the 'continuum' of type I SN, what will be the source of opacity in the inner layers of the shell big enough to give LTE.

L. Searle: Although at maximum light there is no evidence for Hydrogen in type I SN, I think the Balmer series is present after 15 days. So perhaps Hydrogen provides the opacity in type I's just as I suppose it does in type II's.

S. F. Gull: The most important parameters that determine the evolution of SN remnants are the total ejected mass and kinetic energy of the ejected shell. Can you summarize the information that may be obtained about these quantities from the spectra of supernovae?

L. Searle: I think the spectra tell us little about the total ejected mass; the figure that I gave of 0.3 $M_\odot$ referred to the transparent layers above the photosphere in type II SN. Expansion velocities are about 10000–15000 km s^{-1} for this mass.

S. F. Gull: Are the spectra of type II SN then consistent with Arnett's model in which a small fraction (1–2 $M_\odot$) of the mass received the majority of kinetic energy although the total mass ejected can be large as $\sim 10\ M_\odot$?

L. Searle: Yes in a general way. The deeper layers are certainly travelling more slowly. The outer shell, that produces the resonance lines observed in the spectra, probably contains a small fraction of the mass but a large fraction of the kinetic energy.

C. Gordon: The mass determined by Searle (0.3 $M_\odot$ in the outer layers $v > 15000$ km s^{-1} is consistent with my determination of the total ejected mass ($M > 5\ M_\odot$ from $v \simeq 4 \times 10^8$ cm s^{-1} to ∞).

L. Woltjer: Having determined the radius, do you obtain the right epoch for the SN explosion by extrapolating backwards at constant velocity?

L. Searle: Yes, the photometric radii and the observed expansion velocities yield an epoch for the explosion which is about 10 days before maximum light.

G. A. Tammann: Can you say something about chemical differences between type I and type II SNe from your exciting scans?

L. Searle: Not much yet. I can only speculate. Since the lines are stronger in type I's and the mass in the reversing layer is determined by electron scattering it is tempting to conclude that the number of metal ions per free electron is larger in type I's than in type II's. I would speculate that Hydrogen is relatively deficient in type I's.

L. Sartori: You don't exclude the possibility that the 4600 band is He II, do you?

L. Searle: No, the feature drifts in wavelength, as is well known, so is probably a blend. We could find no convincing evidence in our spectra for either He I or He II but we can't exclude it either. If there are He lines in type I SN they carry less than 10% of the flux at all phases observed by us – including my most recent observation obtained 350 days after maximum light.

M. F. McCarthy: In the spectrum of SN 1972e can you explain the variation in time of the feature at 7500–7600 Å? Is there any attribution of this feature to a common element? Why does it disappear about 100 days after maximum?

L. Searle: We have identified only a few of the elevations and depressions in the spectra of type I SN. The variations in the region between Hα and the infrared Calcium triplet are very complicated and suggest that there are several blended contributers to the feature you mention.

THE SPECTRUM OF SN 1972e AND THE FLUORESCENCE THEORY OF SUPERNOVA LIGHT

L. SARTORI
University of Nebraska, Lincoln, Neb., U.S.A.

and

BELLA C. CHIU and P. MORRISON
Massachusetts Institute of Technology, Cambridge, Mass., U.S.A.

Several years ago two of us proposed an explanation for the SN I light curve based on a fluorescence mechanism (Morrison and Sartori 1966, 1969). According to our model, the observed broad spectral features represent fluorescence excited in the medium surrounding the supernova by a pulse of soft x-rays produced in the explosion. The time behavior of the intensity measured by a fixed observer reflects the spatial attenuation of the primary pulse; for a single line and uniform density of absorbing material, the light curve is given by an exponential integral, which behaves asymptotically as e^{-x}/x. With the dominant photographic band identified as He II 4686, the most elementary version of the theory gives an excellent one-parameter fit to the photographic light curve from about 40 days on. The redward shift of the dominant feature is also naturally accounted for. The principal demands of the theory are:

(i) some tenths of a solar mass of helium must have been ejected over the integrated history of the pre-supernova, and must have diffused out to fill a region a light year or two in radius;

(ii) approximately 10^{51} erg is required in the exciting pulse in the energy range 50–54 eV. This leads to an estimate of $\sim 10^{52}$ erg as a lower limit on the energy associated with the explosion.

Additional observational support for the fluorescence theory was provided by recent studies of the Gum nebula (Brandt *et al.*, 1971), which indicate the presence of a very large 'fossil Stromgren sphere'; the theory predicts that such an ionized region ought to surround every SN I remnant, produced by the *u-v* and x-ray burst.

The theory implies also that other strong He II lines ought to be present in the spectrum, among them the infrared line at λ 10124 (5–4 transition). One can calculate the ratio between the intensity of this line and that of λ4686; in the simplest version of the theory, the ratio is independent of time, although that is by no means a general result. Chiu and Shields (1973) have discovered a numerical error in Table I of Morrison and Sartori (1969), and report a revised value of 0.117 for this ratio.

A feature that could well be the He II line in question is present in the infrared spectrum taken by Ciatti (1973) of the recent bright supernova in NGC 5253, SN 1972e (Figure 1). The blue shift of the line is consistent with that of λ4686. The identification is, just as for most supernova spectral features, not unambiguous. Unfortunately, it is not possible from the available data to make a direct determination of the intensity

C. B. Cosmovici (ed.), Supernovae and Supernova Remnants, 131–133.

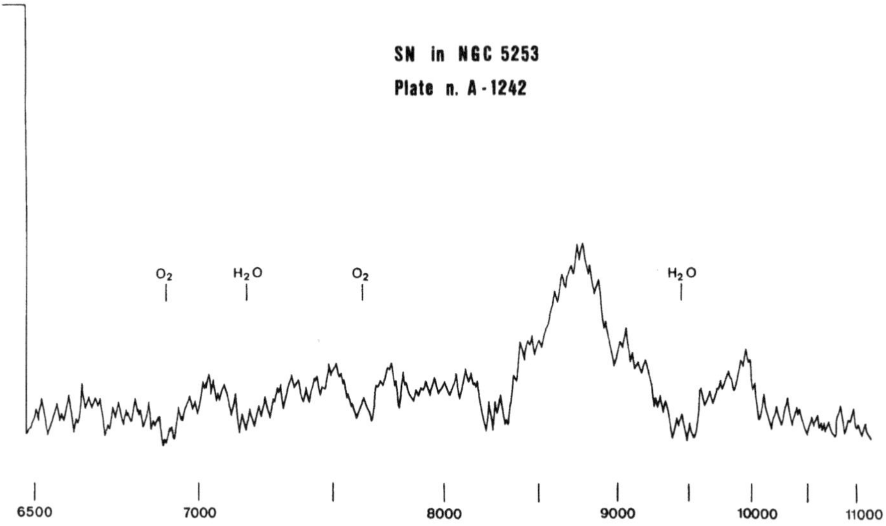

Fig. 1. Infrared spectrum of SN 1972e taken by F. Ciatti at the Asiago Observatory, 6/5/72.

ratio between the infrared feature and the 4686 band in SN 1972e. Chiu and Shields (1973) have made a crude estimate of this ratio by matching Ciatti's spectrum with that obtained by Bloch *et al.* (1964) on the supernova in NGC 4496 at a similar epoch. They report agreement, within a factor of about 2, with the theoretical value quoted above. It would clearly be of interest to obtain a reliable determination of the intensity ratio from data on the same supernova (including time dependence if any).

Spectra of recent supernovae, and particularly SN 1972e, are presented and discussed in some detail by Kirshner *et al.* (1973; see also the contribution of L. Searle at this conference). According to these authors, the principal shortcoming of the fluorescence theory is its failure to account for the strong observed continuum. (The fluorescent continuum contributes only about 8% as much as the lines, whereas according to Kirshner *et al.*, the 'bulk' of the energy is carried by the continuum.) In this connection, we ought to emphasize a remark that was perhaps not made explicitly enough in our original paper. The fluorescence radiation observed at time t emanates from regions at distances $\geqslant \frac{1}{2}ct$ from the explosion site – far outside the expanding shell. In addition, one of course observes any direct radiation originating from the dense expanding matter. The excitation pulse has long since passed, and this material has recombined and is acting like a star surface. Its radiation may be expected to be rich in continuum, and also to contain the P Cygni lines that Kirshner *et al.* discuss. Certainly during the early days, the direct contribution is likely to be dominant. Provided the continuum contributed by the fluorescence mechanism is not *greater* than what is observed, the theory is not in any intrinsic difficulty. It should also be remarked that even if the fluorescent lines and the direct continuum are of comparable intensity, the fluorescent contribution remains the dominant factor in the energy budget of the supernova, because it implies the presence of a much greater amount of excitation energy.

References

Bloch, M., Chalonge, D., and Dufay, J.: 1964, *Ann. Astrophys.*, **27**, 315.
Brandt, J., Stecher, T. P., Maran, S. P., and Crawford, D. L.: 1971, *Astrophys. J. Letters* **163**, L99.
Ciatti, F.: 1973, *Astron. Astrophys.* **22**, 465.
Chiu, B. C. and Shields, M.: 1973, Unpublished.
Kafatos, M. and Morrison, P.: 1971, *Astrophys. J.* **168**, 195.
Kirshner, R. P., Oke, J. B., Penston, M. V., and Searle, L.: 1973, *Astrophys. J.* **185**, 303.
Morrison, P. and Sartori, L.: 1966, *Phys. Rev. Letters* **16**, 414.
Morrison, P. and Sartori, L.: 1969, *Astrophys. J.* **158**, 541.
Searle, L.: 1974, This volume, p. 125.

DISCUSSION

L. Searle: It seems to me that you can't have it both ways. If your model fails, as it does, to explain the main character of the spectrum and you say that this is because there is also a central object that your theory doesn't account for – then you can't, at the same time, claim to explain the light curve.

L. Sartori: That all depends on the relative amounts of energy in the lines and the continuum. If it were true that the 4686 feature contributes at all times a negligible fraction of the photographic light, then indeed our explanation for the light curve would be invalidated. You apparently feel that this is the case. Upon examination of the available data, however, we remain unconvinced that the continuum is that dominant, particularly in the later spectra. Our work was originally motivated by Baade's observation that the main band contains more than half the photographic light.

V. Mansfield: (1) The Gum nebula is not unambiguous evidence for your theory since the blast wave model can account for the required ionization in a very natural way.

(2) Given that your theory forms an H II region at the velocity of light, can you estimate what fluxes of forbidden line radiation would be observed several years after maximum light?

L. Sartori: (1) Your statement is correct. There are in fact still other explanations, for example the cosmic-ray ionization model of Ramaty, Boldt, Colgate and Silk. At the present time, no definitive test exists to discriminate among the various alternatives.

(2) M. Kafatos has studied this problem in his thesis (Kafatos and Morrison, 1971). In the case of galactic supernovae, lines of [O II] and [O III] may be observable.

It is also worth remarking that even though the ionized region is formed instantaneously (i.e. at the velocity of light), its observation is subject to the same kinematic argument which we have invoked to explain the light curve. The observed radiation would come from expanding paraboloidal regions, and the source would appear to expand at a velocity which is initially infinite.

A. Treves. Why should the continuum be peaked at 50 eV?

L. Sartori: It doesn't have to be peaked there. There merely have to be enough photons in the appropriate energy region to provide the necessary number of excitations.

OBJECTIVE PRISM SPECTRA OF SN 1972e NEAR MAXIMUM PHASE

M. F. McCARTHY*
Vatican Observatory, Castel Gandolfo, Vatican City State

In July 1895 W. P. Fleming at Harvard discovered (Pickering, 1897) on a objective prism plate exposed on a field in Centaurus a most peculiar object which was first classified as an R type spectrum and is now identified as SN 1895b in the southern galaxy NGC 5253. This was the first spectrum of a supernova recorded photographically.

Almost 77 years later another supernova occured in NGC 5253; it was discovered by Kowal (1972) at Palomar on May 13, 1972. On 19 May 1972 the news of this discovery reached the Cerro Tololo Interamerican Observatory in Chile. The observations of this supernovae made with the Curtis Schmidt telescope and the University of Michigan objective prisms are reported here.

Two initial reports of the optical observations made at Cerro Tololo have been published (Osmer *et al.*, 1972; McCarthy *et al.*, 1973). The first appeared in *Nature Physical Sciences* for 10 July 1973; it summarized the findings made at optical wavelengths by several different astronomers at CTIO and announced the discovery of new bands in the near infrared at 8650 Å and 7900 Å. In January 1973 the author reported on these infrared observations of SN 1972e at the Las Cruces meetings of the American Astronomical Society and suggested the identification of the 8650 Å band with the Ca II triplet at 8498 Å, 8542 Å and 8663 Å and with the lines of N I at 8629 Å and 8683 Å. A fuller report on these infrared observations is in press at Vatican City.

On the night the news of the discovery reached Cerro Tololo 12 Schmidt plates were obtained of the supernova; four plates were obtained on the following two nights. Table One presents a journal of observations made by the author and by G. Araya of CTIO. Table II presents a summary of the outstanding features to be found on the series of objective prism plates in the spectral range from 3500 Å to 8800 Å.

The present paper discusses these objective prism spectra. As can be seen from Tables I and II many different plate filter combinations were employed as well as three different objective prisms. We shall describe the spectrum of SN 1972e as observed on these objective prism plates from the near ultraviolet to the near infrared near the time of maximum and conclude with a suggestion for the possible detection of optical remnants of supernovae by means of extremely low dispersion objective prism spectra.

* Consultant Astronomer, Cerro Tololo Inter-American Observatory, May 1972.

C. B. Cosmovici (ed.), Supernovae and Supernova Remnants, 135–141.

TABLE I

Journal of observations of SN 1972e

19–20 May 1972					
	11193–6°	IIaO	GG 13	7m	
	11194–6°	IIaO	UG 2	30	
	11195–6°	IN	–	25	No filter; unhypersensitized
	11196–6°	098–02	RG 1	15	
	11197–6°	103aD	GG 14	12	
	11198–6°	IIaO	GG 13	14	
	11199	IIaO	GG 13	10	Direct plate
	11200	IIaO	UG 5	15	Direct plate
	11201	IIaD	GG 14	30	Direct plate
	11202	IIaO	UG 5	15	Direct Plate
	11203–0°.4	IIaO	UG 5	30	
	11204–0°.4	IN	–	30	No filter; hypersensitized
20–21 May 1972					
	11206–10°	103aO	–	10	No filter
	11207–10°	IIaO	UG 5	17	Clouds; 100% humidity
21–22 May 1972					
	11208–10°	098–02	RG 2	60	
	11216–10°	IN	Wr 89B	90	Hypersensitized.

TABLE II

Summary of major emission features and depressions in spectrum of SN 1972e May 19–22, 1972

Wavelength region	Emission feature	Depression
Ultraviolet	3700	
		3800
	3900	
Blue	4200	
		4050
	4600 V. Strong	
		4800
	4900 Strong; V. Narrow	
		5050
Yellow	5250–5340	
		5400 Shallow
	5600 Strong; Broad	
		5800
	6050 Strong	
Red		6175 V. Deep
	6300–7000	
Infrared	7100–8100	
		8250 V. Deep
	8650 Strong; Broad	

1. Ultraviolet and Blue Sensitive Plates

Beginning with the near UV there is a broad and very weak emission feature at 3650 Å to 3750 Å followed by a deep depression near 3800 Å. This is followed by a moderately broad emission band and then a rather weak emission band near 3900 Å. This feature precedes a shallow depression centered near 4050 Å. From 4100 Å to 4300 Å we note a very broad and strong emission band followed by a deep depression near 4450 Å. The most intense emission feature observed on blue sensitive plates is the very strong band at 4600 Å. At 4800 Å there is a rather deep and narrow depression beyond which the strong and sharp feature near 4900 Å is prominent. In the parent galaxy the spectrum shows strong emission features here due to Hβ and [O III]

2. Yellow and Red Sensitive Plates

The 4900 Å emission feature noted above is also observed on the yellow plate and it is followed by a broad shallow depression which runs from 5000 Å to about 5100 Å. From 5150 Å to 5380 Å a moderately strong emission band is seen. Next another shallow but narrow (5400 Å to 5450 Å) depression occurs which is succeeded by the strong emission band which extends from 5500 Å to 5700 Å. Another shallow and weak though quite broad depression extends from 5700 Å to about 5900 Å. At 5950 Å to 6150 Å we note the strongest emission band to be observed in this region. It precedes immediately the deep depression in the red observed first by Ford and Rubin (1968) at 6175 Å and extends from 6170 Å to 6250 Å on our plates. This strong feature can be observed both on the yellow and the red sensitive plates. Directly following this deep depression comes the 6300 Å emission feature which is moderately strong and extends from 6300 Å to the end of the sensitivity of the red plates. There is an emission feature superposed on the weak broad emission band near 6550 Å; it may belong to the H and N II lines which are so easily visible in the spectrum of the parent galaxy NGC 5253.

3. Near Infrared Sensitive Plates

The only feature observable on the IN unhypersensitized plate exposed with no filter for 25 min on 19 May is a very weak feature near 8650 Å; in comparison with the intensity of the strong features noted at 4250 Å, 4600 Å and 4900 Å on the same plate, this 8650 Å feature appeared to be extremely weak. On the long exposure (90 min) IN hypersensitized plate exposed two days later through the 10° prism and using a Wratten 89B filter much more detail is observable in the infrared region. A weak emission band which is very broad extends from 7150 Å to 8100 Å; its maximum intensity occurs about 7900 Å; the terrestrial A band at 7590 Å is seen in the middle of the spectrum and provides a most convenient reference mark. A very deep depression extends from 8100 Å to 8390 Å and is comparable to the very deep depression mentioned earlier at 6178 Å. The strongest Infrared feature noted on our plate extends

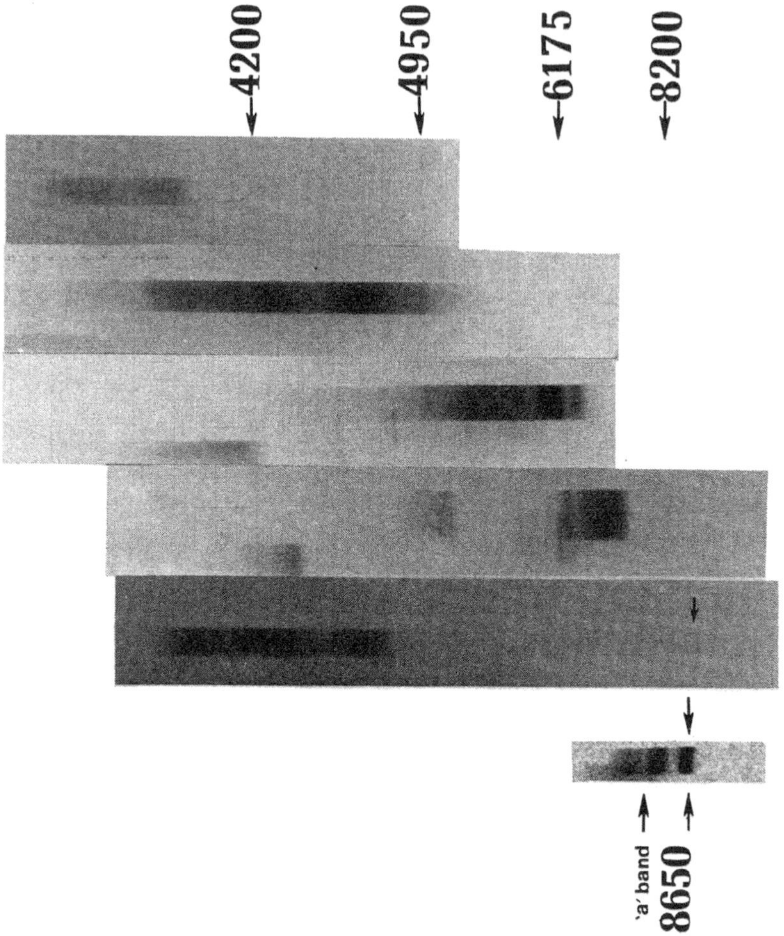

Fig. 1. Objective Prism Spectra of SN 1972e from 3600 Å to 8900 Å. (a) Ultraviolet Spectrum of SN 1972e in NGC 5253. Emission Features; 3700 Å and 3950 Å. Exp. 30^m. 6° Prism. (b) Blue Spectrum: Emission Features: 3900 Å, 4200 Å, 4600 Å, 4950 Å (sharp), Note the emission features near $H\beta$ in the overlapping spectrum of the parent galaxy, NGC 5253; the galactic spectrum is located below and to the left of the brighter SN spectrum. Exp. 14^m. 6° Prism. (c) Yellow Spectrum. Emission Features: 4950 Å (sharp), 5250 Å (broad) 5600 Å (strong), 6050 Å (v. strong); note the prominent and strong depression near 61750 Å. Sharp emission lines seen near $H\beta$ in overlapping spectrum of NGC 5253. Exp. 15^m. 6° Prism. (d) Red Spectrum. Emission features: 6050 Å, 6300 Å, ??$H\alpha$??. Note strong $H\alpha$ in overlapping spectrum of NGC 5253. Exp. 15^m. 6° Prism. (e) Blue and Near Infrared Spectrum; IN plate with no filter; Blue emission features similar to those noted in (b) above. At the extreme right note the faint IR feature separated by a deep depression; this is marked by an arrow on the print and is more readily visible on the original plate. Exp. 25^m. Unhypersensitized. No trace of spectrum of parent galaxy. 6° Prism. (f) Higher Dispersion Infrared spectrum of SN 1972e. Plate exposed for 90^m on hypersensitized IN emulsion with 10° combination prism. Wratten Filter 89B. Arrows indicate atmospheric "a" band and also the new infrared band at 8650 attributed to Ca and O. Note that the prominent emission features in the 6° spectra are indicated in the upper margin. Between (e) and (f) an indicative scale for the 6 degree spectra (a) to (e) is presented. The final spectrum shows the portion from 7000 Å to 8900 Å on the 10° combination prism. For identification the 'a' band at 7590 Å and the new emission band at 8650 Å are marked with arrows. Spectra (a) through (e) were taken on 19 May 1972. The spectrum shown in (e) was obtained on 21 May 1972.

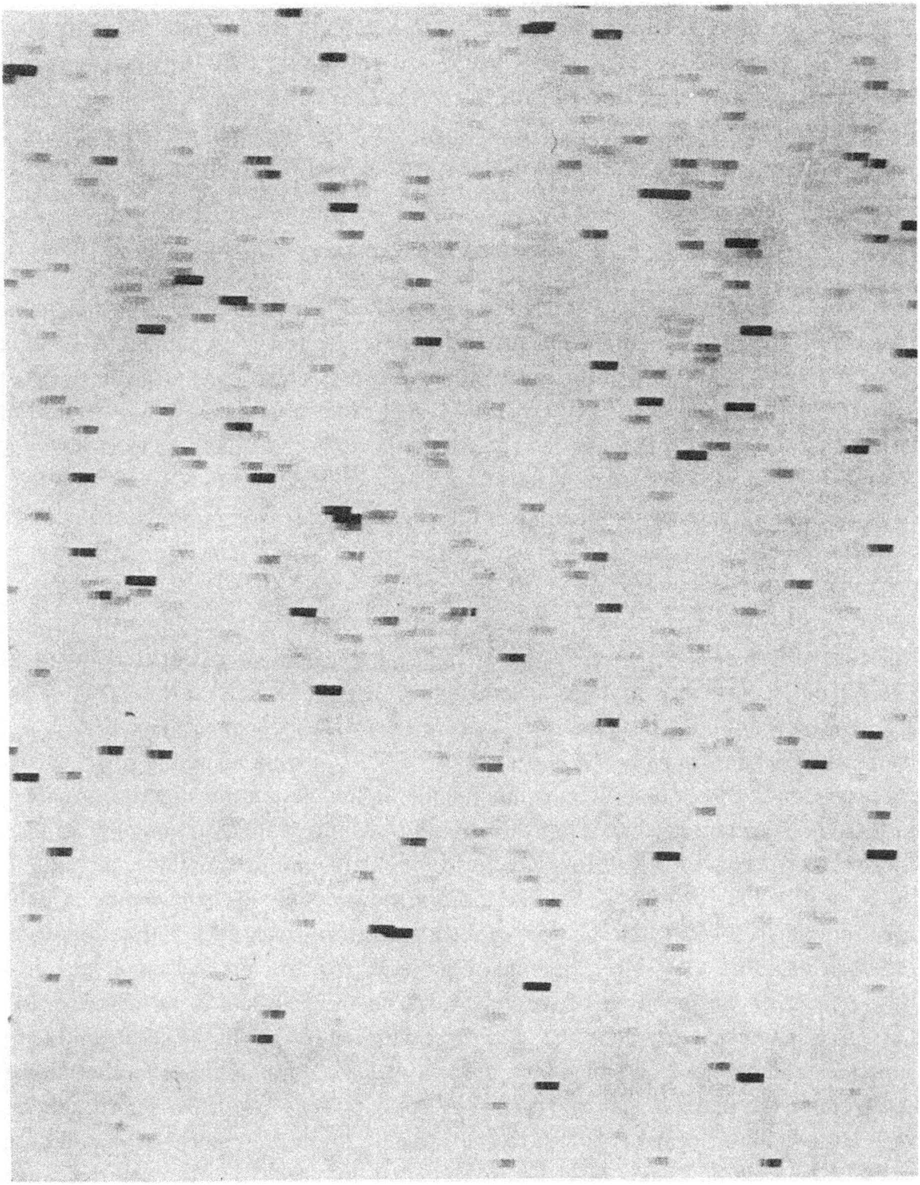

Fig. 2. Objective Prism Plate taken at Cerro Tololo Interamerican Observatory on 21 May 1972 with the Schmidt and its 10° combination prism. Spectrum of SN 1972e The Spectrum extends from 7000 Å to 8800 Å and the atmospheric A-Band is visible in all the Spectra. The new bands are prominent at 8650 Å and at 7900 Å.

from 8400 Å to the limit of the spectral sensitivity of the IN emulsion beyond 8800 Å. This emission feature is broad and strong with an apparent maximum intensity near 8650 Å. As mentioned earlier a possible identification with the Ca II triplet is suggested.

On a plate obtained at Asiago with the image tube spectrograph two weeks after the observations reported here Ciatti (1972) found the maximum relative intensity of this band near 8770 Å. The width of this emission feature is estimated to be approximately 700 Å.

4. Discussion and Summary

In this brief survey of the spectral details observed on three nights near the maximum phase of SN 1972e throughout the spectral interval from 3500 Å to 8800 Å there is good general agreement with the descriptions of type I supernovae as given by Minkowski (1939), by Ford and Rubin (1967, 1968), and most recently by C. Gordon (Pecker-Wimel) (1972). Perhaps the closest similarity is that found to exist between the spectrum of SN 1972e as described above and the spectrum of SN 1960f in NGC 4496 some 20 days after its maximum; the emission and absorption features of SN 1960f have been reproduced in a tracing by Gordon (Pecker-Wimel) in her paper.

Coverage of the same spectral range described above (from 3400 Å to 8800 Å) at very much lower dispersion was achieved by means of the extremely small 0°.4 degree objective prism of the University of Michigan attached to the Curtis Schmidt at Cerro Tololo. This prism yields spectra whose length from 3400 Å to 8800 Å on IN plates is only 0.6 mm. Thus the mean reciprocal dispersion near 6000 Å is approximately 9000 Å mm^{-1}. The appearance of SN 1972e on a 30 min hypersensitized IN plate exposed without a filter is quite unique. In the picturesque descriptions given to these mini-spectra after their introduction by Meinel (1955), Morgan and Schulte, the spectral images of the various spectral types are found to show resemblances to 'tonic bottles', to 'teardrops', to 'exclamation points' and to 'candle pins'. In the spectrum of SN 1972e in NGC 5253 we see a spectrum which does not resemble any other spectrum on the plate; it is much bluer than red. The artifact most suggested by the form of the spectrum of the supernova on the IN plate is an 'arrow head'. On a 30 min exposure made in the ultraviolet with the 0°.4 prism the spectrum of the supernovae resembles a 'double-diamond' and again is quite unlike any other stellar spectrum on the plate. It may be possible by applying very low dispersion techniques with large Schmidt cameras to follow the progress and development of supernovae to much later evolutionary stages and to discern here a new tool for exploring some of the old novae and supernovae remnants in our own and nearby galactic systems.

Acknowledgements

It is a pleasure to thank the Director and Staff of the Cerro Tololo Interamerican Observatory for their hospitality and aids during this observing run, and especially to Gaston Araya for his expert handling of the Curtis Schmidt. That so many different observations could be made of SN 1972e in so short a time is due to him. Our thanks are due also to Dr M. Bappu for profitable discussions concerning the possible identification of the 8650 feature with the Ca II triplet.

References

Ciatti, F.: 1972, *Astron. Astrophys.* **22**, 465.
Ford, W. K. and Rubin, V. C.: 1968, *Publ. Astron. Soc. Pacific* **80**, 466.
Gordon, C.: 1972, *Astron. Astrophys.* **20**, 79.
Kowal, C.: 1972, IAU Circular, 2405.
McCarthy, M. F. and Araya, G.: 1973, *Bull. Amer. Astron. Soc.* **5**, Pt 1, 12.
Meinel, A. B.: 1955, *Mitt. A.G.* (Hamburg Schmidt Dedication), p. 22.
Minkowski, R.: 1939, *Astrophys. J.* **127**, 513.
Osmer, P. S., Hesser, J. E., Kunkel, W. E., Lasker, B. M., McCarthy, M. F., and Landolt, A. U.: 1972, *Nature (Phys. Sci.)* **238**, 21.
Pickering, E. C.: 1897, *Harvard Circ.*, 4.
Rubin, V. C. and Ford, W. K.: 1967, *Publ. Astron. Soc. Pacific* **79**, 322.

A CHECK ON THE SIZE OF THE REGION PRODUCING Fe II EMISSION LINES OF THE 1961 SUPERNOVA IN NGC 1058

M. FRIEDJUNG
Institut d'Astrophysique, Paris, France

A detailed study of the slow 1961 supernova in NGC 1058 (classified as being of type V by Zwicky (1964), has been published by Branch and Greenstein (1971). Line identification for this supernova is much easier than for others, and the presence of Fe II emission and absorption lines is fairly certain. The method of Friedjung and Malakpur (1971) used to find the size of the region producing the narrow emission lines of Nova Delphini, can thus also be applied to the supernova considered here.

For a region producing line emission:

$$F_\lambda H = \frac{R^2}{r^2}(1 - e^{-\tau}) \frac{2\pi hc^2}{\lambda^5} \frac{1}{\exp(1.44/\lambda T) - 1} \tag{1}$$

where F_λ is the continuum flux at the Earth corrected for interstellar absorption, H the ratio of line emission to continuum emission, R a mean radius of the line emitting region in the direction perpendicular to the line of sight, r the distance of the emitting region from the earth, τ its optical thickness, and T the 'excitation temperature', which gives the relative populations of the upper and lower levels of the line considered.

An upper limit to R in Equation (1), can be found using the data of Branch and Greenstein for the 12th November, 1961. $\log F_\lambda H \simeq -6.1$ for a strong Fe II line such as 5018 Å, averaging H over the profile, and $\log r = 25.25$ if Hubble's constant is taken as 100 km s^{-1} Mpc^{-1}. The absence of [Fe II] 4815 Å indicates that $T > 2400$, while τ for Fe II 5018 Å must be near 1, if emission and absorption are supposed produced by the same part of the supernova envelope. Applying Equation (1) to the Fe II 5018 Å line, one finds that $R \leqslant 3 \times 10^{16}$ cm.

This upper limit would have to be somewhat raised if there was more interstellar absorption than that estimated by Branch and Greenstein, or if Hubble's constant was lower.

The envelope of the supernova was ejected at least 4 months (and probably) a year before the date studied here. Using the expansion velocity of 2000 km s^{-1} given by Branch and Greenstein, its radius was at least 2×10^{15} cm, and probably 5×10^{15} cm. These values are consistent with the upper limit to R found above.

References

Branch, D. and Greenstein, J. L.: 1971, *Astrophys. J.* **167**, 89.
Friedjung, M. and Malakpur, I.: 1971, *Astrophys. Letters* **7**, 171.
Zwicky, F.: 1964, *Astrophys. J.* **139**, 514.

C. B. Cosmovici (ed.), Supernovae and Supernova Remnants, 143–144.

DISCUSSION

C. Gordon: At what time after the explosion are you applying your computations?

M. Friedjung: One year after the explosion.

VARIABLE RADIO EMISSION FROM THE EXTRAGALACTIC SUPERNOVA-1970g IN M101*

W. M. GOSS, R. J. ALLEN, and R. D. EKERS
Kapteyn Astronomical Institute, University of Groningen, Groningen, Holland

A. G. DE BRUYN
Leiden Observatory, Leiden, Holland

1. Introduction

The type II supernova 1970g was discovered by Detre (1970) at the end of July 1970 in the southern part of M101 (NGC 5457). The supernova is located about 7″ to the northwest of the H II region NGC 5455. Radio observations were made with the NRAO interferometer at 11 and 3.7 cm starting in August 1970 and continuing to April 1971 (Gottesman *et al.*, 1972). By combining five different base line configurations (with a total of 120 h observing) Gottesman *et al.* (1972) detected a point source at 11 cm of about 4 mfu (4×10^{-29} W m^{-2} Hz) quite close to the position of the optical supernova. In February 1972 the same group using the NRAO interferometer with only three base lines (23 h observing) failed to detect any source exceeding 3 mfu.

In this communication we report measurements of this source made with the Westerbork Synthesis Radio Telescope (WSRT) at 21 and 6 cm and at Effelsberg with the 100 m telescope of the Max Planck Institute at 2.8 cm.

2. Measurements at 21 cm

M101 was mapped in the continuum at 1415 MHz in four 12 h observing periods (November 17, 1970; December 19, 1970; December 31, 1970; and January 9, 1971). The rms noise on the final map is 0.5 mfu and the resolution is 25″ × 30″. In addition to 19 sources in the field (1° × 1°) we see faint (~1.0 K) extended emission arising from the disk of the galaxy. At the position of the optical supernova we detect a point source with an intensity (Stokes I) of 5.4 ± 1.0 mfu (Table I). The errors in flux density are one σ. From an analysis of the four observations separately, there is a slight suggestion of an increase during the four observing dates at a level of 2–3σ.

During late 1971 and early 1972 M101 was mapped in the H I line near 1420 MHz at Westerbork with linearly polarised feeds in P.A. = 90°. The measurements were made in sixteen 12 h observing periods between late November 1971 to late January 1972 (mean epoch December 23, 1971). An average continuum map was made using 11 of the 16 available 129 kHz channels; five of the channels which contain line radiation near the supernova were excluded. The rms noise is essentially the same as in the continuum map since the loss in bandwidth (4.2 MHz to 1.42 MHz) is com-

* This paper is essentially identical to a publication in *Nature Physical Science* **243** (1973), 42.

C. B. Cosmovici (ed.), Supernovae and Supernova Remnants, 145–152.

TABLE I

Radio Source near SN 1970g

λ	Date	α (1950)	δ (1950)	s (mfu)	Reference
optical	Aug. 1970	$14^h01^m14^s.64 \pm 0.06$	$+54°28'54''.7 \pm 0.5$	–	van Altena (1970)
11 cm	Aug. 1970–April 1971	14 01 14.6 ±0.2	+54 28 54 ±2	3.8	Gottesman *et al.* (1972)
21 cm	Dec. 1970	14 01 14.9 ±0.4	+54 28 51 ±5	5.4 ±1.0	This paper, Westerbork
21 cm	Dec. 1971	14 01 14.5 ±0.1	+54 28 54 ±1	11.3 ±1.0	This paper, Westerbork
6 cm	Jan. 1973	14 01 14.7 ±0.2	+54 28 54 ±2	4.0 ±1.5	This paper, Westerbork
2.8 cm	Feb. 1973	–	–	< 6	This paper, Effelsberg

pensated by lower noise receivers. At the position of the supernova a point source of 11.3 ± 1.0 mfu is found. The position of this source is within 1″ of the optical position determined by van Altena (1970) (Table I). Several other sources in the M101 field have also been compared in the two 21 cm maps and their fluxes are constant to within 5%. Thus we may conclude that there are no significant gain differences between the two maps. We have also analysed the line data separately for several of the observing dates. To within the noise the flux of the source is constant at 11 mfu on these dates.

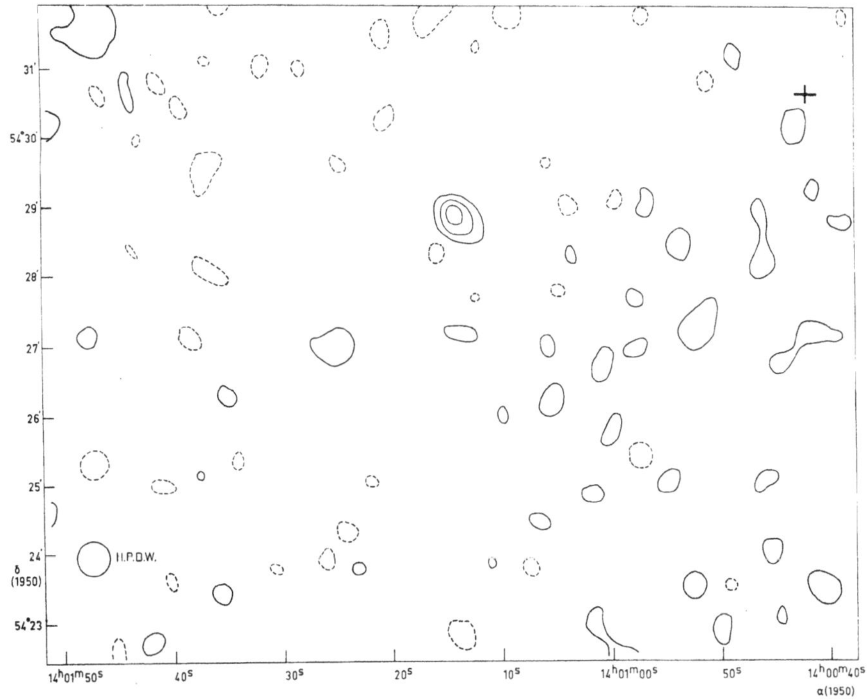

Fig. 1. The region near NGC 5447 and NGC 5455 in M101 at 21 cm as observed with the Westerbork Synthesis Radio Telescope. This contour map shows the difference between two observing epochs approximately one year apart (see text). The contour unit is 1.5 mfu/beam or 1.3 K brightness temperature.

In Figure 1 the 21-cm *difference* map is shown (the second observing period minus the first period). The cross indicates the position of the H II region NGC 5447 which has a total flux of 15 ± 1 mfu and is slightly extended. As expected this source is constant between the two observing dates. The only significant source on this difference map is a point source of ≈6 mfu at the position of the SN 1970g.

3. Measurements at 6 and 2.8 cm

On January 26, 1973, the region of the supernova was observed at 6 cm for a single 12 h period with the Westerbork telescope. The resolution is 6″ × 8″ and the rms noise is 1.2 mfu. A source of ~4 mfu (Stokes I) is found at the position of the supernova (Table I). During the night of 16–17 February we searched for emission at 2.8 cm using the 100 m telescope at Effelsberg (Max Planck Institute for Radio Astronomy). The beam width is 1.3′ and is sufficient to separate the sources known from the 21 cm Westerbork map to be present in the field. No emission greater than 6 mfu (3σ) is found at the position of the supernova remnant. For these observations the beam includes the H II region NGC 5455, and the linearly polarised telescope feed was oriented in P.A. = 20°.

4. Contribution of NGC 5455

At the position of the supernova we have observed an increase of 6 mfu in a one-year period (December 1970 to December 1971). Since radio emission from H II regions is expected to be constant over this time scale we conclude that the variable radio emission is associated with the supernova and may be the first stages of radio emission from the supernova remnant.

The major problem in the interpretation of this source is in determining the contribution of the nearby H II region NGC 5455 (incorrectly called NGC 5445 by Gottesman *et al.*, 1972). This problem was also recognized by Gottesman *et al.* Fortunately absolute Hβ fluxes for NGC 5455 have been measured by Searle (1971). If we assume an electron temperature of 10^4 K then the Hβ flux would imply a 21 cm flux of 1.4 mfu if we assume no optical extinction as implied by Searle's measurement of the ratio Hα to Hβ. However for several H II regions in M33, Israel and van der Kruit (in preparation) have found that the 21 cm flux predicted from Searle's Hβ measurements is typically 2.5 to 4.5 times less than that measured. Possible explanations such as non-thermal radio emission and larger absorption than predicted from Searle's measurements of the Hα to Hβ ratio are being investigated. In M101 Searle has published Hβ fluxes for only two H II regions for which the predicted radio flux is greater than ~1 mfu: NGC 5471 and NGC 5455. For NGC 5471 we measure at 21 cm about 3.5 times the flux predicted. For this source we can rule out any large non-thermal contribution since measurements with the 100 m telescope indicate that the radio spectrum is essentially flat between 21 and 2.8 cm.

We are thus faced with two extreme possibilities for the contribution of the H II region NGC 5455 to the observed radio emission:

(1) We could apply the same correction factor of 3.5 to the predicted 21 cm flux from Searle's Hβ measurements as seems to be necessary for NGC 5471. In this case all of the 5 mfu in December 1970 is due to NGC 5455. However in this case there seems to be significant positional discrepancy between the December 1971 radio source and the centroid of the H II region. In Table II we give the position of the centroid of the H II region as measured by us relative to nearby reference stars. The shift in position between the 200″ telescope photograph (103aO) made by Baade and reproduced in *The Hubble Atlas of Galaxies* (Sandage, 1961) and the plate (103aJ) made by Arp (1966) using the 48″ Schmidt telescope is not surprising since in a private communication to us Tammann has pointed out that the position of NGC 5455 may be color

TABLE II

Measured position of the H II region NGC 5455

α (1950)	δ (1950)	$\Delta\theta$	Description	Reference
$14^h01^m14^s.9 \pm 0.2$	$+54°28'47'' \pm 2$	7".5	Core $11'' \times 5''$ at P.A. $= 0$ with circular halo of diffuse emission $25'' \times 21''$	Sandage (1961)
14 01 14.5 $\pm$ 0.1	+54 28 50 $\pm$ 1	4.1	Dark circular image $24'' \times 24''$	Arp (1966)

and/or exposure time dependent. This H II region is quite overexposed in the Arp atlas while the Hubble Atlas photograph shows a great deal of detail inside the nebulosity. The quantity $\Delta\theta$ in Table II is the angular distance in arc seconds between the centroid of the H II region and the average radio position of $\alpha\,(1950) = 14^h01^m14^s.6$, $\delta(1950) = 54°28'54''$.

(2) The second possibility is that the H II region contributes the 1–2 mfu directly predicted by the Hβ flux and the 5 mfu source observed at the end of 1970 is a composite source consisting of H II region plus supernova radio emission.

Although the positional information favours the second alternative, at present we cannot definitely exclude the first. For the first option the radio emission of the supernova at 21 cm increases from <1 to 6 mfu at some time during an age of 6 to 18 months but is $\lesssim 2$ mfu at 6 cm at an age of 30 months. For the second option the radio emission at 21 cm is already 3–4 mfu at an age of 6 months and is 9–10 mfu 12 months later. At 6 cm it is 3–5 mfu at an age of 30 months.

5. Comments on the Previous 11 cm Observations

It is at present not clear to us how to incorporate the observations of Gottesman *et al.* (1972) into the picture, in particular the disappearance of the source at 11 cm in February 1972 just after we saw the largest flux density at 21 cm. For the first option discussed above, the spectral index, α, ($S \propto \nu^{\alpha}$) between 11 and 21 cm is steeper than -1 in early 1972. In this case the H II region was detected at both 11 and 21 cm at a level of $\sim$4–5 mfu in late 1970–early 1971. For the second option α is steeper than -1.8 in early 1972. If we accept the NRAO measurements at face value the disappearance of the source at 11 cm in early 1972 is inconsistent with the first option since we expect no time variation from an H II region. Certainly the source detected by the NRAO group at 11 cm between August 1970 and April 1971 is at the same position as the source which we detected at the end of 1970 at 21 cm.

6. Interpretation

The radio evolution of SNR (supernova remnants) has been discussed by many authors in recent years. For old SNR an empirical relation has been established between the radio surface-brightness and radius of the SNR (Milne, 1970; Downes, 1971). Since a good physical basis for this relation is still lacking, any predictions about the very early radio luminosity are quite uncertain. Gull (1973) recently proposed a theory for the origin of the radio emission in SNR at a time when the amount of swept up interstellar gas is a reasonable fraction of the ejected shell. This model then predicts that no radio emission should be expected from SNR up to an age of 50–100 yr. On the other hand neutron star theories of SNR (Ostriker and Gunn, 1971), mostly taking the Crab Nebula as a prototype, do predict radiation from a young supernova but unknown details about the electron spectrum make a quantitative prediction concerning the *radio* emission uncertain.

Recently a radio survey of 35 historical supernova has been conducted by one of us (de Bruyn, 1973) with the WSRT at 1415 MHz. The ages of these supernovae range from 0–85 yr. The results of these observations, which were essentially negative, make the present result on SN 1970g a rather unexpected one. Possible explanations of the negative results include free-free absorption in the SNR and the lack of low energy electrons.

To get an idea of the intrinsic strength of the radio emission of SN 1970g we can compare it with the present luminosity of Cas A. For the formerly accepted distance of 3 Mpc for M101, the source is 2 times as strong (based on an assumption of 6 mfu at 21 cm for SN 1970g) as Cas A. If M101 is at 7 Mpc (as has been suggested by Tammann and Sandage (1971)) this becomes 11 times as strong. Assuming the latter distance and a radius of 5×10^{16} cm for the SNR (age ≈ 1.5 yr, velocity of expansion $\approx 10^4$ km s^{-1}), the 6 mfu implies a brightness temperature of 10^{10}K. Hence the mechanism is probably non-thermal.

We would like to explore the two possible origins for the radio emission: (1) the radiation originates from a region surrounded by the ejected supernova material, and (2) the radiation is coming from outside the supernova shell. In both cases we assume that radio emission ≈ 6 mfu was detected at an age of 1.5 yr at 21 cm.

For case (1) we assume that the optical depth for free-free absorption is less than unity at 21 cm at 1.5 yr since the 11 cm flux density is less than that at 21 cm. We can thus place a lower limit on the electron temperature of the ionized envelope. If we assume that $1 M_\odot$ of ionised gas is uniformly distributed inside a sphere with radius 5×10^{16} cm this limit becomes $\approx 10^7$ K (Bahcall *et al.*, 1970). Varying the mass or expansion velocity cannot appreciably decrease this value since the chosen values are probably conservative for type II supernova. (Temperature $\propto V_{\rm exp}^{-10/3} M^{4/3}$). The presence of filaments or a shell-type structure would also increase the temperature although in the very extreme case the radiation could escape through the spaces between the filaments.

How could the envelope reach such a high temperature? From an analysis of the light curves of type II supernova, Rees (1970) suggested that the temperature of the supernova envelope could drastically increase some 50–100 days after the explosion due to the filtering through of soft X-rays from a central neutron star. These in turn heat the envelope to a high temperature. From the work of Tarter and Salpeter (1969) it seems unlikely that the temperature could rise much above 10^6 K (depending on the heavy element abundance) mainly because X-rays harder than 5 keV do not appreciably heat the envelope but escape freely. Even if the temperature could be raised to 10^7 K by X-rays, an X-ray luminosity of $\approx 10^{42}$ erg s^{-1} in X-rays below 5 keV would be needed. The X-ray luminosity in X-rays harder than 5 keV and also the fraction of soft X-rays that escape unabsorbed should be enough to make type II supernova detectable by modern X-ray telescopes out to at least 10 Mpc in the first few years after the supernova explosion. As far as we know no such detections have been reported.

Another heating mechanism is by high energy relativistic particles. In the dense ($\approx 10^7$ particles cm^{-3}) supernova envelope, ionization and collisional losses of relativistic particles are enormous and particles with energies less than a few 100 MeV will lose all their energy within roughly a year. Depending on the available energy in relativistic particles the envelope could be heated to a very high temperature. For example, to heat 1 $M_\odot$ of thermal plasma to a temperature of 10^7–10^8 K requires about 10^{48-49} ergs in relativistic particles. Also in this case we expect about 10^{41-42} ergs s^{-1} in X-rays (free-free emission) from the hot envelope itself. Hence in both cases we expect large amounts of X-rays ($>10^{41}$ ergs s^{-1}) in the range 1–10 keV in the first one or two years after the explosion.

It is interesting to speculate that the particles which heat the envelope may also be responsible for the radio emission by the synchrotron mechanism. Some interesting consequences of a synchrotron model for the radio emission are that for an age of 0.5–1 yr, synchrotron self-absorption could become important around 1 GHz at these flux-levels. Also Razin-Tsytovich absorption could become important at these electron densities if the relativistic electrons are mixed with the thermal plasma.

An alternate possibility which cannot be excluded is that the radio emission we see is the smeared-out pulsed emission from a young, rapidly spinning pulsar.

In case (2) the radiation comes from outside the supernova shell which alleviates the necessity of having a very high electron temperature in the supernova envelope. We would like to mention one model, already quite old, which perhaps does predict radio emission in the early phases. This is the well known cosmic-ray model of Colgate and Johnson (1960). Although this model provides the relativistic electrons, a magnetic field is also required to give radio emission. It is unlikely that an interstellar field of 10^{-5} G would suffice to explain the radio emission since in that case about 10^{51} ergs in radio electrons would be needed according to synchrotron theory.

For further discussion of theoretical models, we need additional information about the spectrum and time variations of SN 1970g. Future observations at 2.8, 6, 21, and 50 cm are planned. An outstanding problem that remains is the contribution of the H II region NGC 5455 to the observed emission.

Acknowledgements

The Westerbork Radio Observatory is operated by the Netherlands Foundation for

Radio Astronomy with the financial support of the Netherlands Organization for the Advancement of Pure Research (ZWO). Goss acknowledges support from ZWO. We thank U.J. Schwarz and R. Sancisi for assistance and the Max Planck Institute for Radio Astronomy for providing observing time at 2.8 cm.

Note added in proof. We have recently made observations at 2.8, 6 and 50 cm of the region near NGC 5455. From these we conclude that option (1) in Section 4 is more nearly correct than option (2): i.e. there is a non-variable component of 5 mfu due to NGC 5455.

References

Arp, H.: 1966, *Atlas of Peculiar Galaxies*, California Institute of Technology.
Bahcall, J. N., Rees, M. J., and Salpeter, E. E.: 1970, *Astrophys. J.* **162**, 737.
Bruyn, A. G. de: 1973, *Astron. Astrophys.* **26**, 105.
Colgate, S. A. and Johnson, H. J.: 1960, *Phys. Rev. Letters* **5**, 235.
Detre, L.: 1970, *IAU Circ.*, No. 2269.
Downes, D.: 1971, *Astron. J.* **76**, 305.
Gottesman, S. T., Broderick, J. J., Brown, R. L., Balick, B., and Palmer, P.: 1972, *Astrophys. J.* **174**, 383.
Gull, S. F.: 1973, *Monthly Notices Roy. Astron. Soc.* **161**, 47.
Milne, D. K.: 1970, *Austral. J. Phys.* **23**, 425.
Ostriker, J. P. and Gunn, J. E.: 1971, *Astrophys. J. Letters* **164**, L95.
Rees, M. J.: 1970, *Astrophys. Letters* **6**, 55.
Sandage, A.: 1961, *The Hubble Atlas of Galaxies*, Carnegie Institution of Washington.
Searle, L.: 1971, *Astrophys. J.* **168**, 327.
Tammann, G. A. and Sandage, A.: 1971, *Carnegie Institution Year Book* **69**, 94.
Tarter, C. B. and Salpeter, E. E.: 1969, *Astrophys. J.* **156**, 953.
Van Altena, W. F.: 1970, *IAU Circ.*, No. 2292.

DISCUSSION

J. R. Dickel: Did you look for the polarisation? (I realize the signal to noise is close to the limit but do you have an upper limit?)

W. M. Goss: No. All of the sources near NGC 5455 are point sources at 21 cm and we expect low polarizations.

J. Heidmann: In answer to Goss' request for accurate absolute positions for SN, I would like to point out that Gallouët and my wife published (in *Astron. Astrophys. Supp.*) the positions of all galaxies in de Vaucouleurs' catalogue with a 3″ accuracy. For each galaxy they give the positions of two small reference stars easily recognizable on any large scale plate and, in case the nucleus of the galaxy is not well defined, these two stars can provide easily the position of any SN.

PART IV

STATISTICS OF SUPERNOVAE

STATISTICS OF SUPERNOVAE

G. A. TAMMANN*

Hamburger Sternwarte, Hamburg-Bergedorf, F.R.G.
and
Institute of Astronomy, Cambridge, England

1. Introduction

The frequency of SNe is strongly dependent on the intrinsic properties of their parent galaxies. The present investigation is an attempt to determine within a clearly defined sample volume the SN frequency as a function of two principal parameters: the type and the luminosity (or mass) of the parent galaxy – which seem to control essentially the intrinsic SN rate. In this respect it is an extension and improvement of an earlier paper (Tammann, 1970), where only Sb and Sc galaxies could be considered. Available distances have become in the mean time sufficiently accurate within the sample volume considered, that it seems possible to use individual intrinsic luminosities for the sample galaxies. The distance scale is based throughout on a value of the Hubble constant of 55 (Sandage and Tammann, 1974), whereas the earlier paper used $H_0 = 75$ km s^{-1} Mpc^{-1}. The other major change in the present paper is that it allows for a newly found selection effect of the observable number of SNe in inclined spiral galaxies.

Throughout this paper the present results are compared with those by other investigators. Wherever discrepancies arise, it is tried to expose the reasons. It is believed that there remain few, if any, major contradictions. However, the statistics of SNe is still confined to small numbers, and several possible selection effects are far from being under control. In this sense the present results are preliminary. Much work along the lines, which sometimes could only be sketched here, remains for the future.

2. Definition of Sample

A. GALAXIES

All galaxies in the Shapley-Ames catalogue (1932) are considered with $v_0 \leqslant 1200$ km s^{-1} ($r \lesssim 22$ Mpc), and $\delta > -36°$. Of 1037 galaxies 755 have known radial velocities. For additional 155 galaxies information is available as to their distances on the basis of cluster and group membership (Holmberg, 1950, 1964; Humason *et al.*, 1956; Karachentsev, 1970; de Vaucouleurs, 1973; Sandage and Tammann, 1974) or of their luminosity classification (van den Bergh, 1960). Decision to include or exclude a member galaxy of a cluster or group rests on the *mean* velocity of the group; in particular all probable Virgo cluster members are included. The velocities used are

* Permanent address: Astronomisches Institut der Universität Basel, CH-4201 Binningen, Switzerland.

C. B. Cosmovici (ed.), Supernovae and Supernova Remnants, 155–185.

weighted means of all available values (e.g. Humason *et al.*, 1956; de Vaucouleurs and de Vaucouleurs, 1964, 1967; Ford *et al.*, 1971; Sandage, 1973b; and many other sources, including 21-cm velocities). Corrections for solar motion were applied; the Roberts correction (Roberts, 1972) was applied to spirals only, as specified elsewhere (Tammann, 1972). From 910 galaxies with distance information 409 galaxies are inside the 'unit volume', 501 are outside. No decision could be reached for 127 galaxies, corresponding to 12 percent of the total. For one galaxy within the sample the type is not known (NGC 1531); it was excluded, leaving 408 sample galaxies.

The Shapley-Ames was intended to be complete to $13\overset{m}{.}2$. Since Holmberg's m_{pg}-magnitude system (1958) is used here and since corrections for galactic and internal absorption are applied. the limit of completeness is brighter, say $\sim 12^m$, with a number of galaxies being fainter. Nevertheless it is believed that the present sample is unbiased as to supernova productivity. There is, of course, the effect that the limiting luminosity of the galaxies increases with distance even within the sample. But this bias does not invalidate the following results. In fact the main merits of this distance-limited (or actually velocity-limited) sample are: (1) very distant, overluminous galaxies are excluded, which are inherent to any sample limited only by the apparent magnitude; and (2) it avoids apparently faint (and on the average more distant) galaxies, for which the present supernova searches are known to be biased and severely incomplete (Sargent *et al.*, 1972; Sargent *et al.*, 1974).

One might think that the member galaxies of a large cluster would represent a more readily defined sample. However, the individual data on cluster galaxies are too meager for analysis, and due to the large distance one would have to cope with the full incompleteness effects of SN searches at faint levels. These two difficulties have become evident from Barbon's (1968) very careful analysis of SN frequencies in clusters of galaxies. The situation would be more favorable for the Virgo cluster; however, its large extension would strongly impair the advantages of a homogeneously searched sample of galaxies – in addition the sample would be small and its definition under debate (de Vaucouleurs and de Vaucouleurs, 1973; Tammann, 1972, 1973a).

In the course of this paper the types of the sample galaxies and their luminosities are needed. The types are taken from the RCBG (de Vaucouleurs and de Vaucouleurs, 1964) and lumped as follows: E, S0, Sa (S0/a, Sa), Sb (Sab, Sb), Sc (Sbc–Sd), Ir I (Sdm, Sm, Im), and Ir II (I0). Apparent magnitudes m_{pg} were adopted from Holmberg (1958), or – where these were not available – B(0)-magnitudes from the RCBG were transformed into Holmberg's system by means of a regression formula. Corrections for galactic absorption and for internal absorption due to inclination follow the precepts by Sandage (1973a) and Holmberg (1958), respectively. Somewhat arbitrarily it was assumed that S0's, which in some cases contain large amounts of dust, have half as much internal absorption as Sa/Sb galaxies. E and Ir I galaxies were assumed to have negligeable internal absorption. Due to lack of statistical data no corrections for internal absorption were applied in Ir II galaxies. Note that at this point the magnitudes are reduced only to face-on magnitudes; the remaining, inclination-independent absorption component is removed only in Section 6. To convert apparent magnitudes into absolute magnitudes distances to individual galaxies or groups of galaxies were adopted from Sandage and Tammann (1974). If these are not available the distances were calculated from recession velocities. Finally, in order to express absolute photographic magnitudes in solar units it was assumed that $M_{pg}(\text{Sun}) = 5\overset{m}{.}37$ (Stebbins and Kron, 1957).

B. SUPERNOVAE

The supernovae of the sample galaxies are taken from an updated version of the list by Kowal and Sargent (1971, 1973). The variable V12 in NGC 2403 (Tammann and Sandage, 1968), which was recently suggested to be a SN (Kowal *et al.*, 1972), was

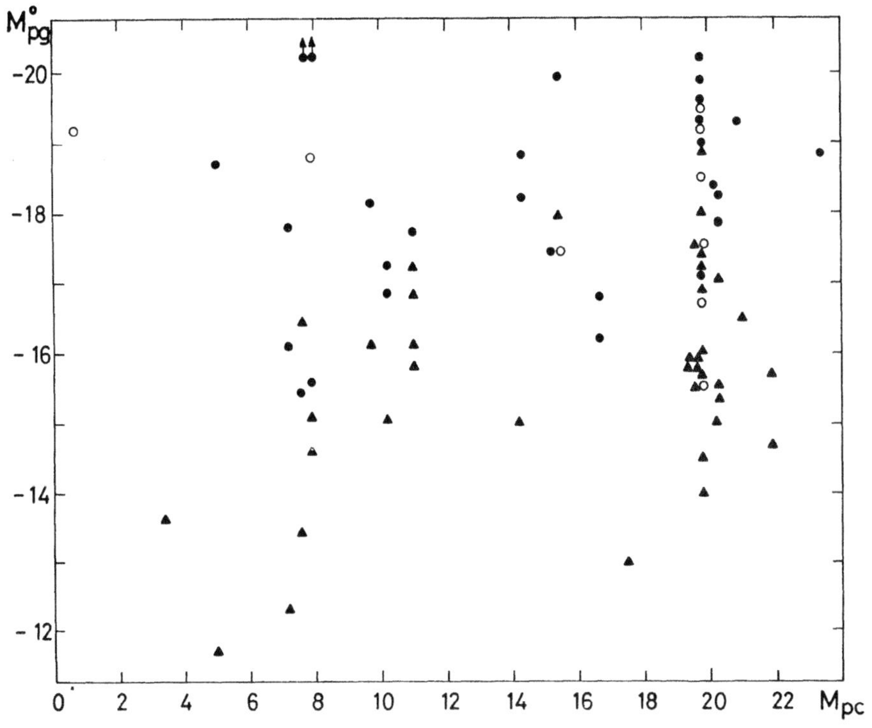

Fig. 1. The absolute photographic magnitude of SNe, corrected for galactic absorption, *versus* distance. Dots are well observed light maxima; open circles correspond to poorer determinations of the light maximum; triangles represent the brightest observed magnitude for SNe whose light maxima were missed.

omitted because there is no evidence that it ever exceeded $M_B^0 = -11\overset{m}{.}3$. The SN 1951 in M101 (Sandage and Tammann, 1974) was included. The probable SN 1969 in NGC 6946 (Rosino, 1971) was given half weight. This brings the number of known SNe for the 408 sample galaxies to 74.5; they appeared between 1885 and the end of 1972.

It has been suggested (Zwicky, 1973) that SN frequency determinations, leading to high rates in some (very luminous) galaxies, are dominated by intrinsically faint SNe ($-11\overset{m}{.}0 > M_{pg} > -14\overset{m}{.}0$). This possibility has been tested by calculating the highest observed absolute magnitude for all SNe in the sample. The observed apparent photographic magnitude maxima were taken from Kowal and Sargent (1973) and a few other sources (only in the minority of cases these magnitudes correspond to the actual maxima); the adopted distances are those of the parent galaxies (see above). The resulting absolute SN magnitudes (corrected for galactic absorption) are plotted *versus* distance in Figure 1. It is reassuring that the absolute magnitudes in Figure 1 depend only slightly, if at all, on the adopted distance. All SNe have $M^0_{pg} \leqslant -15^m$ except nine objects. The true maxima of these nine SNe are not observed; three of them are classified as type I (SN 1940e, 1959e, 1972r) and one as type II (SN 1941c), therefore they must have been much brighter during their actual maxima. The faint SN 1951 in M101 was caught long after maximum as indicated by its very flat light curve. Hence, there is a good likelihood that also the remaining four SNe actually exceeded -15^m at maximum.

From this follows that the present sample of SNe is confined (to at least 95%) to objects with $M^0_{pg}(\max) \leqslant -15^m$, and that the statistical bias from intrinsicly faint SNe is negligeable. The bright lower limit of SN luminosities makes them distinct from all other known stellar objects, which – with the possible exception of η Car – have never been observed to be brighter than $M_{pg} = -12^m$.

3. The Overall Rate of Supernovae

As stated above, the 408 sample galaxies have produced a total of 74.5 SNe, 30.5 of which appeared within the 13-yr interval from 1960–1972. After having procured on the average 1 SN per year from 1885 to 1959, the search for SNe has been strongly intensified on international basis around 1960, resulting since in an increased discovery rate of ~18 fresh SNe per year. If 30.5 sample SNe are found within the last 13 years, a total of 74.5 SNe corresponds to 32 search years at the present intensity level of surveillance. In order to use the full number of SNe for the following statistics, it is assumed throughout this paper, that the 74.5 sample SNe correspond actually to 32 search years. Several tests indicate that this assumption is permissible and does not introduce statistical bias into the sample.

74.5 SNe in 408 galaxies within 32 search years lead to an observed over-all rate of one SN per 175 years per 'average' galaxy, or, with other words, to a mean interval between SN outbursts of $\tau = 175$ yr. for an 'average' galaxy. This value lies between the values of other investigators, e.g. $\tau = 350$ yr. (Zwicky, 1942, 1973), 316 yr. (Barbon, 1968), 100–240 yr. (Chai and van den Bergh, 1970), and 40 yr. (Katgert and Oort, 1967). It has been suggested elsewhere (Tammann, 1970) that the wide range of τ-values is entirely explainable on the basis of *different mean luminosities* of the different samples considered.

In spite of the uncertainty of the mean rate it may be a useful value for practical purposes: one can estimate from it the expected SN yield for a given search field. The mean rate is, however, meaningless and in fact misleading if applied to any individual galaxy. The reason lies in the misconception of an 'average' galaxy. Galaxies exhibit such a tremendous range of their properties (e.g., stellar population, mass, mass-to-light ratio and angular momentum per unit mass), that any compliance of an individual galaxy with the average SN rate would be purely accidental. It is for this reason that an attempt seems worthwile to investigate separately the SN rates in galaxies with different properties.

4. The Distribution of Supernovae of Different Types in Different Types of Galaxies

In Table I the galaxies are subdivided into types as explained above. The number of galaxies for each type is given on line 1. The number of SNe which appeared in them during the last 13-yr interval and the total number of their SNe (the latter in parentheses) are shown on line 2. It is obvious that the SN productivity depends markedly on the type of the parent galaxy. For instance, the 143 E/S0/Sa galaxies have produced a total of 8 SNe, while 49.5 SNe are known for the 156 Sc galaxies.

TABLE I

Distribution of SNe with Galaxian type

	E	S0	Sa	Sb	Sc	Ir I	Ir II	All
n_{Gal}	38	79	36	56	156	36	7	408
n_{SN}	1(5)	2(2)	0(1)	1(10)	22.5(49.5)	3(5)	1(2)	30.5(74.5)
type I	5	1	1	3	13	5	2	30
type II	0	0	0	4	21	0	0	25
Others	0	1	0	3	15.5	0	0	19.5

Lines 3, 4 and 5 of Table I give the distribution of SNe according to their types. The main source for the type I SNe is Barbon *et al.* (1973); the informations on other types were taken mainly from Kowal and Sargent's compilation (1973) and from a few other sources. For the present statistical purpose uncertain types were treated as definite assignments. All SNe with types III, IV etc. and those with unknown types were lumped together under 'others'.

The obvious result of Table I is the fact that type I SNe occur in all types of galaxies, while type II SNe are apparently confined to Sb and Sc galaxies. The same result holds true if one considers *all* SNe with known types. Of course, many authors have commented on this fact.

In particular the very surprising result, pointed out earlier by Reaves (1953), van den Bergh (1959), and Minkowski (1964), is still correct: there are no type II SNe in irregulars. With five known SNe of type I in Ir I's (all in the present sample) the result is now significant at least in the sense that type II SNe are underabundant in irregular galaxies.

If one considers finer subdivisions of galaxian types the present sample suggests a slight, but uncertain dependence of the ratio type I: type II on the subtypes of Sc's: it is high for Sbc's (~2:1) and decreases for Sc's (~1:2) and Scd's (~1:3). This trend does not extend into the Sb's – and there are no data on Sd's. Taken at face value the ratios indicate a sharp change somewhere between Scd's and Sdm's. But in view of the small numbers and the statistical uncertainties this result is at the most suggestive.

Even if one compares the less detailed classes of Sb and Sc galaxies one cannot reach a firm conclusion about the relative frequencies of SN types. The ratio type II: type I was formerly believed to be constant (Tammann, 1970), whereas now we obtain 1.3 ± 1.0 for Sb's and 1.6 ± 0.6 for Sc's. (The indicated errors are always the mean square root errors as determined by the sample size.) These numbers could eventually turn out to be identical or significantly different.

On the whole, type I SNe are slightly more frequent in the present sample, giving type II: type I $= 0.8 \pm 0.2$.

The preceeding remarks depend strongly on the tacit assumption that the discovery chance for different types of SNe does not vary within the sample. A preliminary analysis by Yilmaz and Tammann (1973) gives for type I $M_{pg}(\max) = -19^{m}.7 \pm 0^{m}.3$ and for type II $M_{pg}(\max) = -17^{m}.6 \pm 0^{m}.3$ (neglecting internal absorption and considering only SNe in the present sample); this result is in perfect agreement with the values derived by Kowal (1968), after the latter are reduced to $H_0 = 55$. The considerable luminosity difference between the types could cause a systematic and distance-dependent discrimination against the fainter type. However, this bias does not seem to exist in the present material: for Sb/Sc galaxies nearer than 12 Mpc one finds type II: type I $= 1.5 \pm 0, 8$, for those more distant than this limit 1.6 ± 0.6. There remains, of course, the possibility that one type of SNe is more easily dicovered than the other, due to preferred locations within the parent galaxy.

5. Some Selection Effects

A. THE EFFECT OF SPECIAL SEARCH FIELDS

Perhaps the weakest point of the present sample is that it includes a large part of the sky, with some sections being under frequent and regular control and other sections, for which no information is available at all as to their control times. If the effect of poorly controlled galaxies were pronounced, it would not affect the relative frequencies, because different types of galaxies and SNe seem to be scattered randomly over the sample area – but it would depress the absolute frequencies.

To investigate the influence of different control times of different directions of the sky, the sample galaxies were combined to three different groups (Table II): (1) all 408 galaxies; (2) only the 177 galaxies which are considered to be Virgo cluster members (these objects are expected to be well surveyed since the rich Virgo fields are particularly gratifying for SN searchers); (3) only the 50 galaxies which lie in the Palomar Virgo search fields no. 4, 7, and 9 [these fields are among the best controlled ones (Barbon, 1968; Kowal, 1973)]. Line 2 of Table II gives the number of SNe during the last 13 years and in parentheses the total number and line 3 the number of SNe per galaxy. These numbers show that the SN frequency depends only marginally on the search intensity of the different samples.

TABLE II

Dependence of SN numbers on search fields

	All	Virgo	Fields no. 4, 7, 9
n_{Gal}	408	177	50
n_{SN}	30.5 (74.5)	14 (27)	5
SN/Gal	0.075 (0.183) ±0.013 (±0.020)	0.079 (0.152) ±0.021 (±0.030)	0.100 ±0.045

It shall be discussed in Section 5E that the SN frequency depends directly on the luminosity of the parent galaxy. It is therefore more meaningful to make the above comparison in terms of SN numbers per unit luminosity per unit time, and to restrict it to face-on Sc spirals, because they form the most homogeneous subsample (chapter VD): the five face-on Sc's in search fields no. 4, 7, and 9 have a total luminosity of $1.63 \times 10^{10}\, L_\odot$ and they showed 8 SNe in 32 search years; this corresponds to an observed rate of 0.015 ± 0.006 SN $(10^8\, L_\odot)^{-1}$ $(100\ \text{yr})^{-1}$. The corresponding numbers for the remaining 21 face-on Sc's are $3.34 \times 10^{10}\, L_\odot$, $n_{\text{SN}} = 21.5$ and 0.020 ± 0.004 SN $(10^8\, L_\odot)^{-1}$ $(100\ \text{yr})^{-1}$. If anything, the rate is even higher outside of the particular search fields!

The result is rather surprising that the SN yield of galaxies does not seem to depend on the search intensity of the fields in which they are located. It is hard to believe that this has to be interpreted in the sense that *all* SNe were found in the sample galaxies, but it does seem to suggest that the incompleteness factor depends very little on the position of the galaxy within the sample.

B. THE EFFECT OF DISTANCE

With an adopted distance limit of $\lesssim 22$ Mpc for the present sample, one could expect some distance effect in the sense that fewer SNe were found in the more distant galaxies than in the nearby ones. To check on this possibility two subsamples were defined: (1) galaxies nearer than 10 Mpc, and (2) galaxies more distant than this limit (Table III). Only face-on Sc galaxies are considered, because they form a particularly unbiased sample (Section 5D). Expressing the SN rate in SNe per $10^8\, L_\odot$ per 100 yr, one finds 0.022 ± 0.008 for the near sample and 0.017 ± 0.004 SN $(10^8\, L_\odot)$ $(100\ \text{yr})^{-1}$ for the distant one. The two values agree quite closely, their ratio of 1.3 ± 0.6 deviating only insignificantly from 1. From this one may conclude that in first approximation any distance-dependent incompleteness factor can be neglected for the present sample volume.

This result is seemingly in contradiction with the results by Katgert and Oort (1967), Sargent *et al.* (1972), and Sargent *et al.* (1974), who find a very strong dependence

TABLE III

Dependence of SN numbers on distance (face-on Sc's only)

	n_{Gal}	L (in $10^8 L_\odot$)	n_{SN}	SN $(10^8 L_\odot)^{-1}$ (100 yr)$^{-1}$
$r \leqslant 10$ Mpc	7	1114	8	0.022 ± 0.008
$r > 10$ Mpc	19	3857	21.5	0.017 ± 0.004

of SN numbers on the *apparent* magnitude. At fainter levels ($m_{pg} > 13^m$) the decrease of SN numbers is certainly a strong effect of increasing distance. For the Shapley-Ames galaxies the correlation between apparent magnitude and distance is, however, more complex, and another factor becomes quite dominant, i.e. that the apparently brightest galaxies tend to be also intrinsically overluminous. Among the 28 galaxies known to be brighter than $m_{pg} = 9\overset{m}{.}5$ (van den Bergh, 1965) there are 6 supergiant spirals (Sb I/I–II and Sc I/I–II), whereas only 0.5 would be expected from their average space density (van den Bergh, 1960). As one goes to apparently fainter galaxies, say from $9\overset{m}{.}5$ to $13\overset{m}{.}0$, a larger proportion of less luminous galaxies is included, and since they are poorer SN producers (chapter VE) they lower the overall SN rate. The decrease of the observed SN rate with increasing apparent magnitude seems therefore mainly a luminosity effect for the apparently brighter galaxies, and mainly a distance effect for the apparently fainter galaxies.

C. THE EFFECT OF RADIAL DISTRIBUTION

Due to the frequent overexposure of the central zones of galaxies, it could be expected that many SNe are missed in them. This possible bias can be studied best in the face-on Sc galaxies (inclination $\leqslant 30°$) of the sample. The left upper panel diagram of Figure 2 shows their SN distribution *versus* the relative radial distance r/R [r is the distance in seconds of arc of a SN from the center of its parent galaxy, as determined from data by Kowal and Sargent (1973); R is the optical face-on radius in seconds of arc of the parent galaxy as given in the RCBG (de Vaucouleurs and de Vaucouleurs, 1964)]. Different symbols are used for the different types of SNe in Figure 2.

The question, one has to test now, is: can the observed radial distribution of SNe in face-on Sc's be understood in terms of physical parameters, or are there indications for observational bias against SNe near the inner zones? – There is direct evidence that the SN numbers are proportional to the *mass* of the parent galaxy (Tammann, 1970). It is then reasonable to assume that the radial distribution of SNe reflects the radial mass distribution within spirals. From Schmidt's (1965) model of the Galaxy one can calculate the fraction of mass, which falls within a ring of width $\delta(r/R)$.

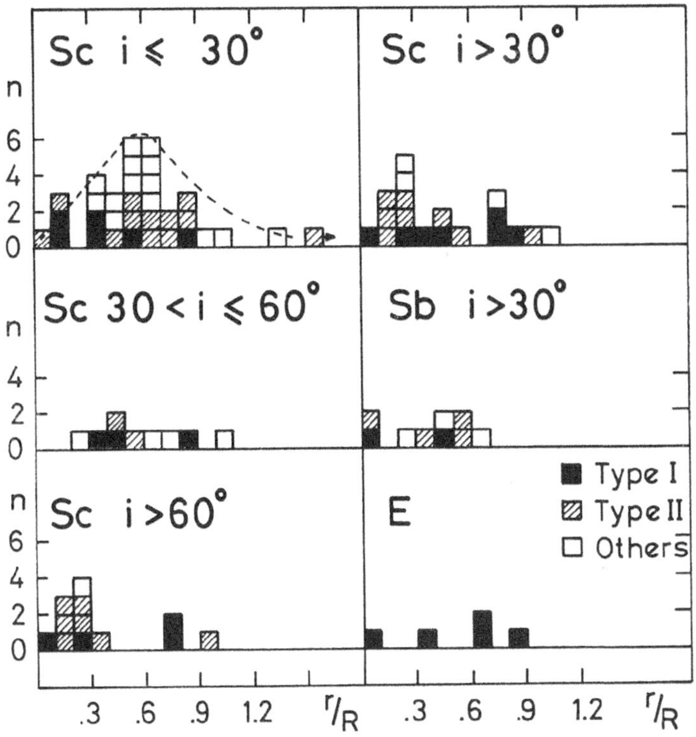

Fig. 2. The relative radial distribution of SNe of different types in Sc galaxies with different inclinations and for Sb and E galaxies. The dotted curve in the upper left histogram gives the relative radial mass distribution according to Schmidt's model of the Galaxy.

For this a value of the 'optical' diameter R of our Galaxy has to be assumed; a value of $R = 14$ kpc was chosen. The ensuing relative mass distribution is shown in function of r/R in Figure 2 as a dotted curve. (Note that the histogram and the curve are not drawn to enclose the same surface, but to facilitate a comparison). The agreement between the SN distribution and the model mass distribution is surprisingly good. In view of the simplifications (our Galaxy can hardly be taken to be typical for a Sc galaxy) and the small numbers of SNe involved, the test is not stringent yet – but taken at face value it (1) gives an independent confirmation for the SN rate being proportional to the parent mass, and, if this is correct, (2) strongly suggests that few if any central SNe were missed in face-on Sc galaxies!

The numbers in the sample are insufficient to make a corresponding analysis for other spirals. Only five SNe have appeared in elliptical galaxies of the sample. This number is, of course, too small for an analysis of their radial distribution. However, the r/R-values of their SNe are also plotted in Figure 2 (lower right panel histogram). In spite of the scarcity of the data, it is noteworthy that all SNe fall within the optical image of the galaxy ($r/R < 1$), and that 4 out of 5 SNe lie within $r/R < 0.7$; one SN was observed near to the galaxian center (SN 1961h in NGC 4564). That *could* mean that the overexposure effect is not pronounced in the sample ellipticals, and that the suspected trend of SNe in E's to concentrate in the peripheral optical parts of their parent galaxies, is exhibited only by more distant objects. But the possibility of considerable selection effects against SNe in the inner parts of ellipticals has to be kept in mind.

D. THE INCLINATION EFFECT OF THE PARENT GALAXY

The Sc galaxies in the sample yield themselves for checking possible effects of the galaxy inclination on the observed SN rate. In Table IV the SN rates (per unit luminosity [10^8 $L_\odot$] and unit time [100 yr]) are calculated for three groups of Sc's with different inclination: $i \leqslant 30°$, $30 < i \leqslant 60°$, $i > 60°$ (line 1, 2, and 3). The SN rates

TABLE IV

Dependence of SN numbers on inclination (Sc's only)

	n_{Gal}	L (in 10^8 $L_\odot$)	n_{SN}	SN $(10^8 L_\odot)^{-1}$ $(100\ \text{yr})^{-1}$
$i \leqslant 30°$	26	4972	29.5	0.0185 ± 0.0034
$30 < i \leqslant 60°$	72	10183	8	0.0025 ± 0.0008
$i > 60°$	58	9952	12	0.0038 ± 0.0011
$i > 30°$	130	20135	20	0.0031 ± 0.0007

of the two latter groups agree within the errors, their combined mean rate being 0.0031 ± 0.0007 SN $(10^8 L_\odot)^{-1}$ $(100\ \text{yr})^{-1}$. The nearly face-on Sc's, however, give a rate which is 6.4 ± 1.8 times higher! Hence, there is a *strong inclination effect* of the observable SN rates.

It is evident that this inclination effect is purely observational. For some reason it is apparently much more difficult to find a SN in an inclined spiral than in a face-on one. At present it is impossible to pin the main reason for this selection effect; one could for instance think of the increased effect of the luminous background, of the merging of bright spots with the SN image, or/and of internal absorption.

The dependence of the observed SN rate with inclination gives a direct indication that SNe do not constitute a spherical population. If they did, one would expect rather more SNe in inclined Sc's than in face-on ones. (In case one wanted to interpret the slight frequency increase from intermediate to strongly inclined galaxies (lines 3 and 4) as being real, it would indicate that the mean z-values of SNe were somewhat larger than the typical value for the disc population.) There is in addition a very slight indication that type I SNe are less affected by inclination than those of type II, because the ratio type II: type I decreases from 2.0 ± 1.0 for nearly face-on galaxies to 1.3 ± 0.7 for galaxies with $i > 30°$. But still, also type I is clearly affected by inclination.

There are too few SNe in sample Sb's to study the inclination effect for this type of galaxy; this is even the more true because of a pronounced and unexplained scarcity of face-on Sb's. Whatever the reason for the observed inclination effect in Sc's is, it seems most plausible to assume that the same selection mechanism works in Sb and Sa galaxies as well. Therefore it is assumed for the remainder of this paper that the true SN numbers are 6 times higher than the observed numbers for *all* inclined spirals and S0's. The size of this correction factor and its statistical error introduce, of course, considerable uncertainties.

For completeness the radial distribution of SNe for two groups of more inclined Sc's and for Sb's with $i > 30°$ is also shown in Figure 2. The pronounced difference of the distribution in these inclined systems from that in face-on Sc galaxies is the consequence both of neglecting the projection effect on the radial distance r, and of the losses due to inclination. Therefore the above decision, to apply the large inclination correction factor to all spirals, may find a slight additional justification in the observation that the character of the SN distribution functions agrees reasonably well for inclined Sb's and Sc's.

E. THE DEPENDENCE ON THE LUMINOSITY OF THE PARENT GALAXY

The first indications that the SN rate depends on the luminosity of the parent galaxy were given by Pskovskii (1961, 1967) and Johnson and MacLeod (1963); Kukarkin (1965) found then that galaxies with multiple SN occurrences are in general overluminous spirals. His result was supported by Ambartsumian (1965), but contested by Katgert and Oort (1967). In the same year as Kukarkin, Zwicky (1965) published an elegant, distance-independent method, which suggested that the SN rate is roughly proportional to the galaxian luminosity. Unfortunately Zwicky (1969) discredited his result later by postulating the existence of 'fast producers', which he believed *not* to be more luminous than average galaxies. Final evidence for a direct correlation between the SN rate and the luminosity class (and hence for a *linear* correlation with the luminosity) of the parent (Sc) galaxy could be given elsewhere (Tammann, 1970).

In order to investigate if the present material supports the result of the SN rate increasing linearly with galaxian luminosity, the unbiased sample of face-on Sc's is subdivided in Table V into three groups with different absolute magnitudes. Their total luminosity is given in column 3 of the table, and their SN rate per unit luminosity and per unit time in column 5. Our expectation is that the SN rate per unit luminosity is constant for all subgroups. Within the statistical errors the individual deviations

TABLE V

Dependence of SN numbers on luminosity (face-on Sc's only)

	n_{Gal}	L (in $10^8 L_\odot$)	n_{SN}	SN $(10^8 L_\odot)^{-1}$ $(100\text{ yr})^{-1}$
$M > -19^m.00$	9	313	3	0.030 ± 0.017
$-19^m.00 \geq M > -20.75$	9	975	5	0.016 ± 0.007
$M \leq -20.75$	8	3684	21.5	0.018 ± 0.004

from the overall mean of $\nu_{SN}=0.0185\pm0.0034$ $(10^8 L_\odot)^{-1}$ $(100\text{ yr})^{-1}$ are indeed insignificant. The agreement is the more reassuring, as the adopted distances enter sensitively into the galaxian luminosities. Possible, small, but systematic inconsistencies in the adopted distance scale for dwarf and giant spirals would have spoiled the comparison.

Unfortunately the data are not extensive enough for a parallel investigation as to the behavior of other types of galaxies. Only by analogy it seems very reasonable to assume that the SN rate *within a given galaxy type* is in general proportional to the luminosity of the parent galaxy.

F. THE DEPENDENCE ON THE SUBTYPES OF SC GALAXIES

The class of Sc galaxies used here comprises four subtypes: Sbc, Sc, Scd, and Sd. Since there is a considerable difference between the two extreme subtypes, it could well be that they differed also in their SN rates. In Table VI the face-on Sc's with known

TABLE VI

Dependence of SN numbers on subtypes (face-on Sc's only)

	n_{Gal}	L (in $10^8 L_\odot$)	n_{SN}	SN $(10^8 L_\odot)^{-1}$ $(100\text{ yr})^{-1}$
Sbc	4	1235	6	0.015 ± 0.006
Sc	8	2139	11	0.016 ± 0.005
Scd	9	1356	12.5	0.029 ± 0.008
Sd	4	211	0	–

subtype are grouped accordingly. The total luminosity of a group is given in column 3 and the SN rate (per unit luminosity and per unit time) in column 5. In view of the considerable statistical errors it is difficult to interpret the different rates. There could be a rate increase from Sbc to Scd by a factor 2 or even 3. The observed low rate in Sb's (Section 6) would conform with the interpretation that the rates increase from Sb (and Sa) to Scd. However, by extrapolation one would expect then 2–3 SNe in the Sd's, whereas none is observed. Therefore the alternative interpretation is still equally plausible, i.e. that the different rates in Table VI are random fluctuations about a mean value.

G. THE DEPENDENCE ON THE COLOR OF THE PARENT GALAXY

Searle *et al.* (1973) have performed model calculation to explain the observed intrinsic color range for galaxies of the same type. They found that the colors could be explained by the assumption of an almost unique initial mass distribution of stars, but with varying decay rates for the star formation rate. The blue galaxies are then those which have maintained a relatively high formation rate, and which therefore should contain an unusual high number of massive stars. Hence, if some or all SNe originate from massive stars, the SN rate should depend on the color of the parent galaxy.

It is very interesting to see whether this predicted color dependence can be confirmed with the present material. The 19 face-on Sc galaxies with known $(B-V)^{\circ}$ colors (corrected for galactic absorption) in the RCGB (de Vaucouleurs and de Vaucouleurs, 1964) were subdivided in three color groups (Table VII). Their SN numbers are given in column 6 of the table, and their distribution over the different types is specified in columns 3–5. Column 7 shows the total photographic luminosity of the galaxies in each color group.

We first consider the SN rates (per unit luminosity and per unit time) for all types of SNe lumped together. The resulting rates and their mean errors are given in column 8. Within the errors the rates agree with one mean value for all 27.5 SNe of 0.019 ± 0.004. A slight indication may be in the data that the SN rate increases for *redder* galaxies, but that is very uncertain, and in fact the error range would still not exclude the possibility that the SN rate is twice as high for the blue galaxies as for the red ones. This large uncertainty is due to the small sample statistics, but also to the narrow color range available.

The preceeding consideration made the tacit assumption that also type I and 'other' SNe are related to massive stars. This may not be correct, and then only the type II SNe should be considered. It happens accidentally that the number of type II SNe is exactly 2.5 times smaller than the total number of SNe for each of the three color groups. Within the groups the relative rate per unit luminosity is therefore the same for type II as for all SNe. However, the statistical errors are now still larger, because the numbers of SNe are smaller. The ratio between the SN rates for blue, intermediate, and red Sc's and the corresponding error range becomes then $1:(0.9 \pm 0.8)$. $:(1.4 \pm 0.9)$. This ratio has to be compared with the predicted one by Searle *et al.* (1973). The mean colors of the three groups are $(B-V)^{\circ} = 0\overset{m}{.}50$, $0\overset{m}{.}59$ and 0.70, respectively; after interpolation in their Table 2 one finds then a predicted SN rate of $1:0.7:0.4$. The agreement between the two ratios is poor, but still inconclusive.

These considerations lead, however, to one firm conclusion: the Ir I galaxies are on the average still bluer than Sc's and their type II SN rate should be quite high. This expectation is definitely not fullfilled, since they have produced (so far) only type I SNe (Section 4). There are apparently only three ways out of the dilemma: (1) The model computations by Searle *et al.* (1973) cannot be applied for some reason. (This possibility would remove in fact only part of the problem, because there is direct evidence in Ir I's for the presence of many very massive stars.) (2) The SN phenomenon of type II (and of type I) is not associated with massive stars. (3) *The SNe of type I* (at least in the case of Ir I's) *belong to the young massive stellar population.* It is this latter possibility which deserves in particular very careful testing.

H. THE DEPENDENCE ON THE BAR OF THE PARENT GALAXY

For the understanding of the SN rate dependence on a given parameter it is essential to know whether or not some other neglected parameter could actually play the key rôle. A last parameter which comes to mind here and which can be tested, is the possible influence of a bar. 26 face-on Sc galaxies of the sample have *A*, *AB*, *B*

TABLE VII

Dependence of SN numbers on galaxian color (face-on Sc's only)

	n_{Gal}	type I	type II	Others	n_{SN}	L (in $10^8\ L_\odot$)	SN $(10^8\ L_\odot)^{-1}$ $(100\ \text{yr})^{-1}$
$(B-V)^\circ \leqslant 0^{m}\!.55$	6	1	2	2	5	910	0.017 ± 0.008
$0^{m}\!.55 < (B-V)^\circ \leqslant 0.65$	7	2	4	4	10	1988	0.016 ± 0.005
$(B-V)^\circ > 0.65$	6	3	5	4.5	12.5	1633	0.024 ± 0.007

TABLE VIII

Dependence of SN numbers on type of bar (face-on Sc's only)

	n_{Gal}	L (in $10^8\ L_\odot$)	n_{SN}	SN $(10^8\ L_\odot)^{-1}$ $(100\ \text{yr})^{-1}$
SAc	7	1588	6	0.012 ± 0.005
SABc	14	3159	22.5	0.022 ± 0.004
SBc	5	225	1	(0.014)

classification in the RCBG (de Vaucouleurs and de Vaucouleurs, 1964). Their distribution is shown in Table VIII. The SN rates (per unit luminosity and unit time) in column 5 seem to scatter reasonably well around the mean rate of 0.019 ± 0.003 $(10^8\ L_\odot)^{-1}$ $(100\ \text{yr})^{-1}$, and therefore to exclude the possibility at least of a strong dependence on the bar type of the parent spiral.

6. The Supernova Rate per Unit Luminosity in Different Types of Galaxies

After the preceding preparatory remarks it seems possible to attack the main question as to the frequency of SNe in galaxies of different types and with different luminosities.

Table IX gives on line 1 the total photographic luminosity for the different types of sample galaxies. For Sc's the sample is large enough, that it can be restricted here to face-on objects ($i \leqslant 30°$). The luminosities are reduced to face-on orientation (Section 2A); the remaining, inclination-independent absorption ($A_{pg} = 0\overset{m}{.}28$ for Sc's, $0\overset{m}{.}43$ for Sa/Sb's (Holmberg, 1958); and $0\overset{m}{.}22$ for S0's (extrapolated); $0\overset{m}{.}0$ for Ir's) is applied on line 2, which gives the absorption-free total photographic luminosities L^0. The total number of SNe, corresponding to 32 search years, are repeated on line 3 for each type of galaxy. Line 4 gives the number of SNe after correcting for inclination; the correction is not applied to E's and Ir's and, of course, not to face-on Sc's. The corrected numbers are calculated by multiplying with 6 ± 2 (Section 5D) each SN, which appeared in a spiral with $i > 30°$, and by adding up the numbers thus obtained. In line 5 the SN rate per (absorption-free) unit luminosity and unit time are calculated.

As to the different SN rates the following remarks are in place:

(1) The best determined rate in Table IX is for the (face-on) Sc's. The other rates are less certain because of small number statistics and because of the adopted effect of inclination on the observable number of SNe.

(2) The rate for Ir I's is slightly *lower* than for Sc's. This is unexpected (Section 5G) and may well be a statistical fluctuation. It is justified then to adopt a mean rate for the two galaxy types. This mean is meaningless, however, if the SNe, which are all of type I in Ir I's, belong to a fundamentally different population than type II SNe, because the rate in Sc's is based on a mixture of all SN types.

(3) Sb's have a SN rate which is about 2 times lower than in Sc's. This difference is very likely to be real, unless the correction factor for inclination is still larger than 6 in Sb's.

TABLE IX

SN frequencies per unit luminosity according to galaxian type

	E	S0	Sa	Sb	Sc[a]	Ir I	Ir II
L (in $10^8 L_\odot$)	6899	9066	8710	16666	4972	1373	628
$L°$ (abs. corr.)	6899	11055	12942	24765	6434	1373	(628)
n_{SN}	5	2	1	10	29.5	5	2
n_{SN} (incl. corr.)		12 ± 9	(6)	55 ± 24			
SN $(10^8 L_\odot)^{-1}$ $(100\ \text{yr})^{-1}$	0.0023 ± 0.0010	0.0034 ± 0.0025	(0.0014)	0.0069 ± 0.0031	0.0143 ± 0.0027	0.0114 ± 0.0051	(0.010)
	E + S0: 0.0025 ± 0.0009				Sc + Ir I: 0.0137 ± 0.0024		

[a] Face-on systems only.

TABLE IXA

Predicted mean intervals τ between Supernova outbursts for some Local Group members

Galaxy	Type	τ (yr)
M31	Sb	20^{-6}_{+16}
M33	Sc	167^{-24}_{+33}
SMC	Ir I	1460
IC 1613	Ir I	8000
M32	E	20000
Leo II	dE	(3×10^6)

TABLE IXB

Predicted and observed Supernova numbers in the 9 most luminous face-on Sc Galaxies of the sample

Galaxy	$L°$ (in 10^8 $L_\odot$)	n_{SN} per 32 search years	
		Predicted	Observed
NGC 3184	330	1.5	3
3938	360	1.6	2
6946	480	2.1	4.5
628	495	2.2	0
5457	500	2.2	3
4254	515	2.2	2
4321	680	3.0	3
4303	715	3.1	3
5236	720	3.2	4

(4) The rate in Sa's is extremely uncertain, but it is definitely lower than in Sc's by roughly one order of magnitude.

(5) The rates in S0's and E's agree within the errors, and it may be best to adopt a mean rate for the two types. This agreement, although rough, is rather interesting, because the two rates are derived with quite different assumptions: the full inclination correction factor of 6 was applied to S0's, whereas it was assumed that actually all SNe in E's were found. If now the true number of SNe in E's were considerably larger, it would seem unavoidable that the SN rate would become higher in E's than in S0's. This conclusion seems unlikely, and one may therefore infer that the observed number of SNe in E's is not too far from the true number. This conclusion is weak and qualitative, but it seems to exclude incompleteness factors of more than about 2 for E's.

E/S0 galaxies have a significantly lower SN rate than Sc/Ir I galaxies. The frequency ratio – excluding the possibility of large selection effects – is 6.0 ± 2.3. Interestingly enough a difference remains if the comparisons is restricted to SN type I only; in that case the rate in Sc/Ir I's is still 2.1 ± 0.9 times higher than in E/S0's.

(6) A surprisingly high rate is found for Ir II galaxies. This rate is very suspect, however. Due to lack of data the luminosities of this galaxy type are not corrected for

internal absorption, although there certainly is a large amount of dust in some of these systems. In addition no correction could be applied for the inclination effect on the observed SN numbers. For these reasons the SN rate in Ir II's is not used in the following discussion.

There is only one case where the SN rates in Table IX can be checked independently: from nine SN remnants in LMC Mathewson and Clark (1973) determined a frequency corresponding to $\tau = 500$ yr. With a luminosity of $L = 19.6 \times 10^8\ L_\odot$ and a mean rate of 0.0137 one would predict $\tau = 370 \pm 55$ yr; using the rate for only the Ir I's of 0.0114 τ increases slightly to 450 yr. In either case the agreement is very satisfactory. This agreement is either accidental or else it has a very important implication: the observed rate in Ir I's rests entirely on type I SNe, whereas the objects of Mathewson and Clark lie in the extreme Population I [also their derivation of τ depends on galactic SNR's, which in turn have a z-distribution characteristic of very young objects (Ilovaisky and Lequeux, 1972a)]. This then is another indication that type I SNe at least in Ir I's are associated with very young objects.

To see what kind of frequencies the SN rates in Table IX imply for individual galaxies, the mean intervals τ between SN outbursts are calculated for some Local Group members in Table IXa, using the appropriate galaxian luminosities and types. The one value in Table IXa which deserves comment is the very high frequency for M 31 ($\tau = 20$ yr). During the last 90 years actually only 1 SN has been observed in that galaxy. The ensuing difficulty has been discussed before (Tammann, 1970). The most probable answer to the difficulty seems now either that the SN rate in Sb's was overestimated here by applying the full inclination correction factor of 6 (which was really derived only for Sc's), – or that the high inclination of M 31 itself ($i = 71°$) suppresses the observable SN rate more than anticipated.

The fastest SN producers in the present sample are the overluminous face-on Sc galaxies, the nine brightest of which are listed in Table IXb with their predicted and observed SN numbers. It is remarkable that this objectively selected subsample contains six cases out of seven sample galaxies, which have produced more than 2 SNe, and two additional cases out of 7 sample galaxies with 2 SNe. (The total number of galaxies with multiple SN occurrences, known by February 1st, 1973, is 19 – only eight of which had 3 SNe or more). If the term 'fast producers', used by some authors, is ment to designate galaxies with 3 or more SNe, one finds now that the vast majority of the 'fast producers' are identical to *nearby* (i.e. within the present unit volume) *face-on overluminous Sc galaxies.*

The galaxies in Table IXb cover a luminosity range of a factor 2; the same expected factor is not exhibited by the observed SN numbers. The most probable explanation for this effect are errors in the adopted distances. In order to list the galaxies in Table IXb in the correct order of their luminosities, distances are required which are determined better than within a factor of $\sqrt{2}$. The expected random errors of the adopted distances are at least half this value, and that shall cause some galaxies in the table to exchange eventually their places.

Only one galaxy, NGC 628, in Table IXb has produced so far no SN. Another nearby supergiant face-on Sc galaxy (not in the Shapley-Ames and therefore not in the present sample) without any known SN is IC 342; due to its low galactic latitude it is disliked by observers. In the light of the present statistics it has to be expected that these two galaxies shall produce in the long run one SN every 10 to 15 yr.

7. The Supernova Rate per Unit Mass in Different Types of Galaxies

It has been discussed in Section 5C that there is evidence for the SN rate being proportional to the *mass* of the parent galaxy of a given type. One therefore would like to express the rate also in SNe per unit mass (10^{10} $\mathcal{M}_{\odot}$) and unit time. For this purpose the total luminosities L^0 from Table IX were transformed into total masses using appropriate mass-to-light ratios for the different types of galaxies. For E/S0 galaxies, which were lumped together here, it was assumed that $\mathcal{M}/L=40$ for $H_0=55$ (Burbidge, 1973; Page, 1973; Smart, 1973); this value is uncertain, because there is increasing evidence that $\mathcal{M}/L$ is for elliptical systems a strong increasing function of the total luminosity (Lynden-Bell, 1973; Smart, 1973). The adopted value, however, agrees closely with Smart's determination, which is based on double ellipticals with 'average' luminosities and which correspond probably closely to the E's of the present sample. The $\mathcal{M}/L$ values for the remaining galaxies are taken from Roberts (1969). Ir II systems are not considered here, as justified in Section 6. The different $\mathcal{M}/L$ ratios are shown on line 1 of Table X, and the resulting total masses for different types of the sample galaxies are given on line 2. The errors given are calculated from the statistical errors in n_{SN} and the errors in $\mathcal{M}/L$ according to Roberts. The numbers of known SNe, corrected for inclination if appropriate, are taken from Table IX, line 3 and 4, respectively. The corresponding SN rates (per unit mass and per unit time) are given in line 5.

The striking result in Table X is the very low SN rate of E/S0 galaxies. Their rate

TABLE X

SN frequencies per unit mass according to galaxian type

	E/S0	Sa	Sb	Sc[a]	Ir I
$\mathcal{M}/L$ (solar units)	40	1.8 ± 0.4	3.7 ± 1.1	2.5 ± 0.3	3.0 ± 0.6
(in 10^{10} $\mathcal{M}_{\odot}$)	7182	233 ± 51	916 ± 272	161 ± 19	41 ± 8
n_{SN} (incl. corr.)	17 ± 9	(6)	55 ± 24	29.5	5
SN $(10^{10}$ $\mathcal{M}_{\odot})^{-1}$ $(100\ \mathrm{yr})^{-1}$	0.007:	(0.08)	0.19 ± 0.10	0.57 ± 0.13	0.38 ± 0.19

[a] Face-on systems only.

per unit mass is about 80 times lower than that for Sc's. Even if one allows for the uncertainties involved and if one assumes that all errors work in the same direction, the difference corresponds to at least one order of magnitude.

Perhaps a more significant comparison is to consider here only SNe of type I, because type II is confined to only two types of galaxies (Sb and Sc). To do this it was assumed that all SNe (also the unclassified or uncertain ones) in E/S0's and Sa's belong in fact to type I, and that one third of the unclassified SNe in Sb's and face-on Sc's are actually also of this type. In that case the rate for type I alone is in E/S0's, Sa's, Sb's, Sc's, and Ir I's as 1:11:11:42:51. Due to the various error sources (possible loss of SNe in some galaxian types, errors in the inclination corrections and in the adopted $\mathscr{M}/L$ values) the relative rate between any two galaxy types in the sequence is quite uncertain, but there is a definite trend of the rates to increase from E/S0's to Ir I's. If it is assumed that type I SNe are inherent to the old population typical for elliptical galaxies, then the rates should not be related to the total mass, but to the old population mass of a particularly type of galaxy. Since the mass fraction of old population decreases from E's to Ir I's, the resulting type I SN rates then differ even more between the different galaxian types. This would lead to the strange conclusion that the old population in Ir I's is much more efficient in producing type I SNe than the population of ellipticals. One way out of this paradox is to assume that type I SNe belong to the young population.

8. The Supernova Rate in Our Galaxy

Assuming the total mass of our Galaxy and its type are known, one can determine the overall SN rate in our system using the data in Table X. It has been argued that the best value of the total galactic mass is $17 \times 10^{10}\ \mathscr{M}_{\odot}$ for the present purpose, and that the type of our Galaxy may be assumed to be Sbc as compromise from the contradictory evidence (Tammann, 1970). In Table X the Sbc galaxies are carried together with Sc's to Sd's under Sc galaxies, for which group a rate of 0.57 ± 0.13 SN $(10^{10}\ \mathscr{M}_{\odot})^{-1}$ $(100\ \text{yr})^{-1}$ was found. From this one derives a mean interval of galactic SN events of $\tau = 10^{+4.2}_{-2.3}$ yr. The corresponding high frequency could become somewhat lower if it were possible to determine the SN rates per unit mass separately for the different subtypes of Sc galaxies. The high frequency could also be caused by an overestimate of the galactic mass or by adopting too late a type for the Galaxy. In fact Schmidt-Kaler and Schlosser (1973) have recently concluded from photometry of the central bulge that our Galaxy is nearest to Sb $(-\text{Sb}^{+})$. Assuming that our Galaxy is an Sb system and that the corresponding rate of 0.19 ± 0.10 SN $(10^{10}\ \mathscr{M}_{\odot})^{-1}$ $(100\ \text{yr})^{-1}$ applies, one finds $\tau = 31^{+40}_{-11}$ yr.

The uncertainty in the galactic value of τ is therefore still considerable. The best compromise is to assume $\tau = 25$ yr and to allow an error range of a factor of 2. With a compromise ratio of 1.5 between SNe of type II and type I (Section 4) and assuming that there are no other types of SNe, one finds then a corresponding rate of 1 type II SN per 40 yr and 1 type I SN per 60 yr, again with a considerable error range.

There are three additional, independent indications concerning the galactic SN rate, i.e. historical SNe, radio SNRs, and pulsars. However, it must be emphasized that all SN rates based on these objects are extrapolations from observations within a galactic subvolume around the Sun, whereas the above rate of $\tau = 25$ yr corresponds to the Galaxy as a whole. Since the SN rate per unit surface varies strongly with the radial distance from the center of a galaxy (as infered from Section 5C), and since in addition the SN rate per unit mass may vary in different regions of a galaxy, the volume correction factors applied in order to extrapolate for the whole Galaxy are necessarily rough and only first approximations.

The results for the SN rate in the wider solar neighborhood, derived from independent observations, are as follows:

(1) Historical SNe. Katgert and Oort (1967) concluded from the four SNe (SN 1006, Crab, Tycho, and Cas A) which appeared within 4.5 ± 0.5 kpc from the Sun during the last 1000 years, that the galactic value of τ is 23^{+27}_{-8} yr. The errors are here calculated from the mean statistical fluctuations of a count rate of only 4, and from the adopted error of the limiting distance. The latter error may be an underestimate; according to Ilovaisky and Lequeux (1972a) the four SNe lie within 5 ± 1.5 kpc, but according to Woltjer (1972) they are within a considerably smaller volume with a radius of only 3 kpc. In any case the indicated error range does not include yet the large uncertainty of the adopted volume correction factor of $(15/4.5)^2$. Finally a systematic error may be introduced by the fact that the four historical SNe considered here have $\langle z^2 \rangle^{1/2} \approx 500$ pc, which is considerably larger than the average value of $\langle z^2 \rangle^{1/2} =$ $= 102$ pc (Ilovaisky and Lequeux, 1972a); additional SNe in lower galactic latitudes may have been missed due to absorption, and in that case the true galactic value of τ would still be smaller.

In summary, the evidence of historical SNe points toward $\tau = 25$ yr with an error range of at least a factor of 2. Although far from being a stringent test in view of the large uncertainties the agreement between the SN rate derived from the galactic mass and from historical SNe is perfect.

(2) Radio SNRs. A number of authors have derived the galactic value of τ from radio SNRs and some model assumptions; some results are: $\tau = 30$–60 (Shklovsky, 1960), $\tau = 50$ (Minkowski, 1964), $\tau = 40$ (supposedly for type II only; Aizu and Tabara, 1967), $\tau = 40$–90 (Kesteven, 1968), $\tau = 75$ (Milne, 1970), $\tau = 40$–80 (Caswell, 1970), $\tau = 34$–45 (Downes, 1971), and τ 50 ± 25 yr (Ilovaisky and Lequeux, 1972b). The latter value and its error range cover quite well all earlier determinations, and hence it may be adopted as typical for the evidence from SNRs. We shall return to the significance of this value.

(3) Pulsars. The SN rate required to account for the observed number of pulsars depends strongly on their adopted lifetime and distance and on their beaming factor. Gunn and Ostriker (1970) found a birth rate for pulsars of one per ~ 30 yr; and they suggested (Ostriker and Gunn, 1971) that pulsars are the cause for the type II SN phenomenon. Their birth rate corresponds so well with the galactic SN rate of $\tau = 25$ yr, as derived here, that it could be possible that *all* SNe (type II *and* type I) have underlying pulsars. However, a recent determination of the pulsar birth rate by Hewish (1973) indicates only one per 100–200 yr. This value would open the possibility, favored by some authors (Dickel, 1973; Gull, 1973), that not even every type II SN event is associated with the birth of a pulsar. It may be significant that Sofia (1972) required independently a rather high minimum mass of $5\,\mathcal{M}_{\odot}$ for the progenitors of

pulsars; this is somewhat higher than the lower mass limit of type II praesupernovae as derived below, implying then that only the more massive type II SN can form pulsars. Also Ilovaisky and Lequeux (1972b) found a marginally significant factor of 2 between the surface density of SNRs and pulsars, which may indicate that not all SNe, or not even all type II SNe turn into pulsars. But the statistics of the galactic SN rate and in particular that of the birth rate of pulsars are still too uncertain to allow a decision whether all SNe or only a fraction of them are connected with pulsars.

The interpretation of the SN rate inferred from galactic SNRs may have a very important consequence. *If* the SN rate of $\tau = 50 \pm 25$ from SNRs is interpreted to agree with the value of $\tau = 25 \pm$ a factor of 2, as derived here from the galactic mass for SNe of all types, then *all* SNe have on the average very low z-values and belong to the very young population. This is because Ilovaisky and Lequeux (1972a) found for SNRs the same z-distribution as for O, B stars. This conclusion can only be avoided by assuming that the catalogues of SNRs are strongly biased against type I SNe. In principle such a bias is possible, because existing searches for radio sources may have discriminated against extended sources in high latitudes (Dickel, 1973; Wendker, 1973), and in addition one expects that (type I) SNRs in high latitudes leave poorer remnants than within regions of higher interstellar density. However, it is not clear whether such a systematic selection effect against type I SNRs actually exists, because the three historical SNe, which are believed to be of type I (SN 1006, Tycho, and Kepler; Minkowski, 1968) *are* contained in the radio lists of SNRs (although two of them have exceptionally large z-values), and because no known SNR is unambiguously identified as the result of a type II SN (Minkowski, 1969).

A clear decision whether the known SNRs represent a reasonably unbiased mixture of type II *and* type I remnants can be expected from statistics only once the birth rate of SNRs and the galactic SN rate are independently so accurately determined that it shall be possible to decide whether the former corresponds better to the expected overall rate of SNe ($\tau \approx 25$ yr) or to the rate of type II only ($\tau \approx 40$ yr). In any case, if one wants to avoid the conclusion that type I SNe belong to the very young population of the Galaxy, one has to assume that existing lists of SNRs contain essentially only type II SNRs.

If type I SNe should eventually prove to exhibit a very flat distribution, the only independent argument to associate pulsars with type II SNe would become invalid. In this connection it is interesting that the Crab SN has always been considered as type I (e.g. Minkowski, 1964) and its reclassification as type II, after a pulsar was dicovered in the Crab nebula, is not very satisfactory (Minkowski, 1969).

If SNe originate from massive stars (and this is uncontested at least for type II) one can compare the SN rate with the death rate of massive stars. Gunn and Ostriker (1970) have estimated that one star with $\sim 4\, \mathcal{M}_\odot$ or more dies every 30 yr in the Galaxy. This result is in good agreement with an independent result by Stothers (1963), who estimated one dying star with 4–5 $\mathcal{M}_\odot$ or more per 30–50 yr. Also Arnett's (1971) result points toward a similar mass limit, because he found that a higher mass limit of ~ 10 (or 5.5) $\mathcal{M}_\odot$ corresponds to too low a SN rate of $\tau = 100$ yr. More recently van den Heuvel (1973) has estimated again the number of O, B stars in the Galaxy and concluded that there are $(2.75 \pm 0.25) \times 10^6$ stars more massive than $3.5\,\mathcal{M}_\odot$. If the observed mass function for the brightest stars has the form $\phi(\mathcal{M}) \propto \mathcal{M}^{-4.5}$ (van den Heuvel, 1973), one can calculate the distribution over mass $\phi(\mathcal{M})$ for these

TABLE XI

The observed and initial numbers of massive stars in the Galaxy and their death rates

Mass ($\mathcal{M}_\odot$)	$\Phi(\mathcal{M})$	$\Psi(\mathcal{M})$	Deaths $(100\ \text{yr})^{-1}$	
4	1.59×10^6	1.52×10^8	1.05	2.75 (4–8)
5	5.85×10^5	8.71×10^7	0.67	
6	2.58×10^5	5.59×10^7	0.46	
7	1.28×10^5	3.84×10^7	0.33	
8	7.06×10^4	2.92×10^7	0.24	
9	4.15×10^4	2.27×10^7	0.18	
10	2.58×10^4	1.84×10^7	0.14	
11	1.68×10^4	1.52×10^7	0.11	
12	1.14×10^4	1.32×10^7	0.09	

stars (Table XI, column 2). The lifetimes until helium exhaustion in the centre (column 3) are taken or interpolated from Paczyński (1971; – the time difference between helium exhaustion and carbon ignition is here negligible). The corresponding initial birth function $\Psi(\mathcal{M})$ is identical to the death rate, because the star formation rate in the Galaxy can be taken constant during the relatively short lifetimes of these massive stars; the death rate per 100 yr is given in column 4. From Table XI one finds that one star from the mass interval $3.5 \lesssim \mathcal{M} < 8.5\,\mathcal{M}_\odot$ dies every 36 yr in the Galaxy in very good agreement with the numbers above. This mass interval is that for which Paczyński and Ziółkovski (1968) and Paczyński (1970) found that carbon ignites violently (this does not necessarily disrupt the core according to Paczyński, 1972). It is pleasing to find that the death rate of these stars corresponds to the galactic type II SN rate of $\tau = 40$ yr. This, however, does not exclude the possibility that type I SNe originate also from massive stars: by lowering the mass limit of 3.5 to $2.5\,\mathcal{M}_\odot$ the resulting death rate would become one death per ~ 25 yr in excellent agreement with the galactic overall SN rate. Such a decrease of the mass limit does not seem unreasonable, because in the absence of mass loss the limit would be as low as $1.4\,\mathcal{M}_\odot$ (Paczyński, 1971).

The death rate of all stars more massive than $8.5\,\mathcal{M}_\odot$ seems from this computation to be smaller than one event per 100 yr. This is apparently in variance with a recent result by Dellaporta (1973); the reason for this unsatisfactory situation is not understood at this moment.

Weidemann (1971) and Sackmann and Weidemann (1972) have cast considerable doubt on the procedure to predict carbon ignition on grounds of mass considerations only. If an evolving star retains a considerable fraction of its original central angular momentum, this parameter influences strongly whether the star can form a degenerate core and therefore whether carbon can ignite. In this case the occurrence of a SN would depend on the original mass *and* angular momentum of the praesupernova. However, Hardorp (1973) has argued that empirical considerations suggest that an evolving star transfers angular momentum from its inner core to the outer surface, a process favored also by Kraft (1970) to explain the slow rotation of cluster giants. Thus it seems still possible, in view of evolutionary as well as statistical arguments, that the ancestors of (type II) SNe are confined to a mass range of $\sim 3 \leqslant \mathcal{M} < 8.5\, \mathcal{M}_{\odot}$.

9. Remarks on the Parent Population of Supernovae of Type I and II

The question as to the parent stars of SNe is not directly connected with the frequency of SNe – however, it might be hoped that their statistics shed some light on their origin. In fact in the course of this paper there were several occasions to comment on the population from which SNe seem to originate. These arguments are briefly compiled here, together with some additional evidences.

The identification of SNe of type II with dying massive stars rests on rather firm grounds, and the multiple evidence shall not be repeated here. In agreement with a large number of authors the identification is assumed to be correct.

The remaining problem concerns the origin of type I SNe (the question whether there are possibly additional types of SNe shall not and cannot be decided here). The conventional assumption is that SNe of type I belong to the old population and that their progenitors are therefore stars with $\lesssim 1\, \mathcal{M}_{\odot}$. The only evidence for this association is the occurrence of type I in E galaxies, which are assumed to be void of massive stars. Ensuing difficulties have been pointed out for instance by Bertola and Sussi (1965), Katgert and Oort (1967), Tammann (1970), and Barbon *et al.* (1973). The list of the main difficulties has become in fact quite long and runs as follows:

(1) The radial distribution of SNe in Sc galaxies shows no separation between type I and II (Figure 2).

(2) The effect of inclination works in spirals on the discovery chance also of type I, which must therefore belong to a flattened population.

(3) According to Johnson and MacLeod (1963) type I SNe in spirals lie preferentially in arms; the result has been questioned, however, by Minkowski (1964).

(4) The very low mean z-values of galactic SNRs are significant, provided one does not assume that the majority of these objects originate from type II SNe (Section 9).

(5) The steady increase of the type I SN rate per unit mass as one goes from E to S to Ir I galaxies is very puzzling (Section 7).

(6) The absence (or underabundance) of type II in Ir I's is totally unexpected (Section 4).

(7) The agreement between the observed and predicted SN rate in LMC would have to be a chance effect, because the former is based on SNRs in the extremely young population and the latter on the frequency of type I SNe in Ir I's (Section 6).

(8) The expected dependence on the SN rate on the color of the parent galaxy predicts a large number of SNe from massive stars for the case of Ir I's – but only type I SNe are found in these systems (Section 5G).

(9) The two SNe which appeared on intergalactic bridges seem to be of type I (Tammann, 1973b) – one expects, however, a large fraction of young population in these bridges because of their blue color (Searle, 1973b).

(10) The overall similarities between type I and type II SNe (for instance in frequency, luminosity, light curves, and expansion velocities) are more striking than their differences. In particular the temperatures of ~12000 K and 10000 K shortly after maximum for type I and type II, respectively (Searle, 1973a), and their agreement in luminosity within factors ~5–10 imply from $L \propto R^2 T^4$ that their radii agree better than within a factor of 3. It seems difficult to understand how an old, low mass star could produce the necessarily very large radiating volume and how it should mimic so closely the SN phenomenon of massive stars.

(11) The proposed existence of transitional SNe between type I and type II (Fairall, 1972) questions severely the possibility of fundamental differences between the origins of the two types.

The weight of every one of these eleven arguments may be judged to be quite low, but their combined evidence strongly suggests that some basic understanding of the origin of type I SNe is lacking. It should be noted that the arguments are selected such as to depend very little on the accuracy of the present statistics; it would take unexpectedly large systematic selection effects to alter only one of the arguments.

How can the dilemma be solved? One possibility would be to assume that SNe of type I come from old stars in E galaxies and from young stars in S and Ir I galaxies. In this connection it is very interesting that Barbon *et al.* (1973) propose a subdivision of type I into two classes: slow and fast ones. The data suggest that the SNe in E's may all belong to the slow class. Since the differences between the two classes are quite subtle, one would, however, still expect that all type I SNe come from the same type of objects, which probably would differ only in age. In this case one could think of exploding white dwarfs (Finzi and Wolf, 1967), or of close binaries (Hartwick, 1972). An ensuing difficulty would be to account for the wide spread in the detonation age: most of these objects would have to erupt while still belonging to the young population (as evidenced in S and Ir I galaxies), but some remaining objects would have to reach very high ages before exploding in E galaxies. Perhaps the least satisfying consequence of this picture is that the underlying phenomenon in type I and type II SNe would still be fundamentally different, and some of the above arguments, especially (6), (10), and (11), would remain unexplained.

There is one, although unorthodox alternative to overcome *all* the above difficulties, i.e. to assume that SNe of type II *and* type I descend from massive, young stars and that E galaxies contain a corresponding number of such stars. One of the most attractive features of this assumption would be that the similarities between different SNe would become plausible. The distinction between type I and type II and the subdivision of type I would then reflect less fundamental differences of the parent stars.

It would be possible for instance that the SN type is a double- (or, in the case of more than two SN types, multiple-) peaked function of the original stellar mass – or the types could be caused by some stars being totally disrupted and others leaving a (neutron star) remnant. In the latter case one might think that the somewhat more energetic type I is correlated with total disruption, and that type II leaves a dense core and possibly a pulsar.

This speculation, which is clearly favored by statistical considerations, raises, of course, one crucial question: is the presence of massive stars possible in elliptical galaxies? That the possibility cannot be rejected off hand is shown in Table XII.

TABLE XII
The relative supernova rates per unit total mass and the relative mass fraction of neutral hydrogen in different types of galaxies

	E/S0	Sa	Sb	Sc	Ir I
SNe per unit mass	1.2	14	33	100	67
relat. $\mathcal{M}_{HI}/\mathcal{M}_T$	10	30	41	100	190

Here the SN rates per unit mass for different types of galaxies from Table X are expressed relative to Sc's (=100) on line 1. The relative neutral hydrogen content on line 2 (Sc=100) is calculated from data by Roberts (1969) and Balkowski *et al.* (1972) for the different galaxy types. There is a clear overall agreement between SN rates and relative gas content in the sense that a higher gas content seems to favor higher SN rates. If this is significant it is interesting to note that the gas content in E/S0 galaxies is eight times larger than their SN frequency would require in the case of a linear correlation. *Prima facie* the presence of a sufficient number of young, massive stars seems therefore quite possible in these galaxies. However, their gas *density* may be so small as to render star formation impossible. This objection can perhaps be overcome by the assumption of metal-enhanced star formation (Talbot and Arnett, 1973), or if the gas in E/S0 galaxies has collapsed into a disc (Lynden-Bell, 1973). Baade's (1951) blue supergiants in NGC 205, an E pec/$S0_1$ galaxy (Sandage, 1961) are, of course, well known. At present it does not seem possible to exclude the possibility of some star formation going on in E/S0 galaxies, and therefore the hypothesis deserves further attention that also SNe of type I descend from young, massive stars.

10. Conclusions

The principal aim of this paper is to demonstrate that the frequency of SNe in individual galaxies can only be understood if the intrinsic properties of the parent galaxies are well known. This and the increasing selection effects with distance confine the statistics of SNe to relatively small volumes. A practicable limit is at present the sample of galaxies within the Local Supercluster. 408 Shapley-Ames galaxies could be considered here, which have produced 74.5 SNe (all or nearly all of which with $M_{pg} \leqslant -15^m$).

Although there remain considerable numerical uncertainties it seems now clear that the SN productivity of a given galaxy is essentially determined by two parameters: its type and its luminosity (or mass). The influence of other intrinsic parameters is smaller, in any case below the reliable detection limit of the present material.

From different observational sources of bias only the inclination of spirals proves to be very important. For some reason only about 15 to 20% of the SNe in spirals with $i > 30°$ are known. Future attempts would be important to detect the missing objects with refined techniques, possibly eventually also by non-optical methods. In the present paper only the SN rates in S0, Sa, and Sb galaxies depend on the adopted inclination correction factor.

No other corrections for incompleteness are applied. *The SN rates obtained*, which range from $\tau \approx 10$ yr for the most luminous Sc galaxies to $\tau \sim 10^6$ yr for dwarf ellipticals – *refer therefore to the present completeness level of SN searches.* Since it is improbable that the present searches yield already the true numbers of SNe, incompleteness correction factors have eventually to be applied. Indications are given that these factors are not large, say < 1.5, but their reliable determination is not possible yet.

Within a factor of 2 the galactic SN rate is found to be $\tau = 25$ yr, in good agreement with historical evidence. The data are insufficient to comment on the completeness of existing lists of SNRs, or to decide whether all type II SNe (or all SNe) become pulsars.

The association of SNe of type II with the death mode of massive stars is not questioned here. However, severe difficulties are encountered when ascribing SNe of type I to the old population. It is therefore suggested that they also originate from young stars. This implies that some star formation is still going on in E/S0 galaxies.

The SN rates derived here are *present* values, which apply to *whole galaxies*. It cannot be expected that any part of a galaxy complies with the overall rate. Moreover, the SN rate changes certainly during the lifetime of a galaxy. Without entering this complex problem, it is a plausible assumption, that galaxies of a given type, which agree *now* in their SN rate per unit luminosity (or unit mass), had coinciding SN rates *at all times*. From this seems to follow that the influence of SNe on the chemical composition of galaxies is the same for a given galaxian type.

Future improvements of the frequency determination of SNe cannot be obtained by just extending the list of known objects. The present sample galaxies shall continue to yield only ~ 2.5 SNe per year (unless new methods are developed to find all SNe in inclined spirals, in which case the overall yield would be three times higher). More reliable data can be expected, however, within a reasonable time along two routes: (1) by improving the statistical sample of parent galaxies (including: additional redshifts to possibly fainter absolute magnitudes, extension of the sample over the Southern Hemisphere, reliable galaxian types, more color measurements, good data on control times of individual galaxies, and a better understanding of the effect of overexposure); and (2) by more detailed information on individual SNe (concerning above all their types and their position within the parent galaxies).

Acknowledgement

It is a pleasure for the author to acknowledge that he received help and stimulation from many sides. Above all he has to thank Dr A. Sandage for the permission to use a very valuable, unpublished list of redshifts for Shapley-Ames galaxies. He owes other unpublished material to Dres R. Barbon, F. Ciatti, C. T. Kowal, L. Rosino, and W. L. W. Sargent. He drew from a number of preprints listed under the references, and he profited in addition from discussions with Dres B. Baschek, J. R. Dickel, W. M. Goss, R. P. Kraft, D. Lynden-Bell, L. Searle, N. C. Smart, V. Weidemann, H. Wendker, P. Wild, F. Zwicky and others. He is very indebted to Prof. B. Batalli Cosmovici for the excellent organization of the present symposium, which prompted this investigation. He sincerely thanks Prof. D. Lynden-Bell for the hospitality at the Institute of Astronomy, Cambridge, and he acknowledges partial financial support from the Swiss National Science Foundation.

References

Aizu, K. and Tabara, H.: 1967, *Progr. Theor. Phys.* **37**, 296.
Ambartsumian, V. A.: 1965, *Astrofizika* **1**, 473.
Arnett, W. D.: 1971, *Astrophys. J.* **166**, 153.
Baade, W.: 1951, *Publ. Michigan Obs.* **10**, 10.
Balkowski, C., Bottinelli, L., Gouguenheim, L., and Heidmann, J.: 1972, *Astron. Astrophys.* **21**, 303.
Barbon, R.: 1968, *Astron. J.* **73**, 1016.
Barbon, R., Ciatti, F., and Rosino, L.: 1973, *Astron. Astrophys.* **25**, 241, see also this volume, p. 99.
Bergh, S. van den: 1959, *Ann. Astrophys.* **22**, 123.
Bergh, S. van den: 1960, *Publ. David Dunlap Obs.* **2**, 159.
Bergh, S. van den: 1965, in H. H. Voigt (ed.), *Landolt-Börnstein*, N. S., VI/1 Springer-Verlag, Berlin, p. 664.
Bertola, F. and Sussi, M. G.: 1965, *Contr. Oss. Astrofis. Univ. Padova*, No. 176.
Burbidge, G. R.: 1973, in A. Sandage, M. Sandage, and J. Kristian (eds.), *Galaxies and the Universe* (=*Stars and Stellar Systems*), vol. IX (in press).
Caswell, J. L.: 1970, *Astron. Astrophys.* **7**, 59.
Chai, C.-S. and Bergh, S. van den: 1970, *Astron. J.* **75**, 672.
Dallaporta, N.: 1974, this volume, p. 187; see also *Astron. Astrophys.* **29**, 393.
Dickel, J. R.: 1973, private communication.
Downes, D.: 1971, *Astron. J.* **76**, 305.
Fairall, A. P.: 1972, *M.N.A.S.S.A.* **31**, 23.
Finzi, A. and Wolf, R. A.: 1967, *Astrophys. J.* **150**, 115.
Ford, W. K. Jr., Rubin, V. C., and Roberts, M. S.: 1971, *Astron. J.* **76**, 22.
Gull, S. F.: 1973, private communication.
Gunn, J. E. and Ostriker, J. P.: 1970, *Astrophys. J.* **160**, 979.
Hardorp, J.: 1973, preprint.
Hartwick, F. D. A.: 1972, *Nature Phys. Sci.* **237**, 137.
Heuvel, E. P. J. van den: 1973, preprint.
Hewish, A.: 1973, private communication.
Holmberg, E.: 1950, *Medd. Lunds Obs., Ser. II.*, No. 128.
Holmberg, E.: 1958, *Medd. Lunds Obs., Ser. II.*, No. 136.

Holmberg, E.: 1964, *Arkiv Astron.* **3**, 387.
Humason, M. L., Mayall, N. U., and Sandage, A. R.: 1956, *Astron. J.* **61**, 97.
Ilovaisky, S. A. and Lequeux, J.: 1972a, *Astron. Astrophys.* **18**, 169.
Ilovaisky, S. A. and Lequeux, J.: 1972b, *Astron. Astrophys.* **20**, 347.
Johnson, H. M. and MacLeod, J. M.: 1963, *Publ. Astron. Soc. Pacific* **75**, 123.
Karachentsev, I. D.: 1970, in S. K. Vsekhsvyatskij (ed.), *Problems of Cosmic Physics*, Vypusk 5, Kiev, p. 201.
Katgert, P. and Oort, J. H.: 1967, *Bull. Astron. Inst. Neth.* **19**, 239.
Kesteven, M. J. L.: 1968, *Aust. J. Phys.* **21**, 739.
Kowal, C. T.: 1968, *Astron. J.* **73**, 1021.
Kowal, C. T.: 1973, private communication.
Kowal, C. T. and Sargent, W. L. W.: 1971, *Astron. J.* **76**, 756.
Kowal, C. T. and Sargent, W. L. W.: 1973, unpublished.
Kowal, C. T., Sargent, W. L. W., Searle, L., and Zwicky, F.: 1972, *Publ. Astron. Soc. Pacific* **84**, 844.
Kraft, R. P.: 1970, in G. H. Herbig (ed.), *Spectroscopic Astrophysics*, Univ. Calif. Press, Berkeley, p. 385.
Kukarkin, B. V.: 1965, *Astrofizika* **1**, 465.
Lynden-Bell, D.: 1973, private communication.
Mathewson, D. S. and Clarke, J. N.: 1973, *Astrophys. J.* **180**, 725.
Milne, D. K.: 1970, *Aust. J. Phys.* **23**, 425.
Minkowski, R.: 1964, *Ann. Rev. Astron. Astrophys.* **2**, 247.
Minkowski, R.: 1968, in B. M. Middlehurst and L. H. Aller (eds.), *Nebulae and Interstellar Matter* (=*Stars and Stellar Systems*), vol. VII, Univ. Chicago Press, Chicago, p. 623.
Minkowski, R.: 1969, in P. J. Brancazio and A. G. W. Cameron (eds.), *Supernovae and Their Remnants*, Gordon and Breach, New York, p. 23.
Ostriker, J. P. and Gunn, J. E.: 1971, *Astrophys. J. Letters* **164**, L95.
Paczyński, B.: 1970, *Acta Astron.* **20**, 47.
Paczyński, B.: 1971, *Acta Astron.* **21**, 271.
Paczyński, B.: 1972, *Astrophys. Letters* **11**, 53.
Paczyński, B. and Ziółkowski, J.: 1968, *Acta Astron.* **18**, 255.
Page, T. L.: 1973, in A. Sandage, M. Sandage, and J. Kristian (eds.), *Galaxies and the Universe* (=*Stars and Stellar Systems*), vol. IX (in press).
Pskovskii, Yu. P.: 1961, *Astron. Zl.* **38**, 656.
Pskovskii, Yu. P.: 1967, *Astron. Zl.* **44**, 1139.
Reaves, G.: 1953, *Publ. Astron. Soc. Pacific* **65**, 242.
Roberts, M. S.: 1969, *Astron. J.* **74**, 859.
Roberts, M. S.: 1972, in D. E. Evans (ed.), 'External Galaxies and Quasistellar Objects', *IAU Symp.* **44**, 12.
Rosino, L.: 1971, *IAU Circ.*, No. 2305; *Inform. Bull. Var. Stars*, No. 515.
Sackmann, I.-J. and Weidemann, V.: 1972, *Astrophys. J.* **178**, 427.
Sandage, A.: 1961, *The Hubble Atlas of Galaxies*, Carnegie Inst. Washington, Washington, p. 3.
Sandage, A.: 1973a, *Astrophys. J.* **183**, 711.
Sandage, A.: 1973b, unpublished.
Sandage, A. and Tammann, G. A.: 1974, in preparation.
Sargent, W. L. W., Searle, L., and Kowal, C. T.: 1972, *Ann. Report, Director Hale Obs.* 1971–1972, p. 677.
Sargent, W. L. W., Searle, L., and Kowal, C. T.: 1974, this volume, p. 33.
Schmidt, M.: 1965, in A. Blaauw and M. Schmidt (eds.), *Galactic Structure* (=*Stars and Stellar Systems*), vol. V, University of Chicago Press, Chicago, p. 513.
Schmidt-Kaler, Th. and Schlosser, W.: 1973, *Astron. Astrophys.*, **29**, 409.
Searle, L.: 1973, private communication.
Searle, L.: 1974, this volume, p. 125.
Searle, L., Sargent, W. L. W., and Bagnuolo, W. G.: 1973, *Astrophys. J.* **179**, 427.
Shapley, H. and Ames, A.: 1932, *Harv. Ann.* **88**, 43.
Shklovsky, I. S.: 1960, *Cosmic Radio Waves*, Harvard Univ. Press, Cambridge.
Smart. N. C.: 1973, private communication.
Sofia, S.: 1972, *Astrophys. J.* **172**, 53.

Stebbins, J. and Kron, G. E.: 1957, *Astrophys. J.* **126**, 266.
Stothers, R.: 1963, *Astrophys. J.* **138**, 1085.
Talbot, R. J., Jr. and Arnett, W. D.: 1973, *Astrophys. J.* **186**, 69.
Tammann, G. A.: 1970, *Astron. Astrophys.* **8**, 458.
Tammann, G. A.: 1972, *Astron. Astrophys.* **21**, 355.
Tammann, G. A.: 1973, in preparation.
Tammann, G. A.: 1974, this volume, p. 215.
Tammann, G. A. and Sandage, A.: 1968, *Astrophys. J.* **151**, 825.
Vaucouleurs, G. de: 1973, in A. Sandage, M. Sandage, and J. Kristian (eds.), *Galaxies and the Universe* (= *Stars and Stellar Systems*), vol. IX (in press).
Vaucouleurs, G. de and Vaucouleurs, A. de: 1964, *Reference Catalogue of Bright Galaxies*, Univ. of Texas Press, Austin.
Vaucouleurs, G. de and Vaucouleurs, A. de: 1967, *Astron. J.* **72**, 730.
Vaucouleurs, G. de and Vaucouleurs, A. de: 1973, *Astron. Astrophys.* **28**, 109.
Weidemann, V.: 1971, *Astrophys. Letters* **9**, 155.
Wendker, H.: 1973, private communication.
Woltjer, L.: 1972, *Ann. Rev. Astron. Astrophys.* **10**, 129.
Yilmaz, F. and Tammann, G. A.: 1973, unpublished.
Zwicky, F.: 1942, *Astrophys. J.* **96**, 28.
Zwicky, F.: 1965, in L. H. Aller and D. B. McLaughlin (eds.), *Stellar Structure* (= *Stars and Stellar Systems*), vol. VIII, University of Chicago Press, Chicago, p. 367.
Zwicky, F.: 1969, in P. J. Brancazio and A. G. W. Cameron (eds.), *Supernovae and Their Remnants*, Gordon and Breach, New York, p. 1.
Zwicky, F.: 1973, *Transactions of the IAU*, **XVA**, p. 344.

DISCUSSION

L. Biermann: Has any attempt been made to derive the incompleteness (inclination) correction separately for type I and type II SN? One might expect a larger correction for type II SN, but I really don't know.

G. A. Tammann: The ratio type I: type II: others is 6:12:11.5 in face-on and 7:9:4 in inclined spirals. Sure, all three categories are affected by inclination. The higher ratio of type I's in inclined spirals might indicate that they are somewhat less affected by inclination and therefore in a less flattened population – but the difference is not really significant. Things would be somewhat clearer if all SNe had known types. From the present material I cannot give a clear answer to your very exciting problem.

F. Zwicky: One group at Palomar attempted to make statistics without any assumptions about absolute quantities, types of populations, star formation etc. If you employ only phenomenological arguments you confirm:

(a) Very rough proportionality of the supernovae frequency $\bar{\nu}$ with the absolute luminosity of the parent galaxy.

(b) Deviations because of the presence of fast producers, and, in clusters of different populations.

(c) $\bar{\nu} \sim$ one in 350 yr per galaxy for differences Δ in apparent magnitudes $\Delta = M_{SN} - M_{galaxy}$ in the range $+3 > \Delta > -3$ (no assumption needed about absolute magnitudes etc.).

G. A. Tammann: We agree completely on (a). As to (b) I don't see from my sample any room left for an important additional parameter acting on the SN rate other than galaxy type (or colour) and luminosity (or mass) – leave aside the observational factors as apparent magnitude, inclination etc. I don't deny that eventually more refined parameters can be found, but at present I don't see a reason to speak of any unexpected fast producers.

Average, overall frequencies, as mentioned in (c), may still be useful for the searcher to indicate how many SNe he should expect within a sufficiently large field of galaxies. But this frequency cannot be applied to any individual galaxy.

L. Woltjer: You demonstrated quite convincingly that there are no substantial incompleteness effects in Sc galaxies as a function of radial position. Should one not still be worried about serious incompleteness in the central regions of ellipticals?

G. A. Tammann: I'm sorry, I skipped this question in my talk. The left lower panel in Figure 2 shows the radial distribution in E's. The data are very scarce, but they do not exclude the possibility that 'no' SNe were missed. This assumption may still be quite wrong, and E's may actually be much better producers than my numbers seem to suggest.

S. D'Odorico: You find supernovae of type I in Magellanic Irregulars but no supernovae of type II. Because of the small number of Irregulars in your sample do you think this result well founded? It would be important because it would confirm the existence of a strong, underlying population II in these galaxies.

G. A. Tammann: From 5 type I SNe in Ir I's one would expect 8 type II's (assuming the same ratio as in Sc's). Zero were found. That is not proof, but one begins to wonder. As to the association of types of SNe with stellar populations I'm unsure – actually confused.

H. Arp: You mentioned that you could predict from your SN statistics what SN numbers should be observed in certain galaxies which are large SN producers. But since these same galaxies dominate your sample of SN is this not a circular prediction? In fact it causes me to wonder how much your sample is dominated by these large producer galaxies and whether your conclusions do not apply very much to galaxies in general, but just this one kind of galaxy.

G. A. Tammann: The division of the galaxy sample according to luminosity (Table V) gives me confidence that the luminous galaxies are not overweighted. If anything, the intrinsically faint galaxies have a (insignificantly) higher SN rate per unit luminosity.

F. Zwicky: In your presentation and in Sargent's paper it is entirely erroneous to assume that, because of the burnt out disks in E's and certain spirals it is more difficult to detect supernovae in the more distant galaxies. The contrary is true because:

(a) the blackening of the disks decreases for more distant galaxies because of the redshifts and because of the characteristics of the HD curves, while

(b) the sizes of the star images stay always greater than 40 μ (for the 48-in.) and therefore survive longer and protrude more from the disks.

G. A. Tammann: My statistics does not depend on that assumption. At this point I would like to leave the question open as for the main reason why so few SNe are found in inclined spirals.

L. Searle: Is it not true that the variation of the SN rate with morphological type is just an expression of the variation with color and that, even quantitatively, it is in reasonable agreement with the prediction by Sargent, Bagnuolo and myself?

G. A. Tammann: My color range is unfortunately too narrow to test your calculations. If I tried to increase the color base to include for instance Sb galaxies uncertain inclination effects and so on would make any conclusions too dubious.

ON THE FREQUENCY OF TYPE I AND TYPE II SUPERNOVAE

N. DALLAPORTA
Osservatorio Astronomico, Padova, Italy

1. From the present evidence it appears that supernovae of type I and II are related to quite different stellar populations.

In fact:

(a) SNII supernovae appear only in spiral and irregular galaxies; moreover, their location is mostly on the outer spiral arms. This is a clear indication of their appertainance to a young population I;

(b) supernovae appearing in Elliptical and So galaxies are only of type I. This testifies that the SNI phenomenon occurs, at least partly, in an old disk population.

I have used the data collected by Tammann (1970), concerning supernovae in Sb and Sc galaxies, supplemented by some other information derived from other authors related to the exact location of events in the spirals, and to relative frequencies of SN in Ellipticals and So's in respect to spirals, in order to deduce from them some speculations concerning the frequencies of SN of both types in relation to the populations to which they belong and to the stellar progenitors of these events, as this may turn out to be an important clue for several theories concerning the advanced stages of evolution.

2. To begin with, Tammann's data related to the frequencies of SN of both types for spiral Sb and Sc galaxies divided into several luminosity groups collected from different tables of his paper are presented in Table I. Column 1 states the names given to the different luminosity groups and the corresponding types and luminosity classes are found in column 2. Column 3 gives then the number of galaxies patroled for each group, column 4 their mean luminosity, and column 5 their mean mass value deduced from the average mass/luminosity ratios of Roberts (1969) for Sb and Sc galaxies; column 6 the total number of SN observed for all galaxies of each group, and column 7 the number of them for the same groups in the fiducial period september 1959–september 1969; columns 8, 9, and 10 yield then the total number of SNI, SNII and SNu of unidentified type; and columns 11, 12, and 13 the same numbers limited to the period september 1959–september 1969. From these data, the average time interval between two SN events per galaxy, the number of supernovae per 10^8 $L_{\odot}$ and 100 yr, and the number of SN per 10^{10} $M_{\odot}$ and 100 yr may be deduced, and are given in columns 14, 15, 16. All these values are deduced from the data for the fiducial period 1959–1969.

If we wish now to obtain separate frequencies for the two supernovae types, owing to the extreme paucity of data relative to events in Sb galaxies in the fiducial period 1959–1969, nothing better can be done than try to use also the unidentified super-

C. B. Cosmovici (ed.), Supernovae and Supernova Remnants, 187—194. *All Rights Reserved*

TABLE I

Group (1)	Type and lum. class (2)	No. Gal. observed (3)	Lpg $\times 10^{10} L_\odot$ (4)	M $\times 10^8 M_\odot$ (5)	No. tot. SN (6)	No. SN 1959–69 (7)	No. tot. SNI (8)	No. tot. SNII (9)	No. tot. SNu (10)
1b	$Sb_I Sb_{I-II}$	11	194	16.4	7	2	2	3	2
2b	$Sb_{II} Sb_{II-III}$	32	77.3	6.5	4	(1.8)	0	2	2
3b	Sb_{III}	22	28.1	2.4	1	0	0	0	1
Total		65			12	2	2	5	5
1c	$Sc_I Sc_{I-II}$	16	149	16.5	19	10	2	7	10
2c	$Sc_{II} Sc_{II-III}$	29	78.0	8.7	9	4	4	4	1
3c	$Sc_{III} Sc_{III-IV}$	25	34.0	3.8	4	2	1	2	1
4c	$Sc_{IV} Sc_{IV-V}$	38	9.4	1.0	2	2	0	0	2
Total		108			34	18	7	13	14
General total		173			46	20	9	18	19

Group	SNI 1959–69 (11)	SNII 1959–69 (12)	SNu 1959–69 (13)	Time Interval (yr) between SN per gal. (14)	No. SN/$10^8 L_\odot$ /100 yr (15)	No. SN/$10^{10} M_\odot$ /100 yr (16)
1b	0	1	1	55	0.0094	0.11
2b	0	(0.9)	(0.9)	(178)	(0.0072)	(0.087)
3b	0	0	0	–	–	–
Total	0	1	1			
1c	2	5	3	16	0.042	0.38
2c	3	1	0	72	0.0177	0.16
3c	1	0	1	125	0.0235	0.21
4c	0	0	2	190	0.0560	0.53
Total	6	6	6			
General total	6	7	7			

novae, statistically divided between type I and type II according to the observed relative ratios of these types. From the totals of columns 8, 9, 11, 12, we get:

(a) $$\frac{\text{SNI}}{\text{SNII}} = \frac{6}{7} = 0.85 \quad \text{if we refer to the period 1959–1969}$$

and

(b) $$\frac{\text{SNI}}{\text{SNII}} = \frac{9}{18} = 0.5 \quad \text{if we refer to the total data.}$$

Such a large discrepancy between two determinations of the same ratio gives an appreciation of the amount of uncertainty underlying all present considerations. As the value of this ratio should not sensitively depend on the random choice of events used to determine it, there should be no strong reason to adopt value (a) rather than value (b), and all following calculations have been carried on with both values. However, as the final results are not much affected by this difference, only the values corrected using (a) are presented in Table II, whose columns 1 and 2 are thus obtained by adding to columns 11 and 12 of Table I, column 13, statistically divided according to (a).

One may now obtain directly the frequency for SNII supernovae which belong only to young pop. I, provided we know the composition of the Sb and Sc galaxies; for our Galaxy one may pick up the following two sets of data:

Authors	% Masses of		
	Halo	Old disk	Young disk
Perek, L. (1962)	17	76	7
Oort, J. H. (1958)	23	67	10

As it is believed nowadays that our Galaxy is of a type intermediate between Sb and Sc, one may tentatively assume the first row as indicative for Sb and the second one for Sc galaxies.

With these figures, the medium amounts of old disk and respectively of young disk population may be calculated from Table I, column 5, for each galaxy group, thus leading to the data of Table II, columns 3 and 4. With these assumptions one then deduces easily mean intervals between two SNII events per galaxy and the frequencies of SNII events per $10^{10}\ M_{\odot}$ and per 100 yr. These figures are given in columns 5 and 6.

3. For SNI events, the analysis is somewhat more involved, as there are some indications that this phenomenon cannot be attributed to the single old disk population. Owing to their presence in Spiral galaxies, from a detailed analysis of Bertola and Sussi (1965) concerning their location in Spirals, it turns out that on a total of

TABLE II (case a)

Group	Correct. $U \times 0.46$ + SNI (1)	Correct. $U \times 0.54$ + SNII (2)	Mass old pop. per gal. in $10^{10}\ M_\odot$ (3)	Mass young pop. per gal. in $10^{10}\ M_\odot$ (4)	Time interval between SNII per gal (yr) (5)	SNII/10^{10} $M_\odot$/100 yr (6)	$y/(y+o)$ SNI (7)	$o/(y+o)$ SNI (8)	No. SNIy /$10^{10}\ M_\odot$ /100 yr (9)	No. SNIo /$10^{10}\ M_\odot$ /100 yr (10)
1b	0.46	1.54	12.5	1.15	71	1.21	0.35	0.11	0.28	0.008
2b	(0.41)	(1.39)	5.0	0.45	(230)	(0.97)	(0.31)	(0.10)	(0.22)	(0.006)
3b	0	0	1.82	0.17	–	–	–	–	–	–
Total	0.87	2.93				Mean 1.09	0.35	0.11	0.25	0.007
1c	3.38	6.62	11	1.65	24	2.50	2.57	0.81	1	0.046
2c	3.0	1.0	5.8	0.87	290	0.40	2.28	0.72	0.9	0.043
3c	1.46	0.54	2.5	0.38	460	0.57	1.11	0.35	1.2	0.056
4c	0.92	1.08	0.67	0.10	350	2.82	0.70	0.22	1.8	0.086
Total	8.76	9.24				Mean 1.57	6.66	2.10	1.2	0.058
G.T.	9.63	12.17					7.01	2.21		

29 SNI events, 16 were situated in the outer arms, 4 in the central parts, and 9 in intermediate regions a distribution completely different from the one of novae, practically present only in the central parts as any typical old disk population product. This strongly suggests that SNI outbursts may occur into two different stellar populations, due to two different explosion triggering mechanisms in different stellar structures and evolution stages, but giving rise to almost identical spectroscopic and luminosity behaviours. If, on behalf of such a situation, and following a first suggestion in this sense of Bertola and Sussi (1965), one may venture to assume that SNI supernovae are a mixture not yet distinguished of two different types of events which may be tentatively indicated as SNIy (young) and SNIo (old), one may try again to statistically divide the figures given by Bertola and Sussi (1965) in order to get a relative frequency of these two types. The simplest and crudest assumption will then consist in attributing the 4 events observed in central parts of the galaxies to the SNIo type, and the 16 in the arm regions to SNIy events. If then one further divides the 9 remaining cases according to the ratio $\frac{16}{4}=4$, one arrives for the relative ratio of the two subtypes at SNIy/SNIo$\simeq\frac{23}{6}$. We may then further divide according to this ratio the SNI frequencies given in Table II, column 1, and thus obtain the numbers of SNIy and SNIo events (columns 7 and 8). Referring then these events to the respective mass fractions of young and old disk populations present in Sb and Sc galaxies (columns 3 and 4), as already done for SNII cases, we obtain for the frequencies of SNIy and SNIo events per 100 yr per $10^{10}\ M_\odot$ of young or old population matter the data of columns 9 and 10. The mean values for all luminosity groups of Sb and Sc galaxies are:

case a			*case b*		
	SNIy	SNIo		SNIy	SNIo
Sb	0.25	0.007	Sb	0.18	0.005
Sc	1.2	0.058	Sc	1.04	0.050

We see that for what concerns the young population, for both SNII and SNIy events, the frequency is higher for Sc than for Sb galaxies. This may be interpreted as giving evidence that the star birthrate function is higher for Sc than for Sb galaxies, in agreement with the general viewpoint of a more intense young population I activity in the Sc than in the Sb cases. Instead, the large difference in the productivity of the SNIo events between Sc's and Sb's galaxies related to their post-birthrate activity, apart again from the strong effect of fluctuations, may be perhaps understood if one keeps in mind that these old population supernovae appear in the central parts of the galaxy, where more of them in the Sb than in the Sc case may be lost, owing to absorption.

These frequencies for SNIo events in Spirals may further be directly compared with those observed in Ellipticals and So galaxies, when we assume they occur in the same type of stars. From the work of Barbon (1968) we then get that 37 supernovae were observed on a total of 1123 Spiral and Irregular galaxies, and 8 Supernovae on a total of 894 Elliptical and So's. Assuming as SNIo events all these 8 supernovae in Ellipticals and So's, and $37 \times 0.105 = 3.9$ among those observed in the Spirals and Irregulars ($0.105 = 2.1/20$ being the average percentage between case a and case b of SNIo's in the statistics on Spirals we have analysed), 0.7 being the average percentage of old disk population in Spirals, we obtain, for the unknown mean masses M_E and M_S of the Ellipticals and Spirals used by Barbon (1968):

$$\frac{M_E}{M_S} \sim 1.8,$$

a not unreasonable figure on the behalf of our general information concerning masses of giant Ellipticals and Spirals, thus justifiying the assumtion of a similar birthrate function in two cases.

4. As a last step, one may try to connect the frequencies we have obtained for all types of SN with the birthrate functions of the different stellar mass ranges in our Galaxy, for the cases in which this function may be surmized to be sufficiently indicative for the corresponding figures in other galaxies, and deduce the mass, or the absolute magnitude range, of the stars undergoing supernovae explosions at the end of their lifes, on the further assumption of equilibrium between deathrates and birthrates.

If $\Psi(M)\,\mathrm{d}M$ is the tabulated Salpeter birthrate function, that is the number of stars created in $T \simeq 6 \times 10^9$ yr per cubic parsec in the magnitude interval $\mathrm{d}M$, and N the total number of stars present per cubic parsec, and $\bar{m}$ the average star mass of thc population considered, then the number n of stars created (or exploded) in 100 yr per $10^{10}\ M_\odot$ between magnitude M_1 and M_2 is given by:

$$n = 100 \frac{\int_{M_1}^{M_2} \Psi(M)\,\mathrm{d}M}{T} \frac{10^{10}\ M_\odot}{(N\bar{m})^2}.$$

We may try to apply this formula in the case of supernovae related to the young populations. It does not appear perhaps too inconsistent as a first approach to extend the data on birthrate functions obtained from the solar neighbourhood to the young populations of the arm regions of both Sb and Sc galaxies. Assuming then (Sandage, 1958):

$$\bar{m} = M_\odot \qquad N = 0.120\ \text{stars/pc}^3$$

one finds:

$$n_y = 1.15 \times 10^4 \int_{M_{1v}}^{M_{2v}} \Psi(M)\,\mathrm{d}M.$$

In Table III we report the mean explosion rates n_y of the two supernovae types of the young disk population (taking a mean value between cases a and b from the Sc data) considered in our previous analysis (columns (6) and (9) of Table II), and the corresponding visual absolute magnitude M_{2v} for which the birthrate function (M_{1v} is assumed as $-\infty$)

$$\int_{-\infty}^{M_{2v}} \Psi(M)\, \mathrm{d}M$$

assumes the value $n_y/(1.15\times 10^4)$. The corresponding values of the bolometric magnitudes (using the bolometric correction for the spectral class B1), and the corresponding mass value (in $M_\odot$ units) obtained from the mass-luminosity relation are then given in the next rows.

TABLE III

	SNII	SNIy
n_y	1.7	1.1
M_{2v}	− 3.7	− 4.0
M_{2b}	− 5.7	− 6.0
Mass	~ 13	~ 15

It may be thus be seen that the birthrates of all main sequence stars brighter than $M_{2b} = -5.7$ and respectively -6.0 provide the death rates of SNII and respectively SNIy events; and thus such absolute bolometric magnitudes correspond to a mass range of about 13 $M_\odot$. Owing to the very steep decrease of the birthrate function with increasing mass, it would thus result that for young populations, only stars with masses higher than this limit should undergo the supernova phenomenon. If the adopted frequencies of SN are still to be considered as lower limits owing to experimental losses, this lower limit of the mass range may perhaps be shifted down to about 10 $M_\odot$. In any case the figures obtained from the present analysis lead to a luminosity and a mass range of the progenitors of young populations supernovae higher than usually assumed or deduced (Stothers, 1963). A more detailed subdivision of the mass ranges respectively responsible for the SNII and SNIy events is of course outside the reach of the present investigation.

For SNIo events, as they occur only in old populations, they must be related either to stars belonging to a mass range not much different from the solar one, or to delayed products of the evolution of larger mass stars, such as residual cores or white dwarfs. In the first case, the very low relative frequency of these events in respect to the abundance of stars corresponding to a solar mass value indicates that either a very restricted mass band or only a small fraction of all the stars pertaining to a larger mass range are subject to such final explosions. In the second case, one could try to deduce the mass range of the progenitors in the same way as for supernovae related to young populations, arriving thus, owing to their much lower frequency, at a still higher luminosity range ($M_{2v} \sim -6$) for the progenitors of these stars than for the young population cases. However, a specific value for these events would be scarcely indicative, because in this case the extrapolation of the birthrate data of the solar neighbourhood to populations older than several 10^9 yr appears very little justified.

References

Barbon, R.: 1968, *Astron. J.* **73**, 1016.

Bertola, F. and Sussi, M. G.: 1965, *Contrib. Oss. Astr. Univ. Padova*, No. 176.

Oort, J. H.: 1958, in O'Connell (ed.), 'Dynamics and Evolution of the Galaxy, in so far as Relevant to the Problem of the Populations', in *Stellar Populations*, North Holland Publishing Company, Amsterdam, p. 415.

Perek, L.: 1962, in Z. Kopal (ed.), *Advances in Astronomy and Astrophysics* **1**, 65.

Roberts, M. S.: 1969, *Astron. J.* **74**, 859.

Sandage, A.: 1958, in O'Connell (ed.), 'Luminosity Functions of Galactic Clusters, Globular Clusters, and Elliptical Galaxies', in *Stellar Populations*, North Holland Publishing Company, Amsterdam, p. 75.

Stothers, R.: 1963, *Astrophys. J.* **138**, 1085.

Tammann, G. A.: 1970, *Astron. Astrophys.* **8**, 458.

SPACE DISTRIBUTION OF SUPERNOVAE IN PARENT GALAXIES

M. F. McCARTHY
Vatican Observatory, Castel Gandolfo, Vatican City State

The systematic study of supernovae is limited both in its observational aspects by selection effects and in its statistics by small numbers. The location of supernovae in parent galaxies has been studied in the past by Whipple (1939), by Zwicky (1958), by Reaves (1953) by Johnson and MacLeod (1963), and by Bertola (1965) and at this conference will be discussed also by Tammann.

In 1958 Zwicky (1958) noted that most supernovae seem to favor the outskirts of stellar system and in 1968 he noted that very few supernovae were to be found close to the nuclei of parent galaxies. Nonetheless Zwicky pointed out clearly both in his 1958 and 1968 studies that supernovae can be easily missed in photographic searches if they flare up in the central luminous discs of galaxies. These selection effects were estimated by him (according to Bertola) to amount to as much as 30% of supernovae exploding in the nuclear regions which pass unobserved because of exposure effects.

In two preliminary studies (McCarthy, 1971, 1973), or this problem the author has investigated the location of supernovae in parent galaxies. Now thanks to the continued success of the supernovae search programs and to the new image tube radial velocity determinations by Sargent and Kowal (1971) and colleagues at Palomar we have much additional information for our study. In the present report we shall consider the space distribution of supernovae in those parent galaxies for which Zwicky's method of indicative distances can be employed. These galaxies number 148 and we divide them into three subgroups. Group A includes those supernovae whose parent galaxies have radial velocities equal or greater than 500 km s^{-1}. There are 109 in Group A.

Group B includes those supernovae whose parent galaxies have radial velocities less than 500 km s^{-1}. There are 18 of these. Group C consists of those supernovae whose parent galaxies do not as yet have individually determined radial velocities but which are considered to be members of a cluster of galaxies and for which the radial velocity of the cluster can be used. There are 21 galaxies in Group C. Supernovae in each of these Groups A, B and C are identified in the Appendix.

We shall consider here the Frequency-Distance Data for the 148 galaxies with radial velocity data. Then we shall consider the Density-Distance Data for these objects. Finally we shall analyze these data according to supernova type and according to the galactic type of the parent galaxy.

We have converted published *angular* distances Δx and Δy of the supernova from the nucleus of the parent galaxy into *linear* distances in kiloparsecs. This was accomplished by using the published radial velocity and an assumed Hubble constant of 100 km s^{-1} Mpsc^{-1}. This calculation yields by Zwicky's Method of indicative

C. B. Cosmovici (ed.), Supernovae and Supernova Remnants, 195–202.

distances R_u, the distance in kiloparsecs of the supernova from the nucleus of the parent galaxy.

This value, R_u, must be corrected for aspect. The correction for this is obtained by measuring on the prints of the Palomar Sky Survey or on discovery photographs of the supernova in the parent galaxy the quantities 'a' and 'b', the semi major and semi minor axes of the image of the parent galaxy and ω the position angle of the major axis with respect to the north point. In addition to these measured quantities we have ϕ the position angle of the supernova with respect to the major axis of the parent galaxy, which is given as the arc tangent of $\Delta x/\Delta y$.

Thus the corrected distance R_c of the supernova from the center of the parent galaxy is given by:

$$R_c = R_v \left(\cos^2 \theta + \frac{a^2}{b^2} \sin^2 \theta \right)^{1/2}$$

where $\theta = \phi - \omega$ as described above.

We have computed R_u and R_c for each of the 148 supernovae for which we have radial velocity data and which are listed in the Appendix. Specific data of these calculations will be published elsewhere.

Table I shows the Frequency-Distance data for the Supernovae of Group A uncorrected for tilt or aspect. Successive columns list the corresponding values as given by Zwicky (1958) and by the author in his 1971 study (McCarthy, 1973). All of the parent galaxies considered here have velocities greater than 500 km s^{-1}. All three Frequency-Distance relations indicate one slight concentration near 7 kps from the center of the nucleus and another at a distance near 3 kps; some care must be taken not to overemphasize these concentrations which as we shall see tend to be smoothed out when the surface densities are plotted.

TABLE I

Frequency-Distance Data for SN in galaxies with radial vel > 500 km s^{-1} (distances uncorrected for aspect)

Distance kpc	N_{SN} (Zwicky, 1965)	N_{SN} (McCarthy, 1971)	N_{SN} (this paper)
0–1	2	16	19
2–3	8	32	37
4–5	4	11	16
6–7	6	15	20
8–9	0	5	8
10–11	0	3	3
12–13	0	4	3
14–15	2	0	0
16–17	0	1	0
18–19	0	0	1
20	0	0	2
Total	22	87	109

TABLE II

Frequency-Distance Data for groups A, B, and C (uncorrected for aspect)

Distance kpc	Group A $RV > 500$ km s^{-1}	Group B $RV < 500$ km s^{-1}	Group C RV from cluster	Totals
0–1	19	9	3	31
2–3	37	4	5	46
4–5	16	1	6	23
6–7	20	2	4	26
8–9	8	2	0	10
10–11	3	1	1	5
12–13	3	0	1	4
14–15	0	0	0	0
16–17	0	0	0	0
18–19	1	0	0	1
20	2	0	0	2
Total	109	19	20	148

TABLE III

Frequency Distance Data for groups A, B, and C (distances corrected for aspect)

Distance kpc	Group A	Group B	Group C	Total
0–1	13	10	3	26
2–3	20	4	2	28
4–5	17	5	1	23
6–7	17	0	4	21
8–9	13	0	1	14
10–11	9	0	4	13
12–13	7	0	0	7
14–15	2	0	4	6
16–17	2	0	0	2
18–19	4	0	1	5
20	5	0	0	5
Total	109	19	20	148

In Table II we see the Frequency-Distance data for 148 supernovae: those shown in Table I above plus 19 whose parent galaxies have radial velocities less than 500 km s^{-1} and 21 which are considered to be members of clusters of galaxies and have been assigned the mean radial velocity of the cluster. The data in Table II are uncorrected for aspect or tilt. Table III lists the Frequency-Distance Data for the same 148 supernovae *after* the correction for aspect has been applied. Now we note that the correction for aspect yields a much smoother distribution and a gradual decrease in number with increasing distances from the central nucleus to the supernova.

Table IV lists the density-distance data for 148 supernovae: all of these distances have been corrected for aspect. This table lists the number of supernovae per square

TABLE IV

Density-Distance Data for groups A, B, and C
(distance corrected for aspect)

Distance kpc	Surface Density ϱ_A	ϱ_B	ϱ_C	ϱ_{A+B+C}
0–1	1.03	0.80	0.24	2.07
2–3	0.53	0.11	0.05	0.74
4–5	0.27	0.08	0.02	0.37
6–7	0.19		0.05	0.24
8–9	0.12		0.01	0.12
10–11	0.07		0.03	0.09
12–13	0.04			0.04
14–15	0.01		0.02	0.03
16–17	0.01			0.01
18–19	0.02		0.004	0.02
20	–[a]			–[a]
Total N_{SN}	109[a]	19	20	148

[a] 5 stars in Group A lie beyond the 20 kpc limit.

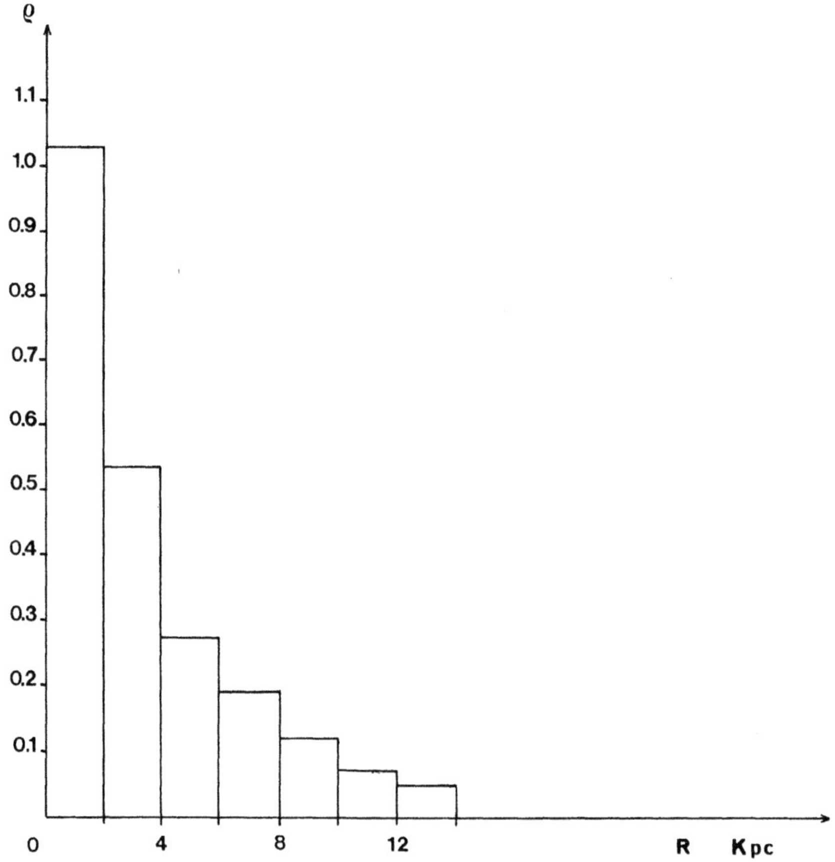

Fig. 1a. Surface Density-Distance Data for 109 SN in parent galaxies which have radial velocities greater than 500 km s^{-1}.

kiloparsec within successive rings or annuli of two kiloparsecs thickness; here one sees a very smooth distribution of supernovae with the density gradient dropping off almost exponentially out to the rim of the galaxy. We can compare in Figure 1 the most homogeneous group of supernovae in our study (i.e., the supernovae whose parent galaxies show radial velocities greater than 500 km s^{-1}) with those SNR, supernovae remnants in our own Milky Way galaxy, as listed by Ilovaisky and Lequeux (1973). Note that the densities are comparable and of the same order of magnitude. There is a marked negative density gradient in both plots. However the Density-Distance plot in the case of the supernovae in external galaxies seems more nearly exponential whereas the drop off in the SNR numbers in our own galaxy is somewhat more gradual.

We look now at the Density-Distance Distribution for type I and type II supernovae. Our sample contains 46 of type I or 31% of the total number of objects studied here. There are 22 supernovae of type II corresponding to 15% of the total studied. We note a sharp drop off in density in both lists as the rim of the galaxy is approached. Here we see another indication that the nuclear regions of galaxies do not lack

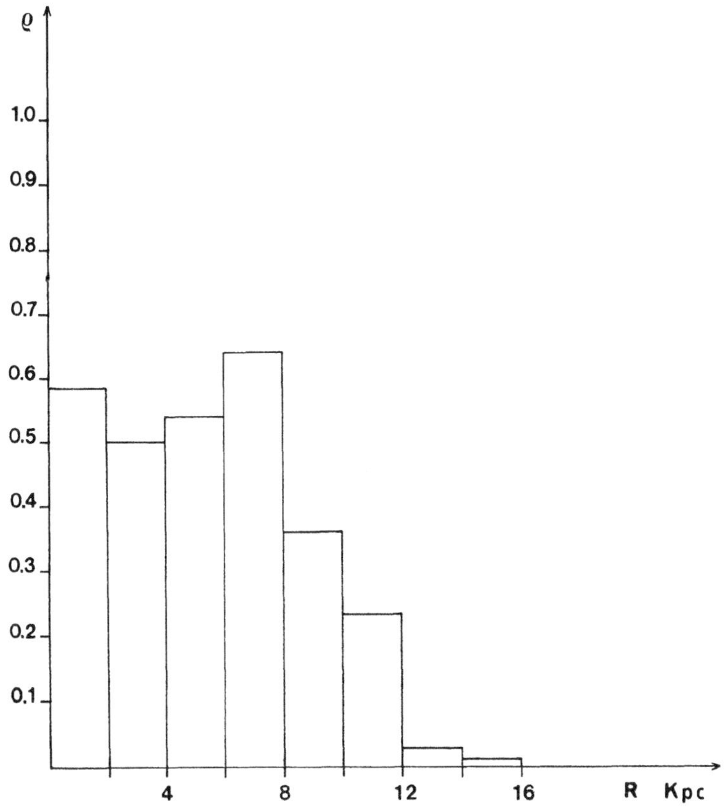

Fig. 1b. Surface Density-Distance Data for SNR, supernova remnants in our own Galaxy as derived by S. Ilovaisky and J. Lequeux (*Astron. Astrophys.* **24** (1973) 143).

TABLE V

Density-Distance Data for SN in parent galaxies of type I and type II

Distance kpc	Surface density $\rho_{\text{type I}}$	$\rho_{\text{type II}}$
0–1	1.03	0.48
2–3	0.19	0.08
4–5	0.14	0.03
6–7	0.09	0.05
8–9	0.03	0.01
10–11	0.01	0.01
12–13	0.01	0.01
14–15	0.01	–
16–17	0.005	–
18–19	0.004	0.004
20		
Total N_{SN}	46	22

TABLE VI

Density-Distance Data for 57 SN in parent galaxies of type Sc

Distance kpc	Surface density $\varrho_{\text{SN}_{\text{Sc type}}}$
0–1	1.11
2–3	0.24
4–5	0.13
6–7	0.11
8–9	0.02
10–11	0.03
12–13	0.02
14–15	0.005
16–17	0.005
18–19	0.008
20	–[a]
Total $N_{\text{SN}_{\text{Sc type}}} = 57$[a]	

[a] 2 SN lie beyond the 20 kpc limit.

supernovae. Indeed if we were to increase, as Zwicky once suggested, the total count in the nuclear regions by 30% (or even by 45% as more recent studies to still fainter limiting magnitudes would indicate), we would find an increase in the central nuclear regions some three or four times the density shown here in Tables IV and V and in Figure 1.

Finally in Table VI we examine the Density-Distance data for 57 supernovae in parent galaxies of type Sc. These are the galaxies which according to Tammann's studies (Tammann, 1970) show the most likelihood of manifesting multiple supernovae and have the highest rate of recurrent supernovae. Here also the density distribution confirms what has been observed throughout this study and is the main observational result of the present investigation: *that while supernovae can and do appear in all parts of parent galaxies they are in no way absent from the central region of parent galaxies.*

Appendix

Here are listed the supernovae discussed in the present investigation. In a subsequent publication we shall list the parent galaxies and their types together with the types of the supernovae and the uncorrected distances from the supernova to the nucleus of its parent galaxy and finally the corrected distance which adjusts for the aspect or tilt. Here are listed the supernovae whose radial velocity data is available from the studies of Karpowicz and Rudnicki (1968) by Sargent and Kowal (1971) and for most recent cases from the *IAU Circulars* (1969–72).

GROUP A

Supernovae in parent galaxies with radial velocities greater than 500 km s^{-1}.

55c, 54e, 37d, 62l, 63p, 68a, 59a, 62q, 01a, 60m,
20a, 12a, 57a, 61f, 47a, 66j, 41b, 67c, 53a, 64a,
65l, 69c, 71k, 61u, 64l, 56a, 21a, 37a, 55a, 65g,
66e, 67h, 36a, 26a, 61i, 64f, 01b, 14a, 59e, 57b,
60j, 60r, 19a, 60f, 41a, 61h, 15a, 39a, 65h, 48a,
40b, 69h, 65i, 56b, 68b, 59b, 50c, 71i, 45a, 23a,
50b, 57d, 68l, 54b, 54c, 40a, 26b, 71 , 60l, 64h,
59d, 65m, 70j, 63e, 50a, 37c, 68o, 68n, 66i, 64j,
60p, 55b, 67e, 61p, 38a, 62o, 68b, 60d, 60n, 65d,
62e, 60c, 60i, 60b, 61a, 68c, 68h, 62a, 59c, 63b,
61e, 63a, 60k, 72 Dec 2841, 72 Jan, 72 Aug, 73 Jan, 72 Dec, 4254

GROUP B

Supernovae in parent galaxies with radial velocities less than 500 km s^{-1}.

40e, 61v, 69l, 54j, 21b, 21c, 37f, 41c, 63i, 54a,
69e, 39b, 95b, 70g, 17a, 39c, 48b, 68d,

GROUP C

Supernovae in parent galaxies with radial velocities taken from the mean cluster velocity.

72f, 95a, 46b, 61b, 61m, 61r, 67j, 67k, 64b, 62f,
62d, 63m, 64d, 62i, 62a, 63h, 63f, 58a, 62c, 64c, +310106.

Acknowledgements

We thank the Director and the Staff of the Carnegie Institution of Washington, Department of Terrestrial Magnetism and especially Dr V. C. Rubin for making it possible to begin this study in the summer of 1970. It has been continued and concluded at the Vatican Observatory.

References

Bertola, F.: 1965, *Contr. Asiago*, No. 176, 6.
IAU Circulars (ed. by B. Marsden), Cambridge, Mass Smithsonian Astroph. Obs/1971 to 1973; see also *Publ. Astron. Soc. Pacific* **81–84** for results of Palomar Supernova Search 1969–72.
Ilovaisky, S. and Lequeux, J.: 1973, *Astron. Astrophys.* **24**, 143.
Johnson, H. M. and MacLeod, J. M.: 1963, *Publ. Astron. Soc. Pacific* **75**, 123.
Karpowicz, M. and Rudnicki, K.: 1968, *Publ Astron. Obs. Warsaw Univ.*, 15.
Kowal, C. T. and Sargent, W. L. W.: 1971, *Astron. J.* **76**, 756.
McCarthy, M. F.: 1971, *Bull. Amer. Astron. Soc.* **3**, 329.
McCarthy, M. F.: 1973, *Ric. Astron. Spec. Vat.* **8**, 411 and 429.
Reaves, G.: 1953, *Publ. Astron. Soc. Pacific* **65**, 242.
Tammann, G.: 1970, *Astron. Astrophys.* **8**, 458.
Whipple, F.: 1939, *Proc. Nat. Acad. Sci.* **25**, 118.
Zwicky, F.: 1958, *Handbuch der Physik* **51**, 766 (ed. by S. Flugge), Springer Verlag, Berlin.
Zwicky, F.: 1964, *Ann. Astrophys.* **27**, 301.
Zwicky, F.: 1965, in L. H. Aller and D. B. McLaughlin (eds.), *Stellar Structure*, Vol. VIII, Univ. of Chicago Press, Chicago, Chapter 7, p. 367.

DISCUSSION

L. Biermann: Could it be that the SN statistics for our own Galaxy are seriously deficient for this region near the nucleus?

M. F. McCarthy: Yes and Ilovaisky and Lequeux (*Astron. Astrophys.* **18** (1972) 169) mention this in their paper from which the figure shown was taken.

K. Johansson: Have you investigated the type II SN distribution according to the density versus the distance from the nucleus in Sc galaxies?

M. F. McCarthy: Yes, the same trend, i.e. a negative density gradient is noted. I have density distance plot for the SN of type II in parent galaxies. Densities are lower and those reflect the smaller number type II SN in our sample. Of the 148 supernovae with radial velocity data, there were 46 of type I (31 % of the total) and only 22 SN of type II corresponding to 15 % of the total. Of the 22 SN of type II 14 are of Sc; those show negative density gradient: 5 are in the 2 kpc ring, 3 in the 6 to 8 kpc-ring and the remaining six are found one apiece in successive rings. The densities of small numbers is not significant.

NOTES ON TYPE II SUPERNOVAE

G. DE VAUCOULEURS

Department of Astronomy, University of Texas, Austin, U.S.A.

1. Type and Maximum of SN 1923 in NGC 5236 (M83)

The supernova of 1923 in M83 has only a partial, but continuous light curve (Figure 1) based entirely on Lampland's assiduous observations at Lowell Observatory from May 5 to July 1, 1923 (Lampland, 1936). There are only three other published observations from Harvard; the SN was invisible on patrol camera plates ($m > 14.0$ and $m > 13.5$) on April 6 and 23 and was at $m = 15.7$ on June 5 (*Harv. Bull.*, No. 910). The luminosity of the SN was stationary at $m = 14.05 \pm 0.05$ for the first ten days (May 5–15), then suddenly started fading rapidly after May 15. Baade (1938) noted that "it is difficult to decide... whether the constant brightness during the first ten days indicates that the nova was near its maximum, or whether it was a halt on the descending branch of the light curve" and remarked that under the first assumption the absolute magnitude at maximum was abnormally faint. Nevertheless, SN 1923 is usually listed in the literature as of type II, IV or V (Payne-Gaposchkin, 1957; Karpowicz and Rudnicki, 1968) with maximum light at $m = 14.0$, which would make

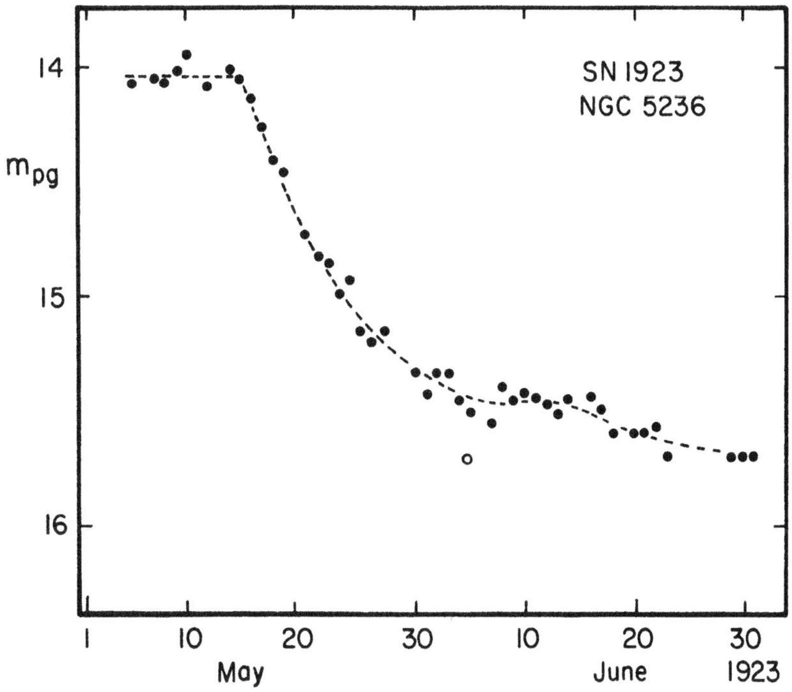

Fig. 1. Light curve of SN 1923 in NGC 5236.

C. B. Cosmovici (ed.), Supernovae and Supernova Remnants, 203–208.

it one of the faintest supernovae on record. Zwicky (1958) goes so far as to suggest it might have been just a bright nova similar to Nova Puppis 1942. However, fast novae have at maximum a bright flash of very short duration which is unlike the initial ten days of the observations for SN 1923. Furthermore, if the distance modulus of M83 is $m-M=28.4$ from its membership in the Centaurus chain (de Vaucouleurs,

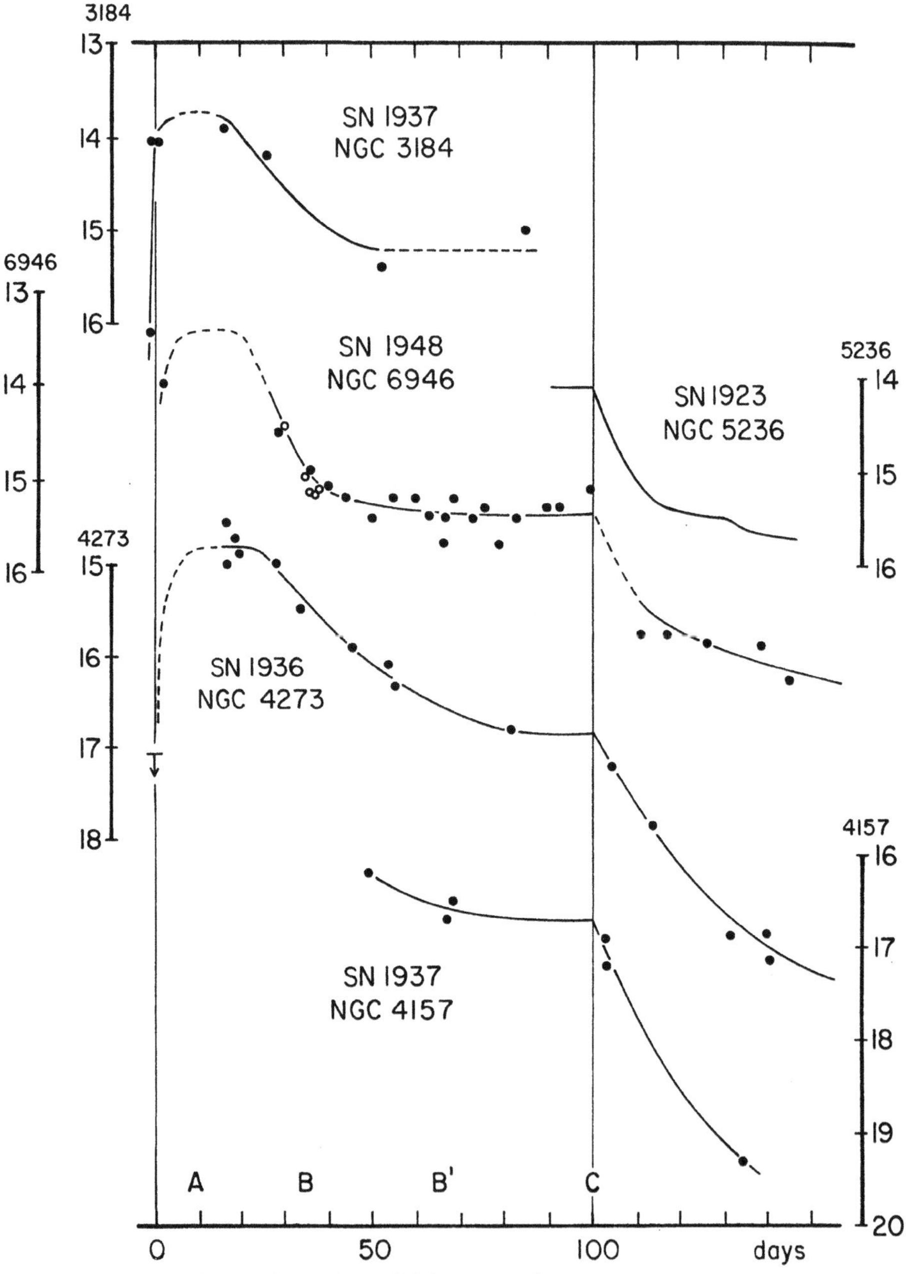

Fig. 2. Comparison of light curves of type II supernovae.

1965), the observed maximum $M \simeq -14.4$ is much too bright for a normal nova. I believe that there is a much simpler and logical interpretation: SN 1923 was a normal type II supernova and the ten-day interval of constant luminosity was the tail-end of the post-maximum plateau lasting for 40 to 60 days which is a common feature of this type. The sudden collapse after May 15 and the shape of the light curve after that date are also characteristic of this phase of evolution of type II supernovae.* The break (point C) takes place about 90 days past maximum in several well-observed supernovae of this type (Figure 2). The rate of decay during the first 10–20 days after point C is about 0.10 mag. day^{-1}, while the initial post-maximum decay after point A is at a slower rate, about 0.05 mag. day^{-1}.

TABLE I

Maximum and post-maximum luminosities of type II supernovae

SN	NGC	m_A	m_{BC}	$\Delta m(BC-A)$
1917	6946	$\leqslant 14.5$	$\geqslant 15.9$	$\geqslant 1.4$
1923	5236	–	14.0_5	–
1935	1511	12.9:	15.3	2.4:
1936	4273	14.8_5	16.7_5	1.9
1937	3184	$\leqslant 13.9$	15.2:	$\geqslant 1.3$:
1948	6946	$\leqslant 13.9$:	15.3_5	$\geqslant 1.4$:
1969	3556	14.3:	15.9:	1.6:
1970	5457	11.1:	13.3:	2.1:
Mean				$\overline{\Delta m} = 2.0 \pm 0.2$

In some supernovae the post-maximum plateau BC is at a very nearly constant level and may last for 50 days or more; examples are SN 1937 in NGC 3184 and SN 1948 in NGC 6946 (Figure 2). In other cases, such as SN 1936 in NGC 4273 and SN 1937 in NGC 4157, only the last 10 to 20 days of the plateau near point C is at a nearly constant level after a longer interval of gradual decay BB′ following the faster initial post-maximum drop AB. In several supernovae (Table I) this level is on the average about 2.0 mag. below maximum light If this interpretation is correct, the unobserved maximum of SN 1923 must have been near $m(\max.) \simeq 14.0 - 2.0 = 12.0$ on or about February 15, 1923. The corresponding absolute magnitude, $M \simeq -16.4$, is about average for a type II supernova; for example, Kowal (1968) finds $\bar{M} = -16.5$ (for $H = 100$) with $\sigma_M = 0.6$.

2. The Post-Maximum Plateau of the Light Curves of type II Supernovae as a Distance Indicator

Most supernovae are discovered at or past maximum light and the magnitude at maximum, often used as a distance indicator (Kowal, 1968, 1969), is rarely well-

* This collapse may correspond to the onset of hydrogen recombination producing a rapid drop of opacity in the expanding envelope when the temperature falls below $T \simeq 6000$K (Falk and Arnett, 1973).

defined. The post-maximum plateau of type II supernovae (Figure 2), and in particular the very nearly constant luminosity 60 to 90 days past maximum, just before the break at point C, is often much better observed than the maximum. If this plateau proves to be at a nearly constant absolute magnitude in different supernovae it may constitute a better distance indicator than the primary maximum; its only limitation is that the SN is then about 2.0 mag. fainter on the average than at maximum (Table I).

TABLE II

Apparent and absolute magnitudes of type II supernovae at end of post-maximum plateau

SN	NGC	$m-M$	m_c	M_c	Source of $(m-M)$
1909	5457	27.7	13.3	−14.4	Sandage (1962)
1917	6946	29.4	15.8:	−13.6:	Ables (1971)
1923	5236	28.5	14.0_5	-14.4_5	Cen chain
1936	4273	31.8	16.8:	−15.0:	Vir W group
1937	3184	30.0	15.2:	−14.8:	N 3184 group
1937	4157	31.1	16.7	−14.4	UMa I (S) cloud
1948	6946	29.4	15.3_5	-14.0_5	Ables (1971)
1969	3556	30.3_5	15.9:	-14.4_5	UMa I (X) cloud
1970	5457	27.7	13.3:	−14.4	Sandage (1962)
Mean			$\bar{M}_c = -14.5 \pm 0.1$		

The apparent and absolute magnitudes of several type II supernovae just before point C of their light curves are collected in Table II; the distance moduli are derived, wherever possible, from group membership, i.e. from the apparent magnitudes and diameters of the five dominant members of group (de Vaucouleurs, 1965) or for isolated galaxies from radial velocity with allowance for local departures from isotropy and linearity (de Vaucouleurs, 1958, 1972) or from the magnitudes of the brightest stars (Sandage, 1962) or from a combination of criteria (Ables, 1971). The average absolute magnitude is $\bar{M}_c = -14.5 \pm 0.1$ (me) and the scatter of individual values about this mean is remarkably small, about ± 0.3 mag. including all nine examples, or ± 0.2 mag. for the six better observed cases. Since the apparent magnitudes and distance moduli have easily uncertainties of this order, the cosmic dispersion may be vanishingly small. This small dispersion makes the type II supernovae competitive with the best extragalactic distance indicators in current use.

3. The-Maximum Halt on the Light Curve of type II Supernovae

Pre-maximum observations of supernovae are rare and the rising branch of the light curve is seldom, if ever, considered in comparing theoretical models of supernovae with observations. A remarkable property of the pre-maximum light curve observed in at least two supernovae of type II does not seem to have received enough attention. The fragmentary light curves of SN 1940 in NGC 4725 and SN 1941 in NGC 4559

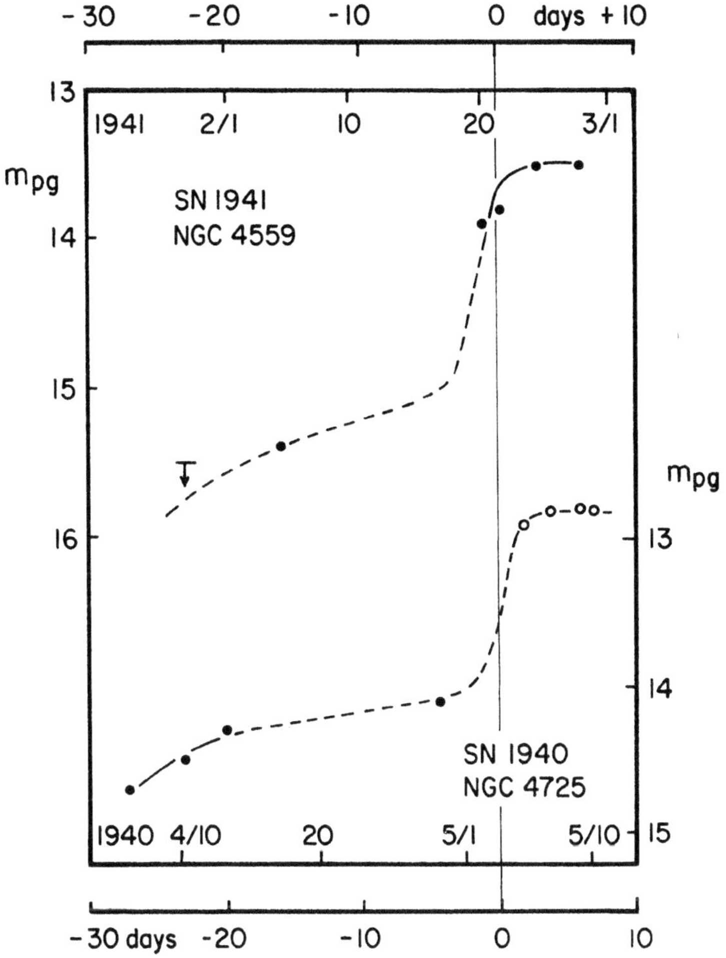

Fig. 3. Pre-maximum halt in two type II supernovae.

TABLE III

Apparent and absolute magnitudes of type II supernovae at pre-maximum halt

SN	NGC	$m-M$	m(max)	M (max)	$m(h)$	M (h)
1940	4725	29.7	12.8	−16.9	14.2	−15.5
1941	4559	29.7	13.5	−16.2	15.2:	−14.5:
Mean				−16.5 ±		−15.0 ±

can be combined in a common mean curve (Figure 3) by matching their maxima, respectively, $m(\text{max})=12.8$ and 13.5. The mean curve shows that during the slow rise to maximum a near halt lasting about two weeks occurs between 5 and 20 days before maximum at a level some 1.7 ± 0.2 mag. below maximum. Both galaxies are members of the same nearby group (Coma I) whose distance modulus is $m-M=29.75$

(de Vaucouleurs, 1965); the corresponding magnitudes for the main maximum and the pre-maximum halt are given in Table III. It may or may not be significant that the pre-maximum halt is at about the same level as the more extended post-maximum plateau (Section 2). This halt is in marked contrast with the near vertical rising branch of normal (fast) novae which take usually less than one day from prenova to maximum. In view of the frequent comments on (spectral) similarities between fast novae and type II supernovae, this striking difference in their initial light curves may be a very significant key to a basic dissimilarity in the triggering phenomenon for the two classes.

References

Ables, H. D.: 1971, *Publ. U.S. Naval Obs.* **XX**, Part IV.
Baade, W.: 1938, *Astrophys. J.* **88**, 285.
Falk, S. W. and Arnett, W. D.: 1973, *Astrophys. J. Letters* **180**, L65.
Karpowicz, M. and Rudnicki, K.: 1968, *Catalogue of Supernovae, Warsaw Univ. Press.*
Kowal, C. T.: 1968, *Astron. J.* **73**, 1021.
Kowal, C. T.: 1969, *Publ. Astron. Soc. Pacific* **81**, 608.
Lampland, C. O.: 1936, *Publ. Astron. Soc. Pacific* **48**, 320.
Payne-Gaposchkin, C.: 1957, *The Galactic Novae*, North-Holland, Amsterdam, p. 269.
Sandage, A.: 1962, in G. C. McVittie (ed.), 'Problems of Extra-Galactic Research', *IAU Symp.* **15**, Macmillan, New York, p. 359.
Vaucouleurs, G. de: 1958, *Astron. J.* **63**, 253.
Vaucouleurs, G. de: 1965, (revised 1973), in A. Sandage (ed.), 'Nearby Groups of Galaxies', *Stars and Stellar Systems*, Vol. 9, Univ. of Chicago Press.
Vaucouleurs, G. de: 1972, in D. S. Evans (ed.), 'External Galaxies and Quasi-Stellar Objects', *IAU Symp.* **44**, 353.
Zwicky, F.: 1958, *Handbuch der Physik* **51**, 775.

DISCUSSION

M. F. McCarthy: Can we associate any spectral changes with the break of the continuum after the horizontal plateau.

G. de Vaucouleurs: I am not aware of spectral data after the 100 day break.

L. Rosino: I fully agree with the conclusions reached by Dr de Vaucouleurs on the brightness at maximum of SN 1923 in NGC 5236 and I agree also, on the basis of the experience made recently on same type II SN, on the importance, as a distance indicator, of the plateau post-maximum and even (as soon as new observations will be collected) of the plateau pre-maximum. I think that de Vaucouleurs' analysis of light curves of type II is a very fine piece of work.

H. Arp: The type II SN in NGC 7331 reached a max absolute magnitude near-19 and fell straight, without a still stand, on the decay. There was some remark earlier that the type II SN in M 101 fell straight also. It is generally known that there is a greater variation in the proporties of type II SN but this now raises the question whether there are two sub classes of type II SN, each sub class having much different light curves and possibly much different absolute magnitudes.

R. Wood: Concerning type II SN in NGC 5236, I would like to remark that the 1968 SN was observed to reach $B \sim 12.0$ near maximum and the plateau in the light curve is at $B = 13.0$; both these magnitudes may be brighter because of absoprtion in the nucleus of NGC 5236. These numbers are, therefore, at least a magnitude brighter than those given by de Vaucouleurs for the 1923 SN in the same galaxy.

J. V. Jelley: Do I understand from this paper that in the second SN you discussed, the time of ignition is really as early as 15 to 20 days before the maximum?

G. de Vaucouleurs: This seem to be the case in the two cases mentioned. (It is even earlier, 30 days at least.) There is as yet no explanation for the pre-maximum pause.

SUPERNOVAE AND EXTRAGALACTIC DISTANCES

DAVID BRANCH
Royal Greenwich Observatory, England

I would like to talk about two questions which are of importance in connection with the use of supernovae as distance indicators. First – what is the dispersion among the absolute magnitudes of type I supernovae at maximum light, and second – what is the mean magnitude difference between type I and type II? Both of these problems were studied by Kowal in 1968; he found a dispersion of $0^m.6$ foı type I (and for type II) and a mean magnitude difference of $2^m.1$, type II being fainter than type I.

The data shown in the following diagrams come primarily from the list of supernovae published by Kowal and Sargent, although a few modifications and additions have been made. The data have been treated in the usual way: supernova magnitudes have been corrected for interstellar absorption in our Galaxy ($A_{pg}=0^m.2 \cos b$) but not for absorption in the supernova parent galaxies, and radial velocities of galaxies have been adjusted to the rest frame of the centre of our galaxy.

Figure 1 shows apparent photographic magnitude at maximum light for type I supernovae in the Virgo and Coma clusters, and in other galaxies having radial velocities of at least 3000 km s^{-1}, plotted against distance modulus. The distances are based for convenience on a Hubble constant of 100 km s^{-1} Mpc^{-1}, but the actual value is not important in this context, since I am talking only about the scatter, and not the absolute distance scale. The small symbols represent supernovae which are not known to have been observed at maximum; many other such limits could have been plotted (higher in the diagram) but only thc interesting ones are shown. Among the nine 'distant' supernovae known to have been observed at maximum, the dispersion is only about $0^m.3$. In this sample the supernovae in spiral galaxies (the square symbols) are $0^m.5$ brighter than the ones in ellipticals. If this difference is real it is probably related to the effects already discussed by Dr Barbon a this conference. But for the time being the difference has been ignored, and a straight line having unit slope has been fitted to the nine distant points. In the Virgo cluster the observed dispersion is a little larger than $0^m.3$, but it is not bad, considering that back-to-front differences among Virgo members are not negligible.

Figure 2 shows two similar diagrams for type I supernovae which have appeared in relatively nearby galaxies (in most cases nearer than the Virgo cluster). On the left, the distances are from de Vaucouleurs' recently revised list of nearby groups of galaxies, and on the right the distances are based on the Hubble law. Whenever possible the mean radial velocity of a group has been used to compute the distance. On both sides the scatter is much larger than in Figure 1, and, compared to the distant supernovae (represented by the solid line), the three nearest ones are too bright and the ones in an intermediate distance range tend to be too faint. The total range in absolute

C. B. Cosmovici (ed.), Supernovae and Supernova Remnants, 209–214. *All Rights Reserved*

magnitude implied by these diagrams is more than $4^m.0$. The big question, of course, is whether this large range is mainly caused by differences in the supernovae or by errors in the adopted distances?

An independent test of the scatter in the intrinsic absolute magnitudes is possible, at least in principle. The luminosity of a thermal source depends only on its tempera-

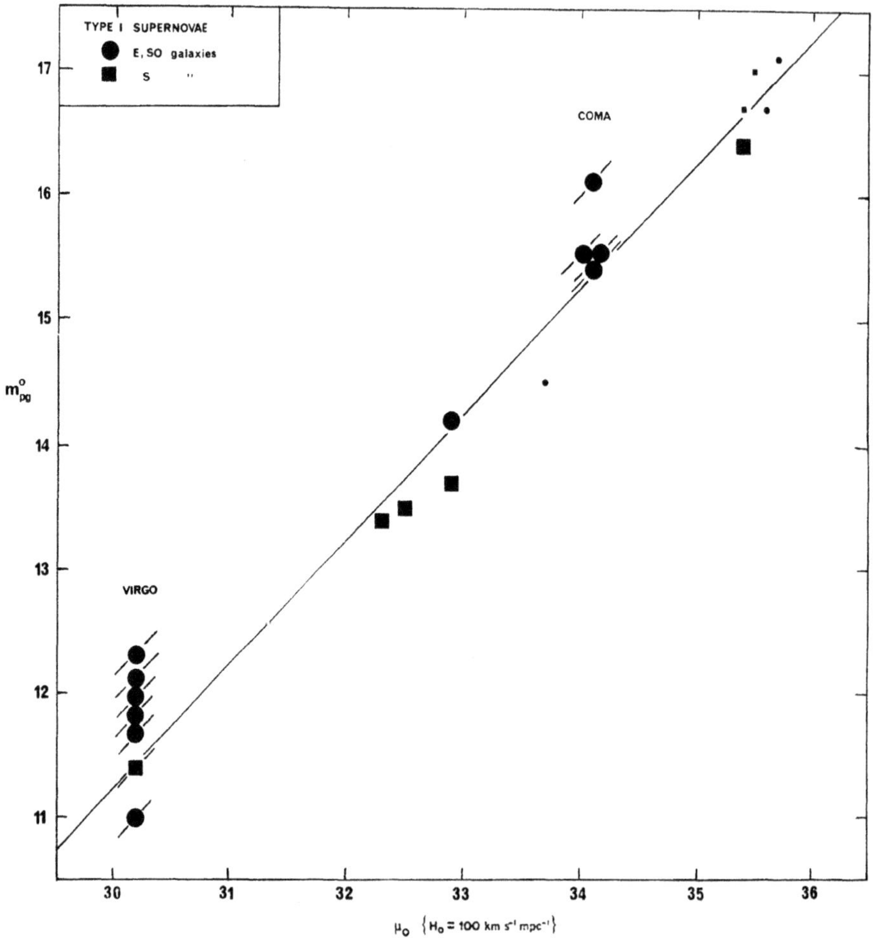

Fig. 1.

ture and radius, and we can look for evidence of a large range in one or both of these quantities. Judging from the available evidence on $B-V$ colors, it seems unlikely that temperature differences at maximum light are extremely important. For instance, Dr Ardeberg has shown us that last year's supernova in NGC 5253, evidently a very bright one, does not seem to have been particularly blue. On the simplest interpretation of the geometry, that of a thin shell expanding at constaut velocity, the radius at maximum light is just a product of the expansion velocity and the interval between outburst and maximum light – the rise time. The expansion velocity is almost certainly not a source of significant scatter; the observed range in blueshifts of absorption lines at maximum light seems to be about 20%, rather than the factor of six which would be required to explain the full four magnitudes. The rise time is difficult to estimate accurately (and will remain so until prompt emissions are detected and identified with optically visible supernovae) but a large range in this quantity should show up as a

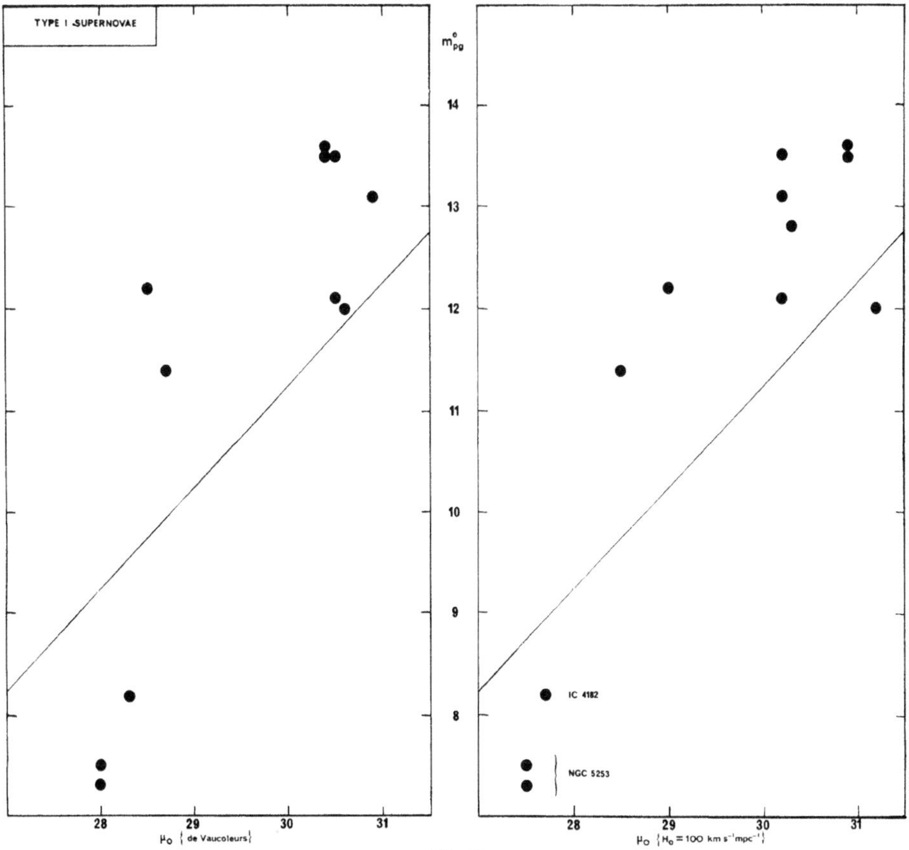

Fig. 2.

large range in the slopes of the light curves during the last few magnitudes of the rise. There is little available data on pre-maximum light curves, but so far there is no evidence for a large scatter.

A further and more direct test is to look for pairs of type I supernovae in the same galaxy or groups of galaxies. The magnitude difference between the two supernovae in NGC 5253 is somewhat uncertain, but it was probably not more than a few tenths. And the published magnitudes of two type I supernovae in the NGC 1566 group differ by only $0\overset{m}{.}1$.

None of the independent evidence supports the large range of absolute magnitudes implied by Figure 2. The possibility that much of the scatter comes from an intrinsic range in rise times cannot yet be excluded, but it does seem likely that a large part of the scatter may turn out to be caused by the distances.

A good comparison between type I and type II would be to use supernovae in galaxies having large radial velocities; unfortunately no appropriate data for type II exist. The Virgo cluster is not a good basis for comparison because of differences of

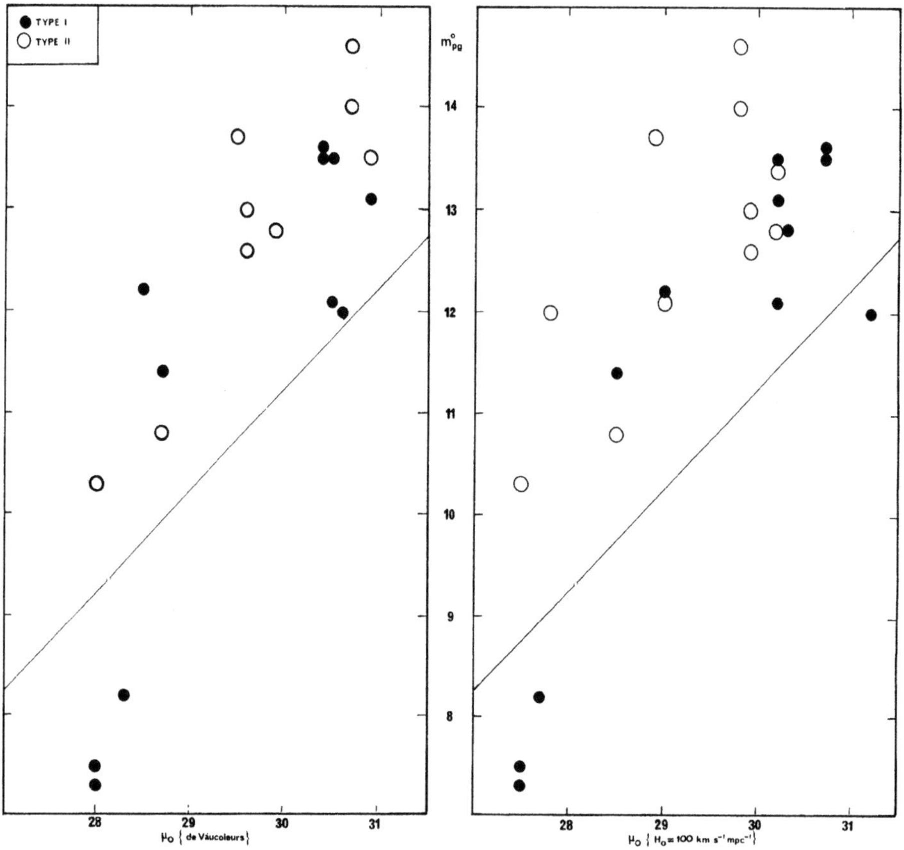

Fig. 3.

opinion regarding its structure. So we are stuck with supernovae in relatively nearby galaxies. The comparison is shown in Figure 3. This is just like Figure 2 except that type II supernovae have been added as open circles. It is true that if the type II's are compared to the distant type I's, the mean difference is almost $2\overset{m}{.}0$. But if they are compared to type I's in the same distance range the mean difference is less than $1\overset{m}{.}0$.

The results of the group test are shown in the Table I. One of each type has appeared in NGC 3938, and the type II was only $0\overset{m}{.}4$ fainter than the type I. In the NGC 1023 group, the type II was apparently as bright as the type I, but the type I may have been affected by absorption in NGC 1003, which is nearly edge-on.

TABLE I

type I–type II pairs in groups of galaxies

group	type I	type II	M (II)–M (I)
Ursa Major I	NGC 3938	NGC 3938	+0.4
NGC 1023	NGC 1003	NGC 1058	0.0
M 101	NGC 5055	M 101	−0.6
NGC 5128	NGC 5253 (2)	NGC 5236	+2.9

In the next two cases the results are discordant, but in both cases doubts have been expressed about the group membership.

The conclusion must be that the mean magnitude difference between type I and type II is very uncertain, but it may be considerably less than two magnitudes.

I would like to conclude by mentioning two examples of the use of supernovae as distance indicators. First, the difference between the mean apparent magnitudes of type I supernovae in the Coma and Virgo clusters is $3\overset{m}{.}9$, in good agreement with the value expected from the ratio of the cluster velocities, and in contradiction to the difference in distance modulus of $4\overset{m}{.}7$ suggested by Abell and Eastmond on the basis of the cluster luminosity functions. Second, Kowal has argued that the apparent magnitudes of supernovae in the Virgo cluster are not compatible with de Vaucouleurs' separation of Virgo into several clouds at various distances. But Kowal's argument depends rather strongly on type II supernovae being several magnitudes fainter than type I. If the mean difference is about $1\overset{m}{.}0$ or less, Kowal's data actually provides some support for de Vaucouleurs. Considering the present uncertainty in the difference, the supernova test as applied to Virgo should be given very little weight.

DISCUSSION

A. G. De Bruyn: Don't you think that the E(lliptical) cloud in the Virgo cluster, which is very compact, is a very good indicator of the intrinsic dispersion in absolute magnitude at maximum?

D. Branch: Yes, it could be. I did not look into it in detail but I thought that the intrinsic dispersion was equal to or less than $0\overset{m}{.}4$ for the type I SN in this 'cloud'.

H. Arp: De Vaucouleurs originally interpreted the redshifts of the different clouds in Virgo as due to differing distances. But now he considers distance difference only one of several possible explanations for this effect. So Kowal's result of all Virgo clouds being at the same distance does not contradict de Vaucouleurs present conception of the possibilities.

M. Friedjung: I was wondering about the relations of the small dispersion of absolute magnitude and interstellar absorption in parent galaxies. A small dispersion would imply a small absorption. Could this be reconciled with Tammann's results?

D. Branch: This is a good point; for the 'distant' type I's the dispersion is comparable to the errors in photometry which must occur, and there is not much left over for variations of absorption in the parent galaxies. It's also interesting that the SN in spirals are, if anything, brighter than the ones in ellipticals, contrary to the expected effects of absorption. But all of this applies to type I's only; there could be large differences of absorption for type II's.

G. A. Tammann: May I mention a certain concern about the scatter in luminosity as derived from distant supernovae. If the true intrinsic scatter were large, one still expects an apparently small scatter at large distances because one discriminates systematically against the intrinsically fainter supernovae.

Your results open up an extremely interesting possibility: there may be a dependence of supernova luminosity on the type of the parent galaxy in the sense that irregular galaxies produce the most luminous supernovae of type I and that the luminosity decreases as one goes toward Sc and E galaxies.

L. Rosino: When you find differences by +0.4, 0.0, −0.6, +2.9 in pairs of SN type I and II in the same galaxies, have you considered the possibility that the apparent magnitude at maximum may have been uncorrect or affected by some indetermination?

D. Branch: The magnitudes of most of the relevant supernovae come from the list of Kowal and Sargent, and they seem to have been careful and conservative in designating the maximum magnitudes as well known.

F. Zwicky: Tammann remarks in the discussion again that we have last supernovae in faint galaxies $m_g \sim 15.0$. This is completely contradictory to the actuality as proved again by my recent search for the predicted 200 supernovae buried on the Palomar Sky Survey Plates and a complete secondary set of comparison plates.

Essentially no supernovae were lost which were intercepted at apparent magnitudes brighter than $m_g = 19.0$ to even 20.0.

R. Wood: In view of Tuesday's discussion about distances to members of Stefan's Quintet, could Dr Branch say how the SN in NGC 7319 fits his line for type I SN?

D. Branch: That SN did not appear on any of these diagrams because its apparent magnitude at maximum was not well determined. But, if one accepts Rosino's extrapolation of the light curve, the SN was just about right, assuming that the redshift of NGC 7319 is cosmological. On the other hand, if NGC 7319 is physically associated with NGC 7320 then the SN must have been wildly peculiar – much fainter than any other known type I SN.

J. R. Dickel: You conclude that the scatter in type I magnitude at maximum is due to distance errors but there is a big range in II/I. Does that imply that there is a large intrinsic scatter in the maximum magnitudes of the II's?

D. Branch: There is a large apparent scatter for the type II's but much of this may also be due to the distances. There are two pairs of type II SN in groups and the respective magnitude differences are 0.4 and 0.6. As for the values of −0.6 and +2.9 for type II/type I differences in groups, Dr Tammann has suggested to me that NGC 5055 may not belong to the M101 group, and of course doubts have been expressed about the physical association of NGC 5253 and NGC 5236.

REMARKS ON OUTLYING SUPERNOVAE AND THE STRUCTURE OF THEIR PARENT GALAXIES

G. A. TAMMANN*

Hamburger Sternwarte, Hamburg-Bergedorf, F.R.G.
and
Institute of Astronomy, Cambridge, England

Some supernovae have appeared at considerable radial distance, outside of the optical boundaries of their parent galaxies. This raises the important question from what kind of stellar population they have originated. This is a first, preliminary report on three typical cases.

1. SN 1909 in M 101 (SS UMa). The supernova was found by M. Wolf in February, 1909, at a distance of $\Delta = 695''$ from the center of M 101. The optical radius of M 101 is given to $R = 736''$ (de Vaucouleurs and de Vaucouleurs, 1964); therefore the relative radial distance is $\Delta/R = 0.94$. This is not yet an extreme relative center distance, but on normal photographs there appears to be practically no background at the position of the SN. The linear distance of the SN from the center corresponds to 24 kpc (Sandage and Tammann, 1974).

The SN is classified as peculiar (Minkowski, 1964), due to its apparently unusual light curve (Himpel, 1942). However, after the correction of an obvious error the light curve becomes reasonably normal (Sandage and Tammann, 1974). Since the SN was caught probably near maximum, only its absolute brightness of $M_B = -15\overset{m}{.}8$ may be unusually low.

On an extremely deep composite photograph taken with the 48-in. Schmidt telescope the optical size of M 101 is so much enlarged that the galaxy extends over the position of the SN. In fact only 1′ from it ends a distinct and deep feature of spiral structure. Hence, it could well be possible that the pre-SN belonged to the normal young stellar population.

The deep picture of M 101 shall be published elsewhere (Sandage and Tammann, 1974).

2. SN 1969l in NGC 1058. This SN was discovered by Rosino (1969) at a very large distance of 220″ from the center of the galaxy. With an optical galaxian radius of 74″ (de Vaucouleurs and de Vaucouleurs, 1964) this corresponds to a relative distance, rarely exceeded, of $\Delta/R = 2.97$. Since the distance of NGC 1058 is 14.5 Mpc (Sandage and Tammann, 1974) the linear radial distance of the SN is 15 kpc. With $m_B(\max) = 13\overset{m}{.}1$ (Ciatti *et al.*, 1971) and a galactic absorption of $A_B = 0\overset{m}{.}24$ the absolute maximum brightness must have been near $M_B = -17\overset{m}{.}9$.

* Permanent address: Astronomisches Institut der Universität Basel, Binningen, Switzerland.

C. B. Cosmovici (ed.), Supernovae and Supernova Remnants, 215–220.

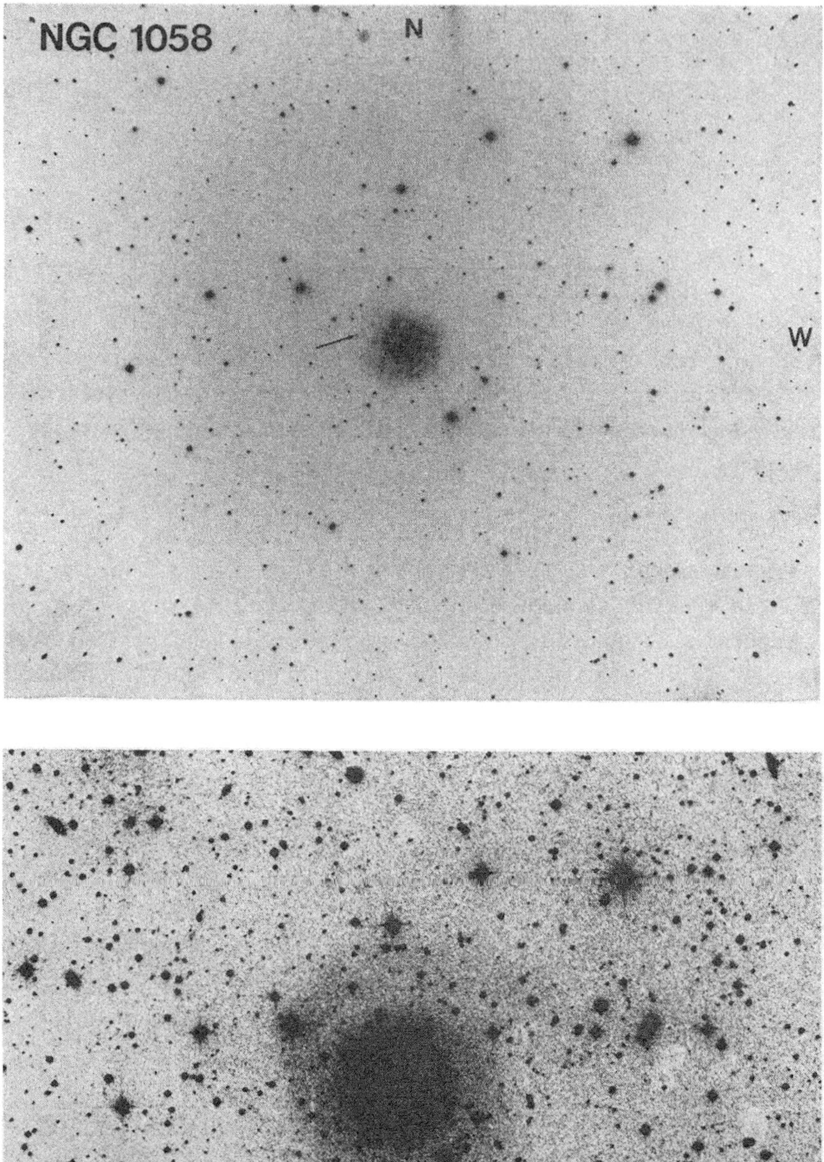

Fig. 1. SN 1969l in NGC 1958. *Upper panel*: A fully exposed 103 a-O plate of NGC 1058 taken with the 200-in. telescope by W. Baade. The SN 1961v is marked. *Lower panel*: A composite photograph of two III a-J plates taken with the 48-in. Schmidt telescope, printed to the same scale(!). The SN 1969l is marked.

Since the SN was classified as type II (Ciatti *et al.*, 1971) the question arises: How can a galaxy have a sufficient amount of young stellar population in its very outskirts to produce a type II SN? To investigate this question a very deep photograph of the galaxy was procured (Figure 1b). It is a composite print of two fully exposed, baked III a-J plates (plus Wratten 4-filter to reduce the sky brightness) taken with the 48-in. Schmidt telescope. The position of the SN is marked on the print.

The surprising fact is that NGC 1058 not only grows considerably in size – to a diameter of about 500″ – on the deep plate (compare Figures 1a and 1b), but that a new generation of a very unusual, faint spiral structure becomes visible: two very broad, smooth spiral arms, traceable over roughly 270° emerge on the north-east and south-west side of the heavily overexposed disc. One arm – the one beginning at the north-east side – winds around to not only reach but actually to exceed the position of the supernova. The arm is not resolved into stars, but at an estimated apparent magnitude limit of 24^m it is still possible that the arms contain stars as bright as -6^m or -7^m. The occurrence of a young population object like the SN 1969 l is therefore not necessarily a surprise.

The case of the faint outer spiral structure in NGC 1058 is reminiscent of LMC, for which de Vaucouleurs (1955) has shown the existence of faint outer arms. However, it is believed that this is not a general feature of spiral galaxies, and that NGC 1058 exhibits a rather rare peculiarity. Final proof has to come from further investigations. The facts that M 101 increases its size on very deep plates to a much lesser extent, and that the other SN in NGC 1058 (SN 1961 v) appeared also outside the conventional optical radius of the galaxy are indicative for NGC 1058 being unusual.

3. SN 1970l in NGC 2968/70. This SN was found by Wild (1970), 140″ from the center of NGC 2968. Since the optical radius of this I0 galaxy is given as 48″ (de Vaucouleurs and de Vaucouleurs, 1964), the relative center distance is $\Delta/R = 2.9$. The mean radial velocity of NGC 2968 is $v_0 = 1533$ km s^{-1} (Kowal and Sargent, 1971; Sandage, 1973), which corresponds to a distance of 27.9 Mpc. At this distance the linear distance of the SN from the center of the galaxy amounts to 19 kpc. According to Barbon *et al.* (1973) the SN is classified as type I and reached a maximum brightness of $m_{pg} = 13\overset{m}{.}3$ ($M = -18\overset{m}{.}9$).

Figure 2 shows a print of a deep baked III a-J (+Wratten 4) plate from the 48-in. Schmidt telescope. The position of the SN is marked. It is striking that the SN lies on a bridge of luminous matter extending from NGC 2968 to the probable E-galaxy NGC 2970 and continuing very markedly beyond the latter. The existence of this bridge has been pointed out before by Wild (1970). The bridge is certainly the result of gravitational interaction between the two galaxies. As the illuminating calculations of Toomre and Toomre (1972) have shown it is not possible to decide from visual inspection to what extent the two galaxies have contributed to the bridge. In spite of the SN being nearer to NGC 2968 it is therefore at present unknown to which galaxy it should be ascribed.

Another case of a SN on an intergalactic bridge has been reported by Rudnicki and

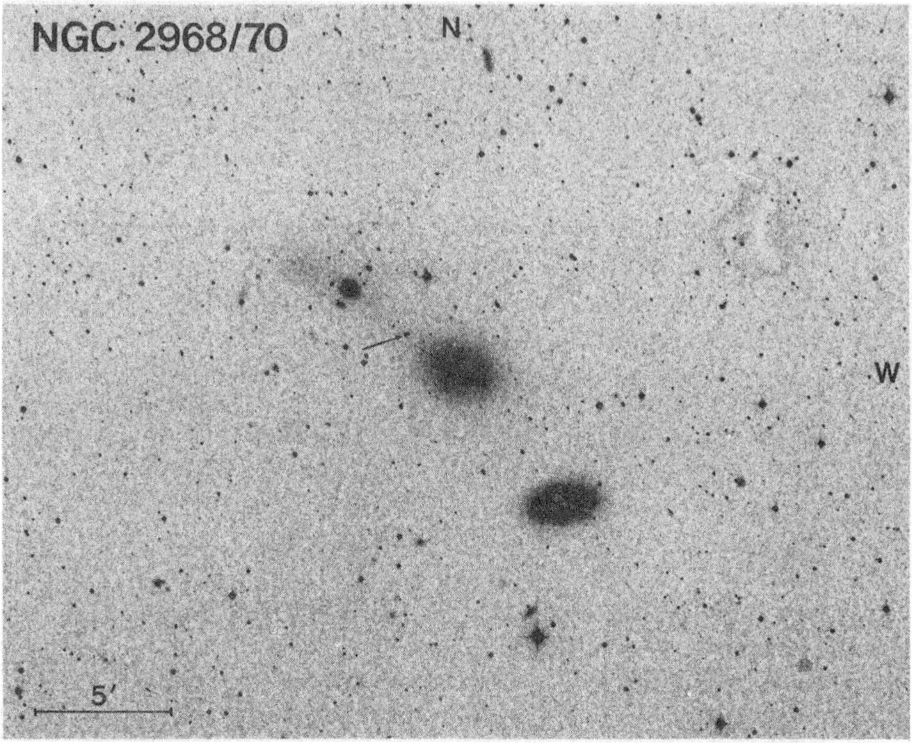

Fig. 2. SN 1970l in NGC 2968/70. The photograph is from a deep III a-J plate taken with the 48-in. Schmidt telescope. The SN is marked.

Zwicky (1967): it is the SN 1962e which appeared between two anonymous elliptical galaxies, the fainter of which being described as compact. It may be hoped that the reality of the bridge shall eventually receive final proof to exclude the possibility of its being only an effect of the extended halo of the brighter galaxy. The SN belongs probably to type I (Rudnicki and Zwicky, 1967).

It is interesting that one and possibly two SNe of type I have appeared on intergalactic bridges, because Searle (1973) has found in several cases that these bridges are blue, and that they must contain therefore a young stellar component. If SNe on intergalactic bridges turned out to be always or predominantly of type I it would be remarkable indeed.

In view of the potential information content of outlying SNe an attempt was made to compile data for similar cases. All SNe in an updated version of the list by Kowal and Sargent (1973) were considered, whose parent galaxies have optical diameters in the RCBG (de Vaucouleurs and de Vaucouleurs, 1964). From 124 cases only 10 have $\Delta/R \geqslant 1.0$. The latter are listed in Table I. This list is very incomplete in the sense that face-on diameters are used and no projection effects are considered. In inclined systems a SN near the direction of the minor axis may be far outside of the optical galaxy

TABLE I

Some outlying Supernovae

Galaxy	Type	SN	Type	Radius of galaxy	Radial dist. of SN	Δ/R	Picture
NGC 1058	Sc	1961 v	(V)	74″	78″	1.1	(1)
1058	Sc	1969 l	II	74″	220″	3.0	(2)
2968/70	I0, E	1970 l	I	48″	140″	2.9	(2)
3177	Sb	1947 a	II	35″	44″	1.3	–
3631	Sc	1964 a	(V?)	117″	118″	1.0	–
3904	E2	1971 c	–	40″	178″	4.5	–
4214	Im	1954 a	I	208″	232″	1.1	(3)
4782/3	E, E	1956 b	–	27″	43″	1.6	(4)
5253	Im pec.	1972 e	I	104″	107″	1.0	(5)
7616	E	1970 j	I	35″	40″	1.1	–

(1) Bertola and Arp (1970).
(2) here.
(3) Wild (1960).
(4) Kowal *et al.* (1970).
(5) *Sky Telesc.* **44** (1972) 16.

image and still yield $\Delta/R<1$. Visual inspection of published SN pictures shall bring additional cases to light. An interesting example is SN 1969n, which appeared in a tiny irregular galaxy (Kowal *et al.*, 1970). On very deep plates it could be that this galaxy turns out to be a peripheral part of a considerably brighter galaxy near to it.

More complete data on outlying SNe shall eventually become valuable for the following problems: (1) they may indicate the presence of unsuspected faint outer structure of the parent galaxy; and (2) they shed light in a statistical sense on the nature of the parent population of SNe of type I and II.

If in addition data on the radial distribution of SNe became available for well-defined samples of galaxies and for the full range of Δ/R they would have bearing on two further questions: (1) Does the radial distribution depend on redshift? If so it would be an independent indication for the existence of observational selection effects in present SN lists. (2) How much stellar mass, beyond the observable rotation curves of galaxies, has to be required at least, to account for the frequency of outlying SNe?

It has been proposed to use radial distances of SNe as statistical distance indicators and to derive a value of the Hubble constant from these data. In view of the very wide range of SN radial distances this would require, however, an enormous and *unbiased* statistical sample, which is far from being available.

Acknowledgements

The author expresses his gratitude to Dr H. W. Babcock for the allotment of observing time and for the hospitality at the Hale Observatories. This paper was written while he was a visitor at the Institute of Astronomy, Cambridge; he extends his thanks to

Professor D. Lynden-Bell for the most pleasant time there. He thanks Messrs J. Bedke and D. Cerrito for the excellent photographic work and Miss Nadine Arber for typing the manuscript. He acknowledges that part of this work was supported by the Swiss National Science Foundation and the Privatdozentenfonds of the University of Basel.

References

Barbon, R., Ciatti, F., and Rosino, L.: 1973, *Astron. Astrophys.* **25**, 241.
Bertola, F. and Arp, H.: 1970, *Publ. Astrophys. Soc. Pacific* **82**, 894.
Ciatti, F., Rosino, L., and Bertola, F.: 1971, *Mem. Soc. Astron. Ital., N.S.* **42**, 163.
Himpel, K.: 1942, *Beob.-Zirk. Astron. Nachr.* **24**, 62.
Kowal, C. T. and Sargent, W. L. W.: 1971, *Astron. J.* **76**, 756.
Kowal, C. T. and Sargent, W. L. W.: 1973, private communication.
Kowal, C. T., Sargent, W. L. W., and Zwicky, F.: 1970, *Publ. Astron. Soc. Pacific* **82**, 736.
Minkowski, R.: 1964, *Ann. Rev. Astron. Astrophys.* **2**, 247.
Rosino, L.: 1969, *IAU Circ.*, No. 2194.
Rudnicki, K. and Zwicky, F.: 1967, *Astron. J.* **72**, 407.
Sandage, A.: 1973, private communication.
Sandage, A. and Tammann, G. A.: 1974, in preparation.
Searle, L.: 1973, private communication.
Toomre, A. and Toomre, J.: 1972, *Astrophys. J.* **178**, 623.
Vaucouleurs, G. de: 1955, *Astron. J.* **60**, 126.
Vaucouleurs, G. de and Vaucouleurs, A. de: 1964, *Reference Catalogue of Bright Galaxies*, University of Texas Press, Austin.
Wild, P.: 1960, *Publ. Astron. Soc. Pacific* **72**, 97.
Wild, P.: 1970, *IAU Circ.*, No. 2287.

DISCUSSION

W. M. Goss: In M 101 the Westerbork HI synthesis map shows a large amount of HI near SN 1909.

G. A. Tammann: That is a good reassuring news.

F. Bertola: The position of the 1969 SN in NGC 1058 in an outermost faint spiral structure was already pointed out in the paper by Ciatti, Rosino and myself from the study of the superimposition of 200″ plates.

G. A. Tammann: Thank you; please forgive my omission.

F. Zwicky: The far extension of all galaxies into intergalactic space was predicted long ago and confirmed without resort to outlying supernovae, for instance by following the H and K absorption lines on overexposed spectra (Humason found that first).

H. Arp: The point of your papers seems to be that III a-J photographs of some galaxies with SN show an enlarged diameter in the longer exposures. But I have been doing a lot of III a-J photography with the 48-in. at Palomar and as you know the diameter of all galaxies enlarge dramatically with the long III a-J exposures. Are there any quantitative measures of the diameter enlargements which would support your claim?

G. A. Tammann: The diameter of NGC 1058 is increased by a factor of about 3. Can you tell me how much diameter enlargement you have on your objects?

H. Arp: No I can't. That is just the point there is no quantitative justification for claiming the galaxies with SN in outer regions have larger than usual faint outer regions.

PART V

SUPERNOVA REMNANTS

OBSERVATION OF X-RAY EMISSION FROM SUPERNOVA REMNANTS

PAUL GORENSTEIN*

American Science and Engineering, Cambridge, Mass. 02139, U.S.A.

1. Introduction

X-ray observations have resulted in much of the recent new information concerning supernova remnants. Observations have been made principally from sounding rockets carrying instruments with various spectral sensitivities over the range 0.1–10 keV. Two satellite instruments have also made important contributions; Uhuru which has provided greatly increased sky coverage and sensitivity above 2 keV, and the small area X-ray instrument aboard Copernicus which has provided greatly improved angular resolution below 1 keV. The X-ray observations have greatly expanded our knowledge of the physical conditions occurring in supernova remnants and have given us new insights into their evolution.

Radio astronomers have defined the criteria for identifying supernova remnants, the existence of an expanding radio nebula with non-thermal emission. Most of the objects defined in this way are less than 10^5 yr old. In this review I shall follow their definition and restrict the discussion of X-ray observations to only those objects which appear in their catalogues (e.g., Milne, 1970 and Downes, 1971). However, there is an increasing amount of evidence that many galactic X-ray sources are very

TABLE I

Catalog of X-ray emitting supernova remnants

Object	l^{II}	b^{II}	Distance (kpc)	Radio diameter (pc)	2U Name
1. Tycho SN1572	120.1°	+ 1.4°	3	5.8	0022 + 63
2. Crab Nebula	184.6	− 5.8	1.7	2.5	0531 + 22
3. PSR0531 + 21	184.6	− 5.8	1.7	–	0531 + 22
4. IC443	189.0	+ 3.0	1.5	19	0601 + 21
5. PSR0611 + 21	189.0	+ 3.0	1.5	–	?
6. Pup A	260.4	− 3.4	1.2	19.4	0821 − 42
7. Vela X-ray Nebula	263	− 3	0.46	27	0832 − 45
8. PSR0833 − 45	263	− 3	0.46	–	0832 − 45
9. Lupus Loop	330	+ 15	0.8	60.7	–
10. SN1006	327.6	+ 14.5	4.9	42.6	–
11. MSH15 − 52A	320.4	− 1.4	5.4	22.0	1509 − 58
12. MSH1811 − 17 (GX13 + 1)	13.4	+ 0.1	9.8	19.0	1811 − 17
13. Cygnus Loop	73.6	− 8.6	0.8	31	–
14. Cas A	111.7	− 2.1	3.4	4.0	2321 + 58
15. North Polar Spur	330	20	0.1 − 0.2	170 − 340	–

* Now at Center for Astrophysics, Smithsonian Astrophysical Observatory, 60 Garden Street, Cambridge, Mass. 02138.

C. B. Cosmovici (ed.), Supernovae and Supernova Remnants, 223–242.

TABLE II

Object	10^9 Hz flux density (Wm^{-2} Hz^{-1})	Radio spectral Index ($-\alpha$)	Surface bright at 10^9 Hz (Wm^{-2} Hz^{-1} sr^{-1})	0.5–2 keV flux (ergs cm^{-2} s^{-1})	2–10 keV flux (ergs cm^{-2} s^{-1})	Confidence of identification
1. Tycho	58	0.6	1. 7E-19		1.8E-10	Certain
2. Crab	1000	0.25	9. 4E-18	–	1. 6E-08	Certain
3. Pulsar				–		Certain
4. IC443	160	0.5	1.2E-20	–	6E-11	Probable
5. Pulsar				–		Possible
6. Pup A	145	0.45	5.7E-21	5.6E-09	1.3E-10	Certain
7. Vela X-Ray	1800	0.3	5.4E-21	7E-09	–	Certain
8. Pulsar				1E-10	1.7E-10	Probable
9. Lupus L	340	0.3	5.6E-22	2E-10	–	Probable
10. SN1006	25	0.6	4.5E-21	< 1E-10	–	Possible
11. MSH15-52A	60	0.6	1.6E-20	–	1.2E-10	Probable
12. MSH1811-77	13	0.5	5E-20	–	5.0E-09	Possible
13. Cygnus loop	190	0.5	6.7E-22	9.2E-09	–	Certain
14. Cas A	2900	0.75	2.3E-17	–	9.1E-10	Certain
15. North polar spur	–	0.7	Brightness temperature $\lesssim 500°K$ at 2.4×10^8 Hz	$\sim$ 1E-08 $ster^{-1}$	–	Possible

compact objects such as neutron stars or balck holes which happen to be members of a close binary system. If they are indeed neutron stars or black holes, and if a supernova explosion is needed for their formation, then it follows that many galactic X-ray sources are supernova remnants in a fundamental sense. However, they are probably much older systems in which the expanding nebula has long since dissipated. Also, the physical process governing their behavior is accretion of matter from a more diffuse and more massive member of the system to the compact member whereas the conventional supernova remnants as found by radio astronomers do not appear to be members of a binarv system or if they are the dynamics of the binary system play no role in their radiative processes.

Tables I and II lists the supernova remnants for which there is evidence of X-ray emission. The 'confidence of identification' column represents the author's own opinion. In all cases, an X-ray source is found within the area of the radio source; however, in some cases the size of the radio source is several square degrees so the association of the two in position may be accidental. This is particularly true in the region of the galactic center where the density of X-ray sources is large. When there is no evidence beyond a positional coincidence, the entry in the 'confidence of identification' column of Table II is 'possible'. The X-ray intensity is given for the energy bands 0.5–2 keV and 2–10 keV. Several objects are also detectible below 0.3 keV but there is a large correction for interstellar absorption. These objects can be subdivided among several categories, remnants younger than 10^3 yr which emit copiously above 2 keV, e.g., Cas A and Tycho, older remnants $\sim 10^4$ yr having a larger intrinsic size and emitting primarily below 1 keV such as the Cygnus Loop and the Vela SNR, young radio pulsars or rapidly rotating neutron stars, and finally the North Polar Spur, a very diffuse object which may be unique.

We can predict how much the catalog is likely to increase with more sensitivity. It appears that only young remnants ($T \lesssim 10^3$ yr) emit significantly above 2 keV. Taking a rate of one born per 10^2 yr we conclude that about half of the young SNR's have probably been already observed. The young objects which remain to be discovered are probably on the other side of the galactic center where their distance, the relatively large opacity of the interstellar medium and the large density of other galactic X-ray and radio source will make it difficult to distinguish and identify them. Among the unidentified galactic X-ray sources now listed in the Uhuru catalog (Giacconi *et al.*, 1972) those which turn out to have finite angular size and non-variable intensity are good candidates for supernova remnants and it would be worthwhile examining their positions carefully for the radio counterpart. However, because of the problems radio astronomers have in looking beyond the galactic center region, it will be difficult to identify the radio counterpart on the basis of their usual criterion, the existence of a non-thermal, expanding nebula. The older remnants in the age range 10^4–10^5 yr are suprising luminous, but they emit primarily below 2 keV. Soft X-rays cannot travel a large distance in the galaxy due to absorption in the interstellar gas. Thus, we can expect only those older supernova remnants within 1000 pc or so of the Sun to be detectible. With a supernova explosion taking place every 10^2 yr and the phase of soft X-ray emission lasting possibly 10^5 yr, there are 10^3 objects in the entire galaxy. However, only a few percent of these would be within 1000 pc of the Sun. Thus, the list of older supernova remnants with detectible soft X-rays fluxes may grow ultimately to as large as 20 or 30.

2. Young Supernova Remnants

Five supernova remnants are less than 10^3 yr old. Their intrinsic radio diameter is about 5 pc or less, and so is the X-ray diameter in the two cases where they have been measured, the Crab Nebula and Cas A. X-ray emission has been detected from four. Kepler's Nova or SN1604 has not been detected to date. However, considering its large distance, if SN1604 has about the same intrinsic luminosity as Tycho, it is not surprising that not even the Uhuru satellite has detected it.

2.1. The Crab Nebula and NP0531

The Crab Nebula was the first visible or radio object to be identified with a cosmic X-ray source. Following the identification, observers have studied the Crab Nebula in X-rays in great detail from rockets, balloons, satellites and even ground-based locations. The identification of the pulsar in the Crab Nebula, NP0531, as the first X-ray pulsar was another important step which stimulated new observational activity. There are a larger variety of observational results available from the Crab and NP0531 than from any other X-ray object. Nevertheless we are still far short of achieving a complete understanding of the dynamics of the neutron star-nebula system existing in the Crab. A great deal more can be learned from future programs which will offer large improvements in sensitivity and angular resolution.

2.1.1. *Angular Structure*

A lunar occultation of the Crab Nebula that took place in 1964 was observed from a rocket by Herbert Friedman and his group at the Naval Research Laboratory (Bowyer *et al.*, 1964). They detected a slow variation rather than an abrupt change of intensity with the apparent movement of the Moon across the Nebula. This observation proved that the X-rays were originating from an extended rather than a point source. The numerical values of the size and centroid as presented in the original paper by NRL were not corrected properly for the down range motion of the rocket. Corrections at a later date showed that the size of the X-ray emitting region was about 2′ and the centroid was close to the star that was later identified as the optical counterpart of the pulsar NP0531.

The angular structure of the Crab Nebula was observed in two dimensions in 1966 with a modulation collimator by a combined MIT/AS & E group (Oda *et al.*, 1967). Their results were consistent with the revised NRL result regarding the finite extent of the source and its location. In Figure 1 the X-ray source in the energy range 2–10 keV as observed in these two experiments is compared with an optical picture of the Crab Nebula.

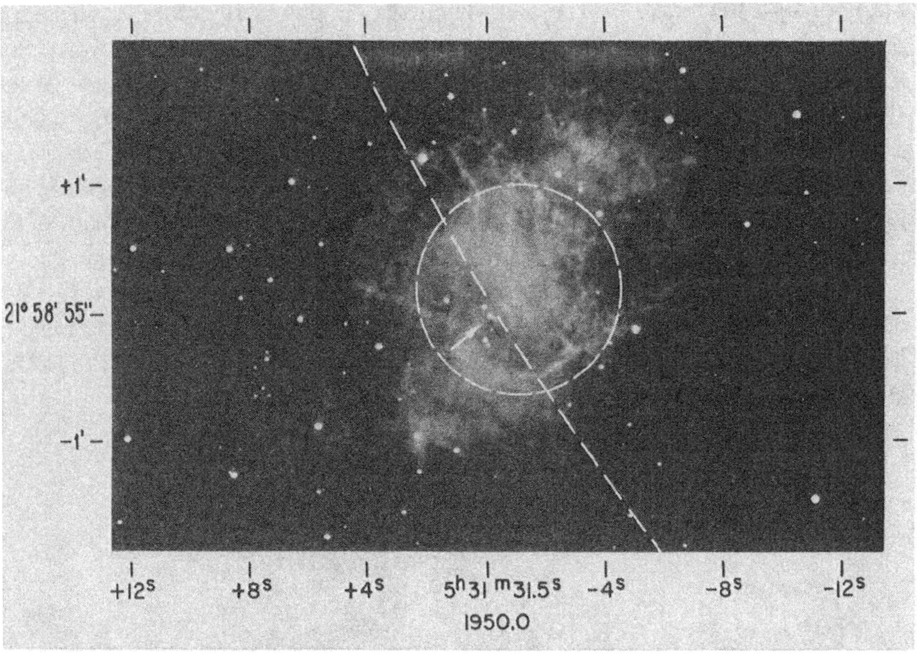

Fig. 1. Location of X-ray source in Crab Nebula. Curved line is from lunar occultation experiment (Bowyer *et al.*, 1964); circle is result of modulation collimator experiment (Oda *et al.*, 1967). Photograph in courtesy of Hale Observatories. The pulsar is indicated by an arrow.

In neither of these two observations were there sufficient angular resolution to detect any features beyond overall size and location. Thus, they were not able to detect the 10% or so point source contribution from the pulsar near the center of the nebula. New lunar occultations of the Crab Nebula will take place during 1974 and a number of X-ray observations are being planned to study the object in finer detail.

2.1.2. *X-Ray Spectrum and Polarization*

It has been demonstrated rather convincingly that the spectrum of the Crab Nebula over the range 2–100 keV can be represented as a power law with an almost constant spectral index. In fact, the Crab Nebula is the only cosmic X-ray source where it has been established that the spectrum is a power law. Exponential functions characteristic of thermal spectra cannot fit the data unless we arbitrarily assume a multi-temperature model. A power law spectrum fits in neatly with the idea that synchrotron radiation is the emission mechanism. In the range $2 \leqslant E \leqslant 20$ keV, obtained mostly from rockets, the spectrum can be described by the function (e.g., Gorenstein *et al.*, 1970a),

$$\mathrm{d}N/\mathrm{d}E = 9E^{-2.0 \pm 0.1} \text{ photons cm}^{-2}\,\text{s}^{-1}\,\text{keV}^{-1}.$$

The integrated intensity of the Crab Nebula is 3×10^{-8} ergs cm^{-2} s^{-1}, 2–20 keV. Taking the distance to be 2 kpc, this intensity corresponds to an intrinsic X-ray luminosity of 1.5×10^{37} ergs s^{-1}. The X-ray luminosity in this energy range is still several times smaller than the energy lost by a rotating neutron star whose rate of slowing down is given by the secular variation of the period of the pulsar NP0531. Between 20 and 100 keV the Crab has been observed from balloons. Fitting the higher energy data to a power law results in a slightly more negative spectral index but not so different as to suggest a sharp break. The Crab is also detectible above 100 keV, but the counting rate is low compared to background and it becomes difficult to determine the shape of the spectrum. The spectral observations in the X-ray region are shown in Figure 2.

At the very high energies, above 2.5×10^{11} eV, a γ-ray flux of $4.4 \pm 1.4 \times \times 10^{-11}$ photons cm^{-2} s^{-1} has been detected in ground-based observations of atmospheric Cerenkov radiation. The γ-flux appears to increase significantly following a major spin-up of the pulsar NP0531 (Fazio *et al.*, 1972).

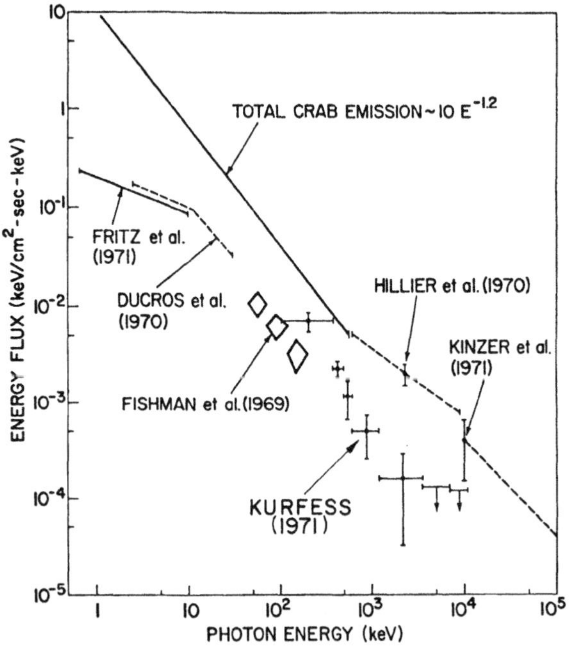

Fig. 2. X-ray spectra of Crab Nebula and NP0531. Figure is from paper by Kurfess (1971).

Aside from the pulsar there is no evidence for time variability of the Crab Nebula X-rays. Unless there are additional compact objects associated with the Crab, we expect no variability. There is also no evidence for line emission in the spectrum at either X-ray or gamma-ray energies.

R. Novick and co-workers at the Columbia Astrophysics Laboratory have measured the X-ray polarization of the Crab Nebula. They report a linear X-ray polarization of ($15 \pm 5\%$) integrated over the whole nebula (Novick *et al.*, 1972). This result represents another convincing argument for the predominant role of synchrotron radiation in the X-ray emission. The Crab Nebula is the only cosmic X-ray source where we are certain that the X-ray emission process is synchrotron radiation and not thermal radiation.

2.1.3. NP0531, *the Crab Pulsar*

Following the discovery of radio and optical pulsation from NP0531 with a period of 33 ms, there were several observations from rockets at about the same time to look for this effect in X-rays as well. The NRL group (Henry *et al.*, 1969) was the first to report positive results followed by MIT (Bradt *et al.*, 1969), Goddard (Boldt *et al.*, 1969), and Saclay (Ducros *et al.*, 1970). The MIT observation included nearly simultaneous optical and X-ray observations of NP0531 and the Crab Nebula. Their data shows that to within an experimental precision of a few tenths of a millisecond the X-ray and optical pulses are in phase. Also at X-ray frequencies the interpulse is much stronger relative to the main pulse than at optical frequencies. It appears that there is a finite residual X-ray intensity between the main and interpulses. At the peak of the main pulse the instantaneous intensity of the pulsar is approximately half as large as the continuum intensity which is dominated by the contribution from the Nebula. The time average X-ray flux of the pulsed component beamed at the Earth is about 8% of the intensity of the Crab Nebula in the range 1–10 keV. The MIT group reports that the pulsed fraction relative to the Nebula appears to be significantly smaller at the lower end of this range. With interstellar absorption likely to be the same for the Nebula and the pulsar, their result suggests a bend in the pulsar spectrum at ~ 1 keV. The intensity of the pulsed component relative to the continuum increases at higher energy to about 20% at 20 keV (Fritz *et al.*, 1971) and does not increase significantly up to about 1 MeV (Kurfess, 1971). The observations are unclear as to whether the fractional intensity of the pulsar continues once again to increase relative to the continuum above 1 MeV. More results on the subject will be reported during this conference.

2.2. Cas A

An early rocket flight by the NRL group identified Cas A as an X-ray source (Friedman *et al.*, 1967). A subsequent flight by our group at AS & E gave a firm basis to the identification by providing a precise location (Gorenstein *et al.*, 1970b). Cas A has a much softer X-ray spectrum than the Crab, for a power law fit to the intensity vs. energy the spectral index is -2.3 ± 0.6, for an exponential fit $kT = 1.3$ keV with a corresponding uncertainty. The source is severely attenuated below 1 keV by the interstellar medium. No X-ray pulsar has been found in Cas A (Gorenstein *et al.*, 1970a; Holt *et al.*, 1973).

A recent report by Serlemitsos *et al.* (1973), cites observational evidence for broadened iron line emission in the X-ray spectrum of Cas A. If this is correct the strength of the iron line exceeds what one would expect for an equilibrium thermal spectrum because the shape of the continuum suggests lower temperatures. The small X-ray telescope aboard Copernicus has observed a finite X-ray size for Cas A and a structure that is suggestive of a shell. More details on the observation will be presented by Adams and Zarnecki at this conference.

2.3. Tycho (SN1572)

The X-ray source associated with the remnant of Tycho's Nova was first observed by the NRL group (Friedman *et al.*, 1967). The identification was firmly established later by AS & E (Gorenstein *et al.*, 1970b). The source is observable above 1 keV and its spectrum of intensity vs. energy can be fitted by either an exponential function with $kT = 2.2 \pm 1.4$ keV or by a power law with a spectral index of $\alpha = -1\ 3 \pm 0\ 3$. A rocket observation by the Wisconsin group finds evidence for greater complexity in the spectrum, possibly line emission (McCammon *et al.*, 1972). Tycho is not easily observable below 1 keV presumably because of interstellar absorption. Searches for time pulsations in the flux from Tycho have not been successful (Gorenstein *et al.*, 1970a; Holt *et al.*, 1973).

3. Puppis A and IC443

Two SNR's in the middle size range, about 15 pc intrinsic diameter and an age probably in the range 10^3–10^4 yr have been identified as X-ray sources, Pup A and IC443. Both have similar radio characteristics in terms of size, surface brightness, and spectral index. However, they are distinctly different in the optical and X-ray bands; IC 443 is a weak X-ray source but a rather bright optical nebula, while the reverse is true for Pup A.

3.1. IC443 and PSR0611+21

A weak X-ray source has been found within a few degrees of IC443 by the Uhuru satellite. IC443 is located in a region of the galaxy (the anti-center) where the density of X-ray sources is comparatively low. Thus, although the X-ray position is rather crude, we can accept the identification. In a rocket flight the Wisconsin group finds enhanced X-ray emission in their 0.5–0.9 keV channel about the position of IC443 (Bunner *et al.*, 1973). Neither of these observations had sufficient angular resolution to distinguish between extended nebular-type emission and a point source at the position of PSR0611+21, a radio pulsar with a comparatively short period that has been associated with the SNR. Based on an analogy to the Crab and Vela SNR's, X-rays from the pulsar, if there are any, are probably much less intense than X-rays from the nebula.

3.2. Puppis A

Pup A has been observed in the soft X-ray range 0.5–1 keV where it is intense, from sounding rockets and by Copernicus and also above 2 keV where it is rather weak, by Uhuru.

The most detailed picture of its angular structure has been obtained by Copernicus and will be reported upon in more detail at this conference by Adams and Zarnecki. Several features were also seen during a rocket flight. The object is extended in X-rays about as much as it is extended in radio; however, the two maps are not identical. In particular the X-ray distribution is located further north than the radio and a

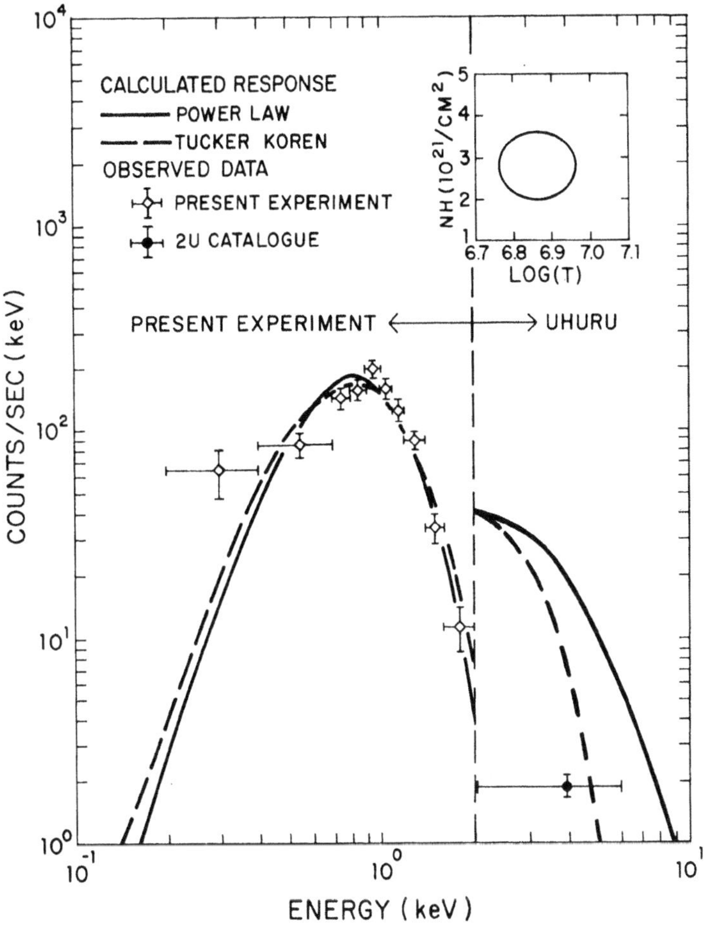

Fig. 3. Proportional counter spectrum of Pup A based on recent rocket data (0.1–1.5 keV) and Uhuru data (2–5 keV) in 2U catalog (Giacconi *et al.*, 1972). Theoretical fits of thermal (Tucker-Koren) and power law spectra are shown.

bright X-ray feature is definitely not coincident with the brightest radio feature. Spectral data from rocket experiments favor a thermal spectrum for Pup A as compared to a power law. Spectral data below 2 keV obtained during a rocket flight in combination with spectral data above 2 keV obtained by Uhuru are shown in Figure 3. The rocket data by itself could be explained as either a power law spectrum or thermal radiation from a hot plasma at a temperature of about 8×10^6 K. The level of X-ray emission above 2 keV seen from Uhuru excludes the power law interpretation. An independent rocket flight also favors thermal emission but at a lower temperature (Burginyon *et al.*, 1973).

4. Older Supernova Remnants

These objects have an intrinsic diameter of about 40 pc or more and their age probable exceeds 10^4 yr. Their X-ray spectrum is soft, essentially all of the emission being below

2 keV. Their large size and their age suggest that the radiation process is probably not synchrotron radiation.

4.1. The Cygnus Loop

4.1.1. *Angular Structure*

The Cygnus Loop was first identified as a soft X-ray source by the Lawrence Livermore group (Grader *et al.*, 1970) who also noted that its size was extended. A more detailed observation with a focusing collector was achieved by a combined group from American Science and Engineering and the Columbia Astrophysics Laboratory (Gorenstein *et al.*, 1971). Among the results the AS & E/CAL group obtained was a one-dimensional picture of the distribution of X-ray intensity in the Loop which showed that the X-ray structure has the same angular size, about 2.8°, as the outermost boundaries of the optical filaments. Furthermore, like the optical filaments, the X-ray structure was suggestive of a shell with a relative absence of intensity at the center. Subsequent observations provided more details concerning the Loop structure (Borken *et al.*, 1972, and Stevens and Garmire, 1973). We can conclude from these observations that the X-ray structure of the Cygnus Loop may be as complicated and as full of details as the optical or radio picture. Although the X-ray size is the same as the optical, there is not necessarily a one to one correspondence between components of the Loop that are bright optically and intense in X-rays. The Cygnus Loop envelopes a mass of interstellar material that greatly exceeds the mass ejected by the supernova explosion. Thus, it would not be surprising if the initial conditions in the interstellar medium are a predominant influence in the present X-ray structure of the Loop.

4.1.2. *X-Ray Spectrum*

The X-ray spectrum of the Cygnus is quite soft, about as soft as any discrete cosmic X-ray source found so far. In fact the Cygnus Loop can be easily detected at energies below 0.3 keV. It has not been detected above 2 keV, not even by the Uhuru satellite. If an exponential function is fit to the data, a kT in the range 0.25–0.4 keV is obtained, the value depending on the amount of interstellar matter to the Loop one assumes in the analysis of the spectral data. The Loop is so large that there is no reason to expect the spectrum to be constant over its entire volume.

The most outstanding result concerning the X-ray spectrum of the Cygnus Loop is the claim of Gorenstein *et al.* (1971) of the existence of lines of highly ionized oxygen, O VIII and/or O VII. The presence or absence of these lines provides a rather unambiguous test of whether or not the spectrum is thermal. The Cygnus Loop is so large in mass that its matter content is primarily interstellar. All the elements of the universal cosmic abundance should be present. Thus, if the X-rays are indeed due to radiation from a hot plasma then line emission from heavier elements cannot be avoided. The density in the X-ray plasma is low, $\lesssim 1\ \mathrm{cm}^{-3}$, so there is no possibility of lines being self-absorbed or scattered in the Cygnus Loop as they might be in

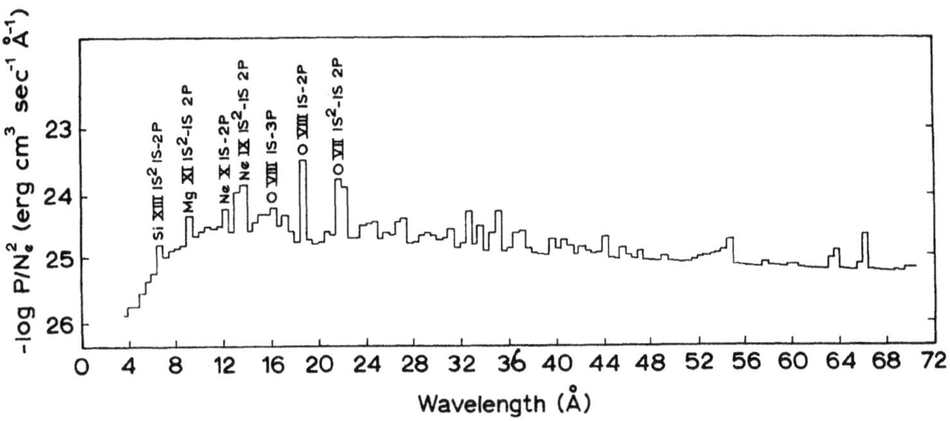

Fig. 4. Theoretical X-ray spectrum for thermal plasma of interstellar matter at 4×10^6 K (Tucker, 1971).

compact objects. Figure 4 shows a theoretical emission spectrum from a plasma at a temperature of 4×10^6 K (Tucker, 1971). For the one angstrom resolution shown in the figure, the oxygen lines lie about a factor of ten above the continuum. The resolution of gas proportional counters used in rocket observation is not better than 10 Å so the lines are difficult to identify. In fact, Figure 3 illustrates approximately how a gas proportional counter responds to the input spectrum of Figure 4. A thermal spectrum was given some support by a collaborative Leiden-Nagoya University rocket experiment (Bleeker *et al.*, 1972) whose data favored an exponential-type spectrum. However, they did not make a positive detection of line emission. In summary, the evidence seems to favor a thermal spectrum for the Cygnus Loop but measurements with high energy resolution are needed to make the case totally convincing.

4.2. The Vela Supernova Remnant

The SNR in Vela may turn to be as important as the Crab Nebula as a laboratory for studying the transformation of energy into radiation. The Vela SNR contains the pulsar PSR0833-45, an X-ray nebula, and a complex radio source. It is older than the Crab and probably younger than the Cygnus Loop, so it still retains several attributes of the Crab but has evolved considerably beyond it and towards the condition of the Cygnus Loop. In addition it has been suggested that the effects of higher energy photons produced in the initial outburst can still be seen in the form of the giant HII region known as the Gum Nebula (Alexander *et al.*, 1972).

4.2.1. *The Vela X-Ray Nebula*

Soft X-ray emission from the Vela SNR was first discovered by the Lawrence Livermore group (Palmieri *et al.*, 1971; Seward *et al.*, 1971). They observed X-ray emission over a broad region whose size was about 5.6° × 4°. They noted that the center of

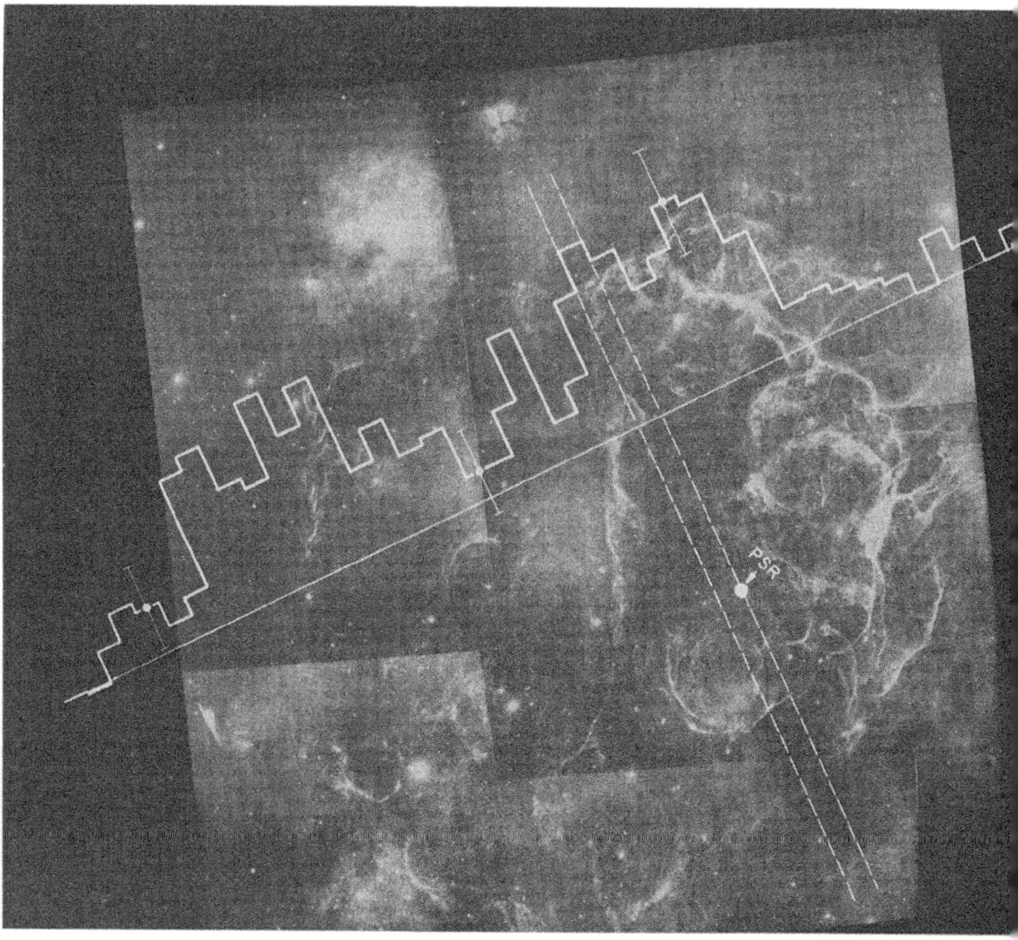

Fig. 5. Scan across Vela SNR and Puppis A with focussing collector having angular resolution in one dimension superimposed upon a composite photograph of region. X-ray data refers to energy band 0.1–0.3 keV. Puppis is not seen. Photograph was taken in ultra-violet light by, E. W. Miller of the Steward Observatory and the Cerro Tololo Inter-American Observatory.

the X-ray source does not coincide with the Vela X radio complex nor with the position of the radio pulsar PSR0833-45. A higher angular resolution survey of the region using a focusing X-ray collector was made by us recently at AS & E. Low energy data (>0.28 keV) obtained during one of our scans is shown in Figure 5, superimposed upon a composite photograph of the filaments. Above 0.5 keV we find the X-ray source in the Vela SNR to consist of a nebula plus a point source at the position of PSR0833-45 (Harnden and Gorenstein, 1973) as shown in Figure 6. The X-ray structure of the Vela SNR has about the same overall size as the radio structure

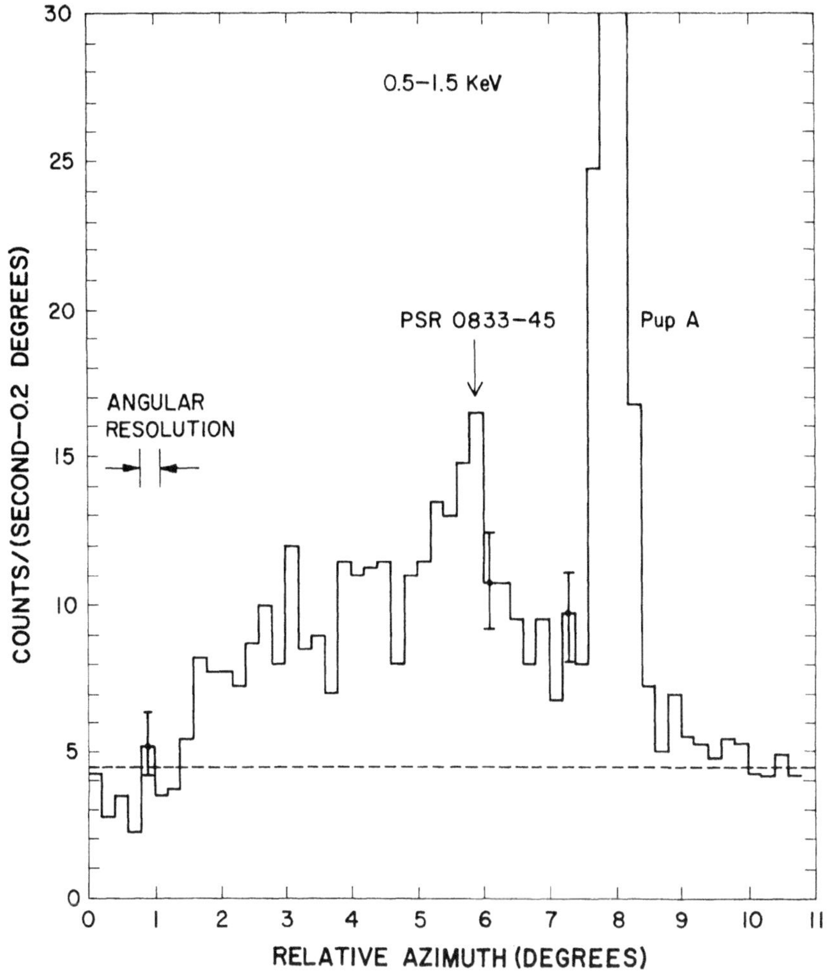

Fig. 6. Scan across Vela SNR and Pup A with focusing collector having angular resolution in one dimension for energy band 0.5 – 1.5 keV. Peak occurs at position of PSR 0833-45 and Pup A is very intense in this energy band.

but very definitely differs from it in detail and in the location of the center. This is also true when a comparison is made between the X-ray and the optical picture. The bright optical filaments that appear in Figure 5 are not the regions when the X-ray intensity is greatest.

The spectrum of the Vela X-ray Nebula is soft, like that of the Cygnus Loop. It is intense below 0.3 keV and no longer detectable above 2 keV except perhaps for restricted regions. Fitting the Lawrence Livermore spectral data of the entire nebula to an exponential spectrum gives a $kT = 0.23$ keV (Seward *et al.*, 1973) but our own data is better fit by a temperature that is twice as high. At the present time it is not known whether or not thermal emission is favored over a power law.

4.2.2. PSR0833-45

Our observation gave evidence that the radio pulsar PSR0833-45 associated with the Vela SNR is also a point source of X-rays. The angular distribution of counts peaks at the expected position of the pulsar over and above the more or less slowly varying contribution from the nebula as shown in Figure 5. An analysis of the phase structure of the data contained in this peak at the frequency of the radio pulsar showed evidence for the existence of an X-ray pulse. The two effects, a peak in the angular structure and the indication of a pulse at the correct frequency, were present in the 0.5–1.5 keV data. However, both effects were absent from the data below 0.3 keV, while the Vela X-ray Nebula was stronger than in the higher energy range. Thus, the apparent point source at PSR0833-45 has a harder spectrum than the nebula. The X-ray spectrum of the Vela pulsar could be bending over at 0.5 keV just as the Crab pulsar's spectrum bends at 1 keV. In observations of the Uhuru satellite of the Vela region, X-rays above 2 keV are seen localized about the direction of PSR0833-45 while the Vela X-ray Nebula is probably not seen (Kellogg *et al.*, 1973) supporting our conclusion that the pulsar is point source of X-rays having a harder spectrum than the nebula. At this time one would say that the evidence in favor of PSR0833-45 being an X-ray pulsar is strong but that we lack final proof needed to make the case complete.

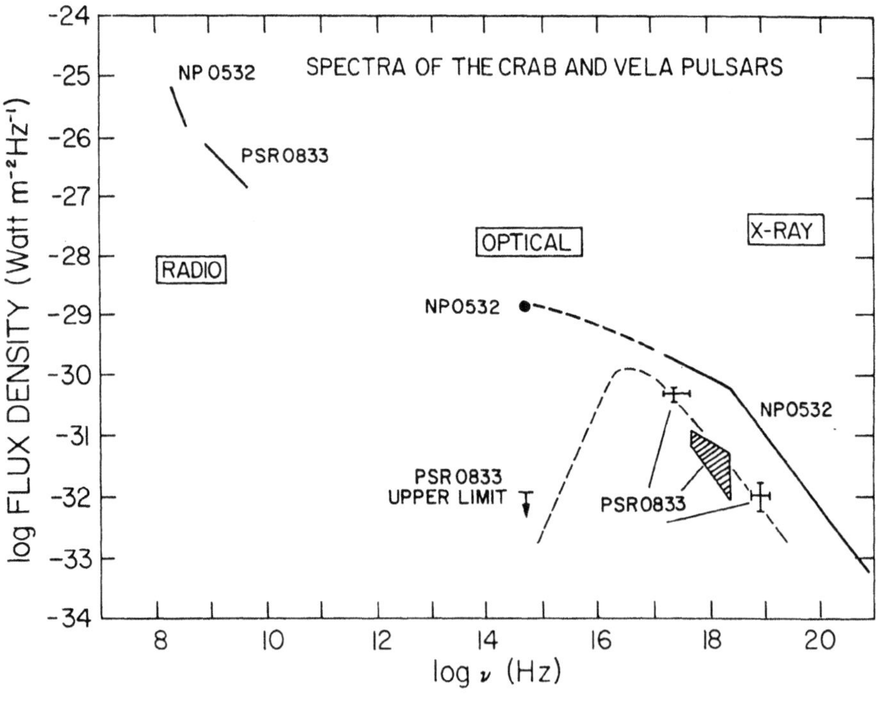

Fig. 7. Data for PSR 0833-45 from the following sources: Optical, Kristian (1970); Low energy X-ray, Harnden and Gorenstein (1973); Medium energy, Kellogg *et al.* (1973); and High Energy X-ray, Harnden *et al.* (1972).

Figure 7 shows the comparative spectra of the Crab and Vela pulsars over a broad region of the electromagnetic spectrum.

5. Other X-Ray Emitting Supernova Remnants

There are a number of other SNR's that are not as well identified as X-ray emitters as those I have already discussed. There is a positional agreement between an X-ray source and the radio object but other observations are needed to make the association convincing. Objects in this category are given brief individual descriptions.

5.1. The Lupus Loop

The Lawrence Livermore group finds excessive soft X-ray emission in a broad region $\sim 5°$ that includes the radio distribution of the Lupus Loop and possibly extends beyond it (Palmieri *et al.*, 1972).

5.2. SN1006

A large region of soft X-ray emission extending beyond the broad radio features of the Lupus Loop also encompasses the more compact radio source SN1006 (Palmieri *et al.*, 1972). There is evidence for excess emission at the position of SN1006, but instruments with better angular resolution are needed to clarify the question of the existence of a discrete X-ray source.

5.3. MSH15-52A

A weak X-ray source has been found by Uhuru within the region of the radio source. The X-ray source does not appear to be variable. The center of the radio source is not coincident with the position of the X-ray source. Since there are examples of SNR's where the X-ray structure differs appreciably from the radio, this is an allowable possibility.

5.4. MSH1811-77

The position of the strong X-ray source GX13+1 falls within the region of the radio source MSH1811-77, which is listed as a supernova remnant in the catalog of Downes. The X-ray source does not seem to be variable. The X-ray source is clearly smaller in size than the radio nebula. If the association is correct the comparatively large distance to the object implies that the intrinsic X-ray luminosity in the 2–10 keV band is very large, much larger than any of the other SNR's. In the case of this object and the one which precedes it, there is a possibility that the X-ray source is indeed a supernova remnant, but not the one with which it is identified. The X-ray source appears to be non-variable and the region of the sky where it is found contains a large density of radio sources possibly including one that has not yet been recognized as an SNR.

5.5. The North Polar Spur

The North Polar Spur is a radio feature of very large angular dimensions. Hanbury-

Brown *et al.* (1960), suggested that it is the remnant of a fairly nearby supernova. However, radio astronomers are by no means unanimous in this opinion. The Wisconsin group has reported the observation of soft X-ray emission from this large radio feature above the soft X-ray background (Bunner *et al.*, 1972). It could be said that the detection of the X-ray emission adds a certain amount of credibility to the idea that the NPS is a supernova remnant because X-ray emission is often a feature of SNR's. However, the NPS is so large that it is difficult to account for its X-rays by the same mechanisms that we invoke for the other remnants; thus, if the NPS is an X-ray emitting SNR, it represents the first member of a new class of SNR's.

6. Blast Wave Interpretation of X-Ray Emission from Supernova

SUPERNOVA REMNANTS

No one seriously doubts that synchrotron radiation is the mechanism accounting for the production of X-rays from the Crab Nebula. The data indicating a power law spectrum and the evidence foı polarization are by themselves convincing. Synchrotron radiation is also consistent with the observed loss of rotational kinetic energy from the rapidly rotating neutron star, NP0531. For the other SNR's we are not so confident that synchrotron radiation is responsible for the X-ray production. In fact, there are four cases, as cited above, where the spectral data is more suggested of thermal radiation from a hot plasma; the Cygnus Loop, Pup A, Cas A and Tycho. Hence, there is a possibility that the X-ray production of all SNR's except the Crab Nebula is thermal, arising from the propagation of a blast wave into the interstellar medium. Tucker (1971) considered the Cygnus Loop as an example of this process. The physical model he derived for the Cygnus Loop appears to be quite plausible except for one difficulty, the high velocities needed to explain a shock wave with a temperature of a few million degrees is not seen in the optical filaments. However, since the spectral data does favor thermal emission it is worthwhile to see how Tucker's model applies to the other remnants. For five SNR's there exists data on three observables; luminosity, temperature, and X-ray diameter (except Tycho where the X-ray diameter is assumed to be equal to the radio diameter). Tucker, in a recent private communication,

TABLE III

Supernova shock wave model parameters

Object	T_e(K)	D_{pc}	n_1(cm^{-3})n	$M_{sn}/(M_o)$	W_o(erg)	t(yr)	V_s(km s^{-1})	T_i
Cas A	1.5×10^7	5.5	3	6	4×10^{51}	200	6000	2×10^8
Tycho	2.9×10^7	5.8	1.6	4	4×10^{50}	400	3000	7×10^8
Pup A	8×10^6	16	0.6	30	3×10^{50}	4000	800	
Cyg L	3×10^6	38	0.15	100	4×10^{50}	15000	460	
Vela SNR[a]	3×10^6	40	0.07	50	2×10^{50}	20000	500	
	6×10^6	40	0.03	25	2×10^{50}	13000	600	

Note: M_{sn} = swept up mass $\simeq (4\pi/3) r^3 n_1 M_p$

[a] Two temperatures are used in the calculation to allow for uncertainties in the observations.

to me, has calculated several other quantities from the three observable within the context of the blast wave model; these are: n_1, the initial density in the vicinity of the supernova; M_{sn}, the swept up mass; W_0, the energy released; and where independent observations do not exist, the age and expansion velocity. Table III summarizes the preliminary results of Tucker's calculations. It seems that the calculated quantities are in reasonable agreement with independent observation or deductions. For example, the calculated initial density is high in the Cas A and low in the case of the Vela SNR in agreement with judgements based on totally different ideas. The values for the age and energy release that come out of the model are also quite acceptable. According to the model, the ejecta of Tycho and Cas A are too young to have come to thermal

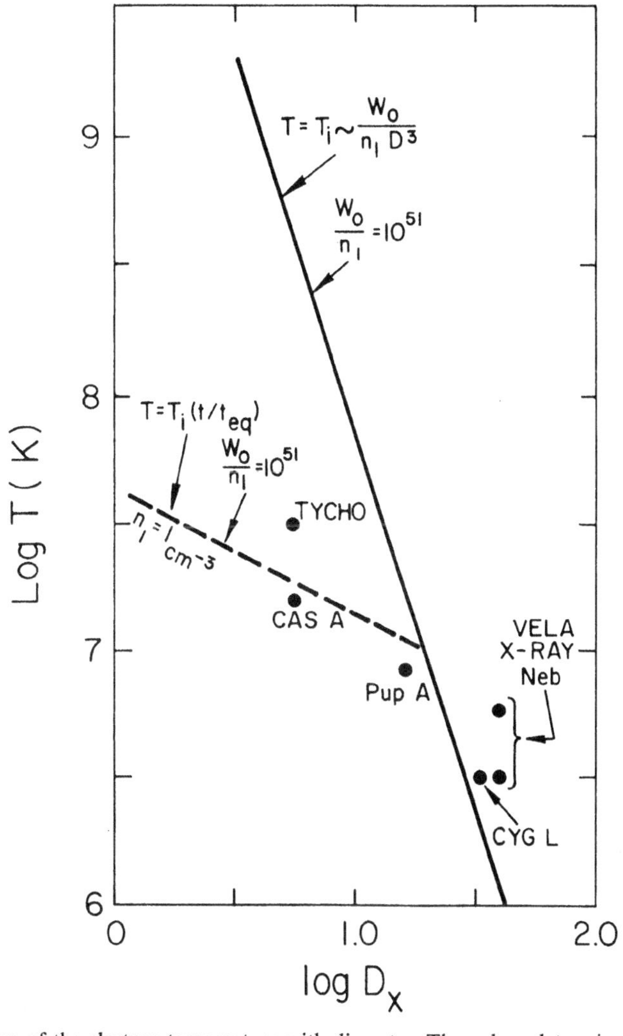

Fig. 8. Variation of the electron temperature with diameter. The values determined from observations are shown, together with a theoretical curve for the case where $W_{51}/n_1 = 1$. The dashed curve shows the electron temperature if the electrons are not as hot as the ions, and are heating up by means of Coulomb collisions with the ions.

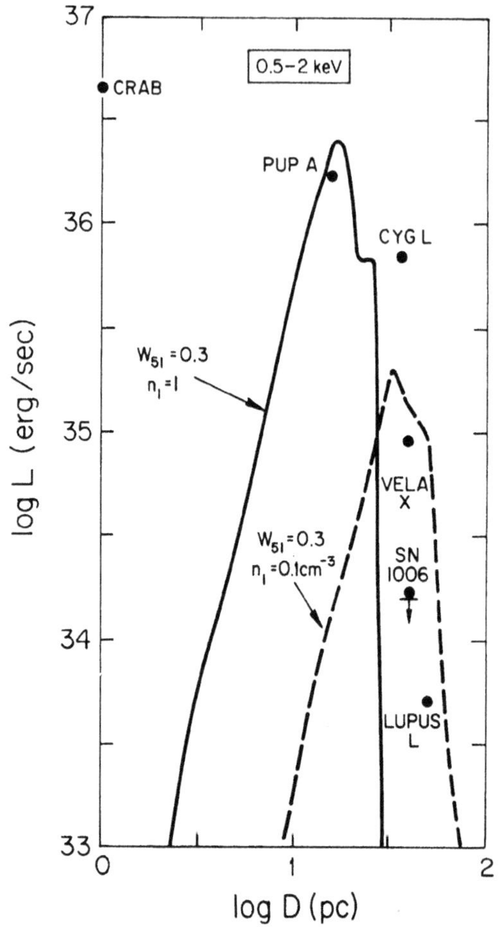

Fig. 9. Variation of the 0.5–2 keV luminosity with diameter. The observational points are shown, together with two theoretical curves.

equilibrium; the ion temperature may differ appreciably from the electron temperature. Thus, it would not be surprising if their thermal spectra contained lines that were either too strong or too weak compared to what one might expect from the shape of the continuum. Figure 8 shows the theoretical variation of temperature with intrinsic diameter compared to the observations. In Figure 9 and 10 the dependence of the 0.5–2 keV and 2–10 keV luminosity with the diameter is shown together with the data points.

In conclusion recent X-ray data from SNR's gives additional support to the idea that X-rays are produced by the propagation of a blast wave into the interstellar medium. However, the Crab Nebula is a notable exception because synchrotron radiation is the X-ray production mechanism. It would not be surprising if marked differences exist between the radio and X-ray structure because the radio is produced by residual low energy electrons that do not enter into the X-ray production. Thermal

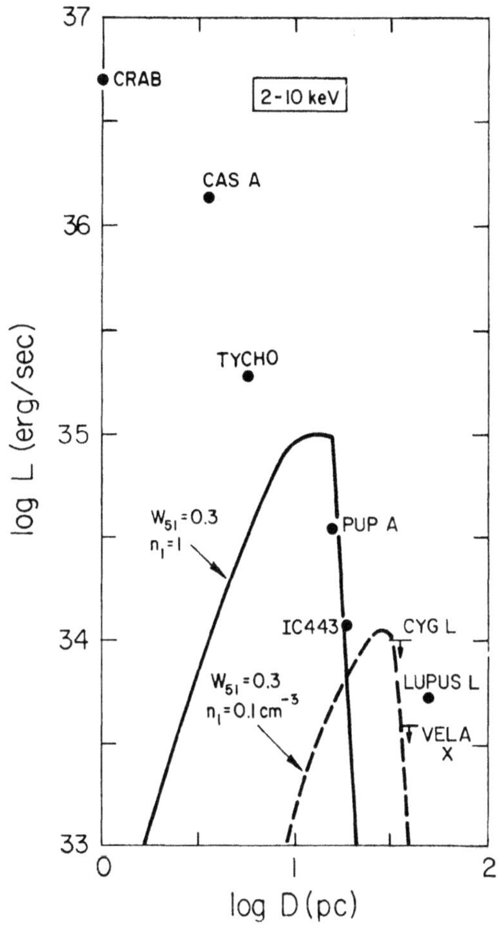

Fig. 10. Same as Figure 2, for the energy band 2–10 keV.

radiation results in line emission which offers us great possibilities of studying the physical processes in occurring SNR's in great detail in future experiments.

Acknowledgements

I would like to thank W. Tucker and F. R. Harnden, Jr. for their help in preparing this manuscript.

References

Alexander, J., Brandt, J., Maran, S., and Stecher, T.: 1971, *Astrophys. J.* **167**, 487.

Bleeker, J., Deerenberg, A., Yamashita, K., Hayakawa, S., and Tanaka, Y.: 1972, *Astrophys. J.* **178**, 377.

Boldt, E., Desai, U., Holt, S., Serlemitsos, P., and Silverberg, R.: 1969, *Nature* **223**, 280.

Borken, R., Doxsey, R., and Rappaport, S.: 1972, *Astrophys. J.* **178**, L115.

Bowyer, S., Byram, E., Chubb, T., and Friedman, H.: 1964, *Science* **146**, 912.

Bradt, H., Rappaport, S., Mayer, W., Nather, R. E., Warner, B., MacFarlane, M., and Kristian, J.: 1969, *Nature* **222**, 728.

Bunner, A., Coleman, P. L., Kraushaar, W. L., McCammon, D., and Williamson, F. O.: 1973, *Astrophys. J.* **179**, 781.
Burginyon, G., Hill, R., Seward, F., Tarter, C., and Toor, A.: 1973, Submitted to *Astrophys. J. Letters*.
Downes, D.: 1971, *Astrophys. J.* **76**, 305.
Ducros, G., Ducros, R., Rocchia, R, and Tarrius, A.: 1970, *Nature* **227**, 152.
Fazio, G., Helmken, H., O'Mongain, E., and Weekes, T.: 1972, *Astrophys. J.* **175**, L117.
Friedman, H., Byram, E. T., and Chubb, T. A.: 1967, *Science* **156**, 374.
Fritz, G., Henry, R., Meekins, J., Chubb, T. A., and Friedman, H.: 1969, *Science* **164**, 709.
Fritz, G., Henry, R., Meekins, J., Chubb, T. A., and Friedman, H.: 1971, *Astrophys. J.* **164**, L55.
Giacconi, R., Murray, S., Gursky, H., Kellogg, E., Schreier, E., and Tananbaum, H.: 1972, *Astrophys. J.* **178**, 281.
Gorenstein, P., Kellogg, E. N., and Gursky, H.: 1970a, *Astrophys. J.* **160**, 199.
Gorenstein, P., Gursky, H., Kellogg, E. M., and Giacconi, R.: 1970b, *Astrophys. J.* **160**, 947.
Gorenstein, P., Harris, B., Gursky, H., Giacconi, R., Novick, R., and VandenBout, P.: 1971, *Science* **172**, 369.
Grader, R., Hill, R., and Stoering, J.: 1970b, *Astrophys. J.* **161**, L45.
Hanbury-Brown, R., Davies, R. D., and Hazard, C.: 1960, *Observatory* **80**, 191.
Harnden, F., Johnson, W., and Haymes, R.: 1972, *Astrophys. J.* **172**, L91.
Harnden, F. and Gorenstein, P.: 1973, *Nature* **241**.
Holt, S., Boldt, E., Serlemitsos, P., and Brisken, A. F.: 1973, *Astrophys. J.* **180**, in press.
Kellogg, E. M., Harnden, F., Tananbaum, H., Gursky, N., Murray, S., Giacconi, R., and Grindlay, J.: 1973, *Astrophys. J.*, in press.
Kurfess, J.: 1971, *Astrophys. J.* **168**, L39.
Kristian, J.: 1970, *Astrophys. J.* **162**, L103.
McCammon, D., Bunner, A. N., Coleman, P. L., Kraushaar, W. L., and Williamson, F.: 1972, Presented at *AAS Meeting High Energy Astrophysics Division*, Pasadena, October 23–26.
Milne, D. K.: 1970, *Aust. J. Phys.* **23**, 425.
Novick, R., Weisskopf, M., Berthelsdorf, R., Kinke, R., and Wolff, R.: 1972, *Astrophys. J.* **174**, L1.
Oda, M., Bradt, H., Garmire, G., Spada, G., Sreekantan, B. V., Gursky, H., Giacconi, R., Gorenstein, P., and Waters, J. R.: 1967, *Astrophys. J.* **148**, L5.
Palmieri, T., Burginyon, G., Grader, R., Hill, R., Seward, F., and Stoering, J.: 1971, *Astrophys. J.* **164**, 61.
Palmieri, T., Burginyon, G., Hill, R., Scudder, J., and Seward, F., 1972, *Astrophys. J.* **177**, 387.
Serlemitsos, P. J., Boldt, E. A., Holt, S. S., Ramaty, R., and Brisken, A. F.: 1973, *Astrophys. J.* **184**, L1.
Seward, F., Burginyon, G., Grader, R., Hill, R., Palmieri, T., and Stoering, J., 1971, *Astrophys. J.* **169**, 515.
Stevens, J. C. and Garmire, G. P.: 1973, *Astrophys. J.* **180**, L19.
Tucker, W.: 1971, *Science* **172**, 372.

X-RAY OBSERVATIONS OF SUPERNOVA REMNANTS BY COPERNICUS

D. J. ADAMS
Department of Astronomy, University of Leicester, England

and

J. C. ZARNECKI
Mullard Space Science Laboratory, University College London, England

1. The Instrument

The observations reported below were obtained using two focussing X-ray systems carried on board OAO-C (Copernicus). Each system comprises a grazing incidence paraboloidal reflector to focus soft X-radiation onto a miniature proportional counter detector. The circular field of view of each telescope is approximately 10′ in diameter, but may be reduced by the use of smaller apertures mounted on a remotely controlled wheel in front of the detectors. After including the efficiencies in the reflecting optics and in the detector response, the geometrical aperture of 12 cm^2 is reduced to an effective aperture of approximately 1 cm^2 for each telescope. One system responds to photons in the energy range 0.7–1.5 keV, the other to 1.5–4.2 keV photons. A third non reflecting system (not subsequently referred to) is collimated to 2 deg by 5 deg FWHM and responds to 4–12 keV photons. The X-ray telescopes are aligned to the spacecraft axis to an accuracy of about 1′, and spacecraft drift errors are smaller than this over periods of several hours. Thus exposures of 20000 cm^2 s are possible on accurately selected regions of sky. The equipment was built at Mullard Space Science Laboratory.

The spacecraft was launched in September 1972, and the X-ray observers have priority for 10% of the spacecraft time. The observations reported here have been made jointly by the Leicester and MSSL groups, with assistance from U.K. and U.S.A. operations personnel. Work has also been done using Copernicus to observe X-ray stars and extragalactic objects.

2. Puppis A

This supernova remnant shows up as a strong radio source of roughly 1 deg in diameter (Milne, 1971). Faint optical wisps are seen, and the remnant is believed to lie at a distance of 1.2–2.2 kpc (Woltjer, 1972). The remnant coincides in position with an extended region of X-ray emission at energies below 2 keV (Palmieri *et al.*, 1971) and is regarded as being associated with the X-ray source.

The circles in Figure 1 show the elements into which the nebula was dissected by the Copernicus observations using the 10′ field of view (Zarnecki *et al.*, 1973). Each pointing direction was maintained for approximately 30 min, and the numbers in the

C. B. Cosmovici (ed.), Supernovae and Supernova Remnants, 243–250.

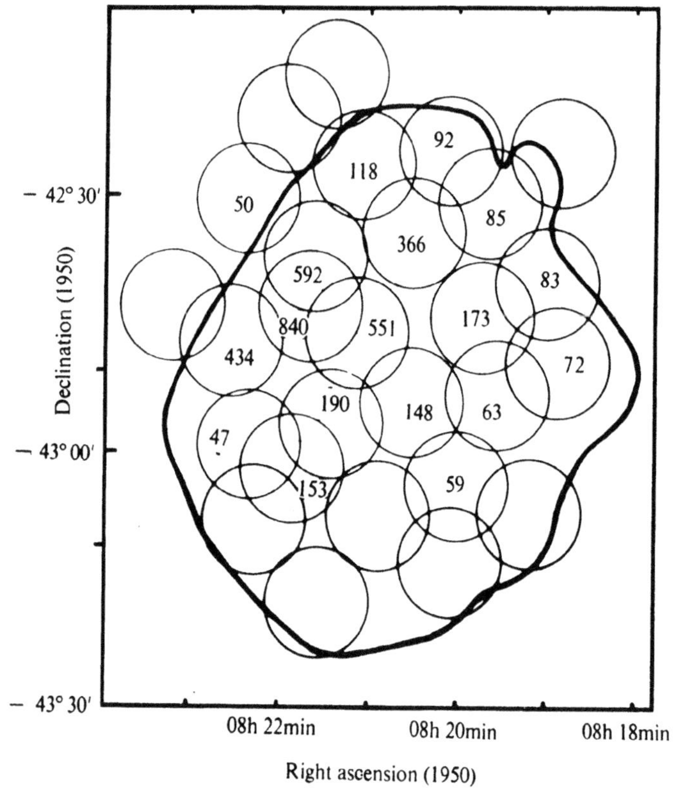

Fig. 1. 0.7–1.5 keV X-ray observations of Puppis A, superimposed upon the outline of the radio map due to Milne (1971).

circles represent the corresponding number of incident photons in the 0.7–1.5 keV system. The cosmic ray background counting rate of approximately 90 per 30 min has been subtracted from the raw data to obtain these signals. Fluctuations in the background rate are such that the 3 sigma level for detection of an X-ray source amounts to 40 counts per 30 min. Signals below this level are not regarded as significant, and when this occurs the circles in Figure 1 have been left blank. In Figure 2 the same X-ray data is presented. In this case a computer program has been used to fit contours around the observed data points. There has been no attempt at deconvolving the data to remove the effect of the instrumental beamwidth. It is apparent from Figure 2:

(a) That there is a distinct peak in the emission of the nebula at RA = 125.4° Dec. = −42.3°.

(b) That this peak of emission lies on the Eastern side of the nebula, but that it fails to coincide with the peak in the radio contours (which are shown in various shades of grey).

(c) That the extent of the X-ray emitting region is smaller than that of the radio

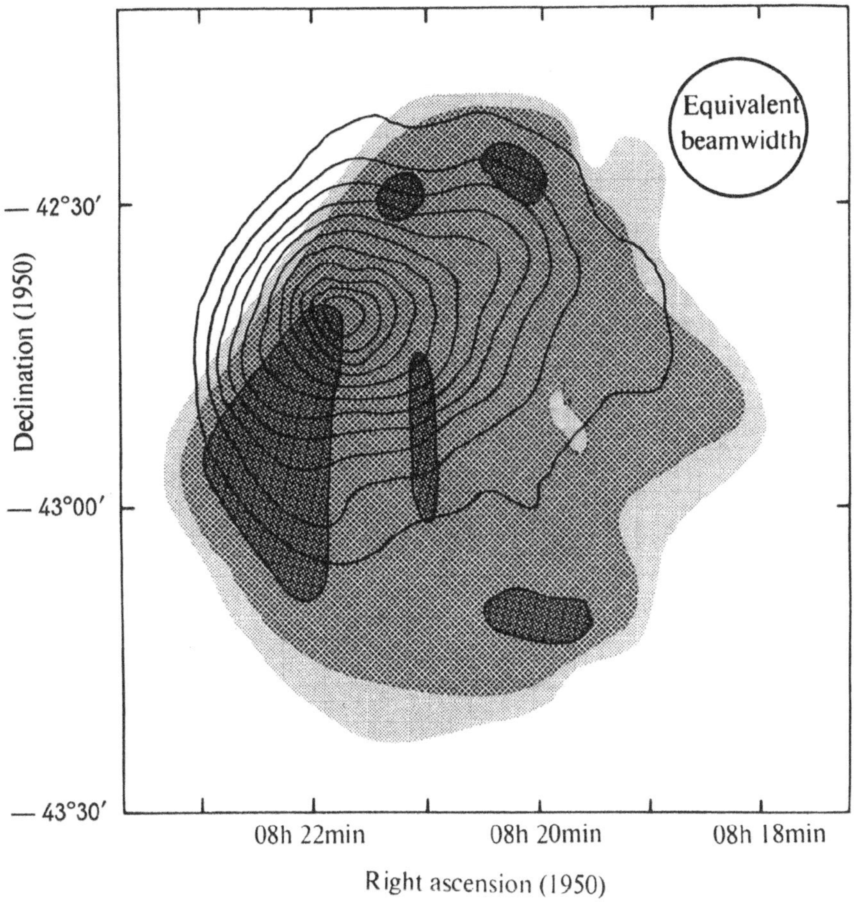

Fig. 2. Puppis A X-ray contour map, derived from the data of Figure 1. The shading represents selected contours from Milne (1971).

emitting region. Because the beamwidth has not been taken into account in constructing the X-ray contours, the size of the region is likely to be exaggerated in Figure 2.

A second set of observations was undertaken by Copernicus to study in more detail the peak of the X-ray emission. Figure 3 shows the results of observing this region using the 6′ field of view. Each observation lasted approximately 1 h. Caution should be exercised in comparing the counts with those of Figure 1. Not only is the acceptance solid angle of the telescope reduced, but scattering effects in the reflector lead to a loss in effective area of the system. From this set of observations it is clear that the central 6′ field has a mean surface brightness 2.5 times greater than that of the region surrounding it. An attempt to unfold the data using a minimum chi-squared technique has shown that the X-ray emission is not consistent with a model comprising a point source superimposed upon a uniform disc. The assymmetry in the contours is real.

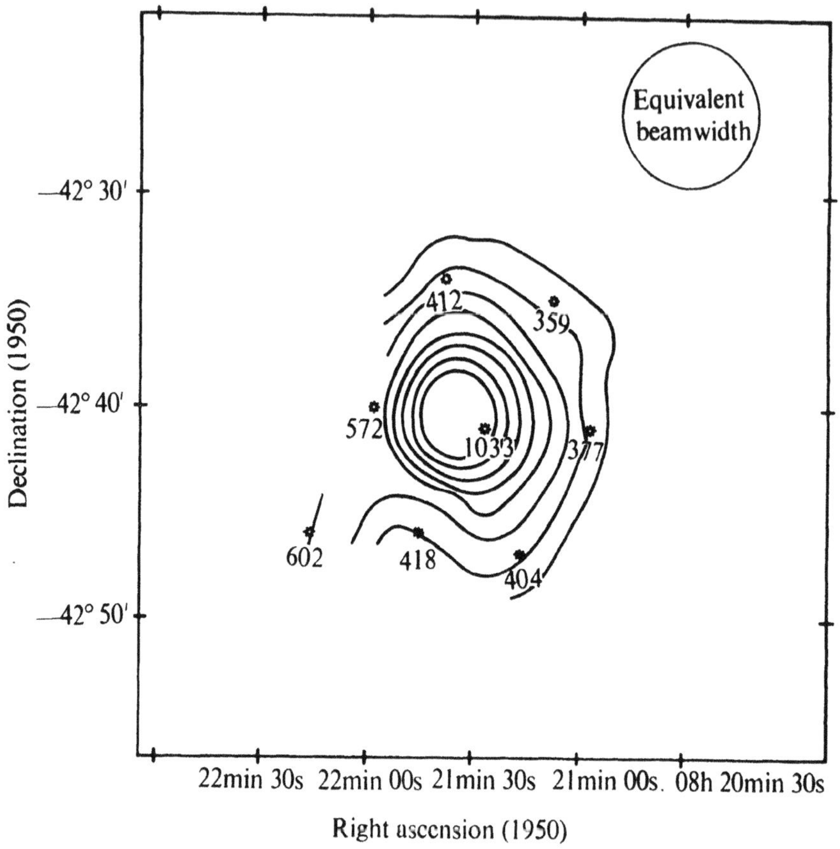

Fig. 3. More detailed observations of the bright X-ray region of Puppis A. The observed data points are shown as stars, and the contours have been fitted to these data points.

3. IC 443

Observations were attempted on this supernova remnant which has an angular diameter similar to that of Puppis A (Woltjer, 1972). A set of pointing directions similar to those of Figure 1 were selected. Using integration periods of 30 min, no significant signals were observed. Two adjacent bins in the part of the nebula which is brightest in the visible region of the spectrum showed up at the 2 sigma level, so that longer observations are needed before any claims can be made.

IC 443 differs from Puppis A in that it is a much more conspicuous object in optical emission. Its distance, however, is uncertain. If it should be 3 kpc or further away, then it is likely that X-ray emission with a spectrum similar to that of Pup A would be obscured by absorption in interstellar matter. The Uhuru identification of IC 443 as an X-ray source is based upon an error box of 7.5 deg^2 in area (Giacconi *et al.*, 1972) and requires confirmation.

4. Cassiopeia A

This supernova remnant is of a much younger class than either Puppis A or IC 443. Studies of the expansion velocities of the optical filaments put the age at about 300 yr (Van den Bergh and Dodd, 1970). The remnant lies at a distance of about 2.8 kpc and shows up as a radio source approximating to a shell structure (Woltjer, 1972). It was identified as an X-ray source by Gorenstein *et al.* (1970) and was shown to have a harder spectrum than Pup A, i.e. to be relatively brighter at the higher X-ray energies. The object was detected both above and below 1.5 keV. Only the higher energy data is reported because the field of view was more favourable.

The X-ray observations (Fabian *et al.*, 1973) are shown in Figure 4. As the radio diameter of the remnant is only some 5′, a small (3′) aperture stop was chosen for the observations. There was an appreciable overlap between the various 30 min. observations. A 3 sigma level of detection corresponded to about 40 counts per 30 min. Inspection of Figure 4 reveals that a single point source is inconsistent with the data, because the effect of such a point source could only be seen in adjacent observing bins. Attempts were made to deconvolve the data by calculating the response of the

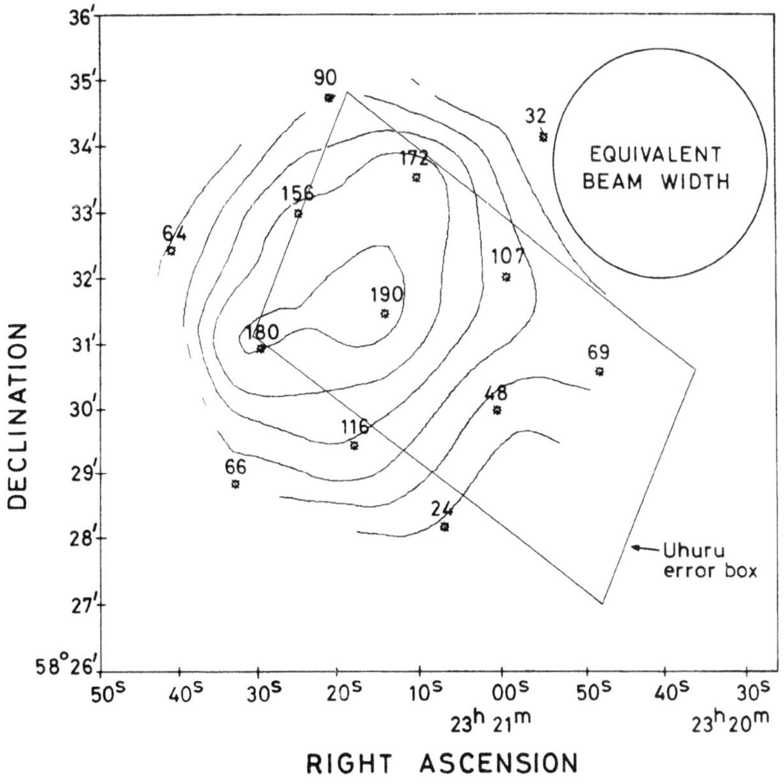

Fig. 4. 1.5–4.2 keV observations of Cas A. The observed data points are shown as stars, and contours have been fitted directly.

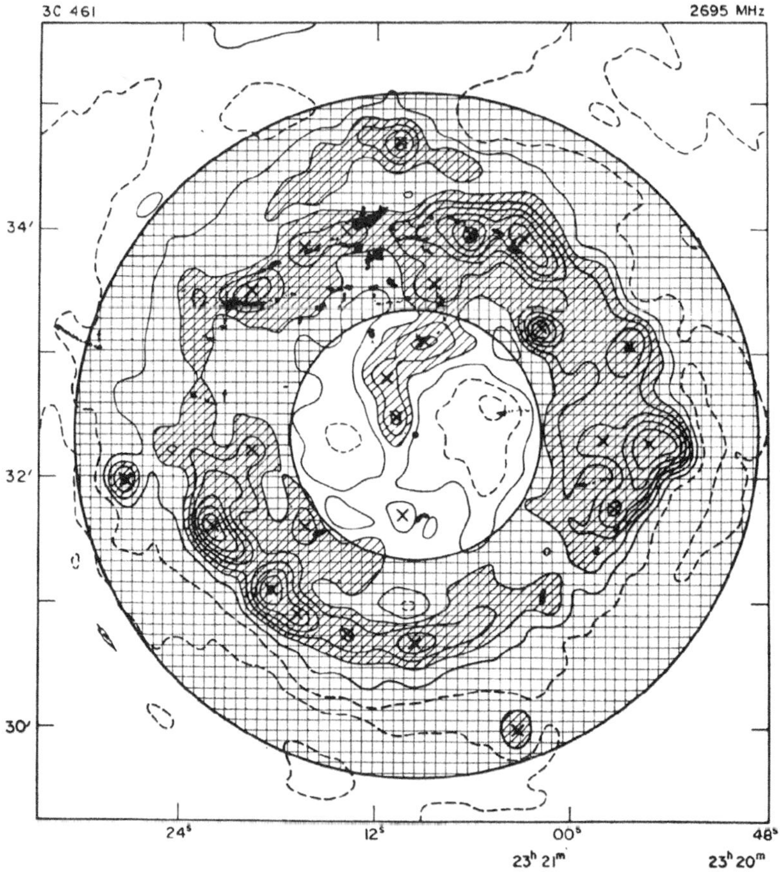

Fig. 5. The best fitting X-ray annulus to Cas A, shown superimposed upon the radio map of Rosenberg (1970).

instrument to various models with circular symmetry, and by comparing the result with the observations using a chi-squared technique. The best fits were obtained with either a uniform disc of 5.5′ diameter, or with an annulus of 5.5′ outside diameter and 1.5′ inside diameter. The chi-square value did not increase significantly for changes of $\pm 1'$ about these values. The high minimum values for chi-square suggest that circular symmetry is an unsatisfactory approximation. The conclusions are as follows:

(a) The X-ray source does indeed coincide with Cas A.

(b) The X-ray source is extended by $5.5 \pm 1'$.

(c) There are indications of complex structure.

Figure 5 shows one of the best fitting models superimposed upon a radio map of the shell source.

5. Cygnus Loop

This old supernova remnant has an X-ray diameter of about 3 deg (Gorenstein *et al.*, 1971) and is therefore difficult to scan with the 10′ field of view of Copernicus. Garmire and Stevens (1973) have shown that two 0.5 by 0.5 deg regions of the nebula are

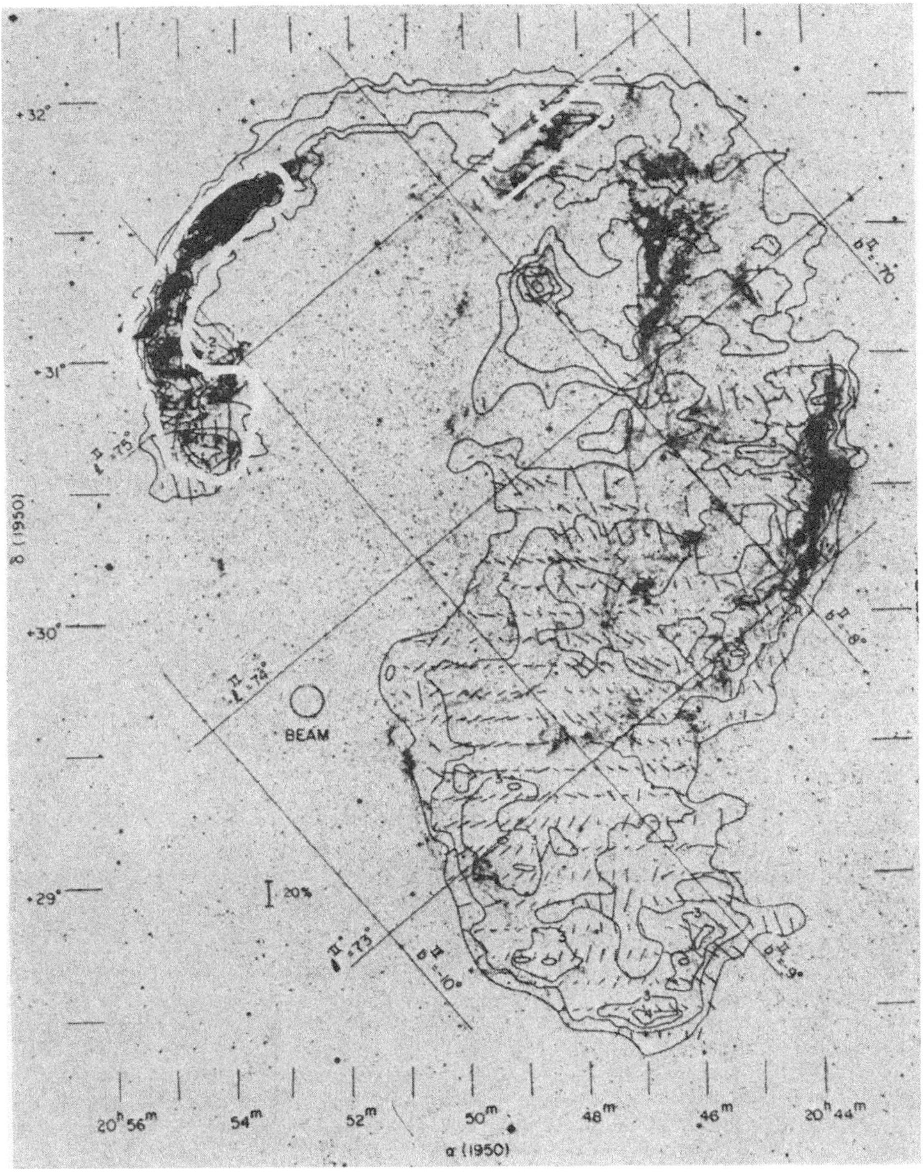

Fig. 6. The regions of the Cygnus Loop observed by Copernicus are shown. The radio map of the supernova remnant is due to Kundu and Becker (1972).

brighter than the mean. A search of these regions was performed with Copernicus using the 10′ field of view in conjunction with 30 min integration periods. The regions scanned are shown in Figure 6. Although weak signals were detected, it is only possible to place upper limits upon the brightness of any fine structure which may be present.

6. Conclusion

Of the supernova remnants observed at high spatial resolution in this work, few similarities in the X-ray structure emerge. It seems that in general the X-ray diameter approximates to that in the radio region, but that a detailed correspondence is not found between X-ray and radio structure. Theories are required which can explain the displacement of the peak in X-ray emission from the radio peak in Pup A, as well as the general diversity in X-ray structure.

References

Fabian, A. C., Zarnecki, J. C., and Culhane, J. L.: 1973, *Nature Phys. Sci.* **242**, 18.

Garmire, G. P. and Stevens, J. C.: 1973, *Astrophys. J.* **180**, L19.

Giacconi, R., Gursky, H., Kellogg, E. M., Murray, S., Schreier, E., and Tananbaum, H.: 1972, *Astrophys. J.* **178**, 281.

Gorenstein, P., Gursky, H., Kellogg, E. M., and Giacconi, R.: 1970, *Astrophys. J.* **160**, 947.

Gorenstein, P., Harris, B., Gursky, H., Giacconi, R., Novick, R., and Vanden Bout P.: 1971, *Science* **172**, 369.

Kundu, M. R. and Becker, R. H.: 1972, *Astron. J.* **77**, 459.

Milne, D. K.: 1971, *Aust. J. Phys.* **24**, 429.

Palmieri, T. M., Burginyon, G., Grader, R. J., Hill, R. W., Seward, F. D., and Stoering, J. P.: 1971, *Astrophys. J.* **164**, 61.

Rosenberg, I.: 1970, *Monthly Notices Roy. Astron. Soc.* **147**, 215.

Van den Bergh, S. and Dodd, W. W.: 1970, *Astrophys. J.* **162**, 485.

Woltjer, L.: 1972, *Ann. Rev. Astron. Astrophys.* **10**, 129.

Zarnecki, J. C., Culhane, J. L., Fabian, A. C., Rapley, C. G., Silk, R., Parkinson, J. H., and Pounds, K. A.: 1973, *Nature Phys. Sci.* **243**, 4.

DISCUSSION

J. R. Dickel: Do Gorenstein and Adams agree on the extent of the Puppis X-ray source? I had the impression that Gorenstein went outside the radio outline but Adams didn't.

P. Gorenstein: The two results are consistent. The Copernicus results had a greater resolution and a better positional accuracy.

R. M. Duin: Is it correct that the peak in the X-ray emission does not correspond to the center of your annulus for Cas A?

D. J. Adams: The center of the annulus is in between the two counting rate peaks in the X-ray picture. But the position of the centre cannot reliably be quoted to better than 1.5′.

S. F. Gull: You showed a value of the shock velocity of Cas A of 2000 km s^{-1}. This seems inconsistent with the observation of filaments behind the shock moving with velocities of 6500 km s^{-1} since these filaments must be moving slowly than the shock. Also the radio structure remnant is not consistent with it having swept up 6 times its own mass of interstellar medium.

W. M. Goss: Have any of the X-ray astronomers attempted to observe extragalactic SN?

P. Gorenstein: Yes, with negative results. The limits are 10^{-11} ergs cm^{-2} s^{-1}, for the range of about 2 to 10 keV.

PRELIMINARY RESULTS ON THE EVOLUTION OF SUPERNOVA REMNANTS AND THEIR X-RAY SPECTRUM

VICTOR N. MANSFIELD* and EDWIN E. SALPETER
Center for Radiophysics and Space Research, Cornell University, Ithaca, N.Y. 14850, U.S.A.

1. Introduction

Many authors have treated the problem of the evolution of supernovae remnants in an analytical or semianalytical fashion (see the review paper by Woltjer, 1972 and the recent work of Cox, 1972a, b, c). Recently, Rosenberg and Scheuer (1973) have presented a numerical solution to the problem. In general, the broad features of the evolution of supernovae remnants are fairly well understood. In the present paper we will discuss some preliminary results of a detailed numerical study of the evolution of supernovae remnants and their associated thermal X-ray spectrum. A more complete description of our methods and results will be given in a paper that is in preparation for publication.

2. Calculation and Results

In the present calculation spherical symmetry has been assumed and the fluid dynamic approach is used. The effects of a possible pulsar, cosmic rays, and an interstellar magnetic field have been neglected. The ejected shell is assumed to expand into a fully ionized, uniform medium of constant temperature

$$T_0 = 2 \times 10^4\,\mathrm{K}.$$

The radiative cooling is accounted for by using the equilibrium cooling rates calculated by Cox and Tucker (1969) and Cox and Daltabuit (1971). Although the gas behind the shock is not strictly in equilibrium since it is cooling, Kafatos and Morrison (1971) have shown that this is an adequate approximation for our purposes for $T > 3 \times 10^4$ K. For $T < 3 \times 10^4$ K it is necessary to calculate the detailed ionization balance of hydrogen between recombinations and photoionizations. When $10^4\,\mathrm{K} \lesssim T \lesssim 3 \times 10^4$ K hydrogen recombinations provide the dominant cooling. For $T < 10^4$ K the equilibrium rates calculated by Jura and Delgarno (1972) are used. The photoelectric heating and ionization are approximated by assuming that all of the energy radiated from the gas with temperature $T > 4 \times 10^4$ K is radiated at a single frequency found by an appropriate averaging process. The effects of helium are also approximately accounted for.

The equations are solved by a second order difference scheme with a quadratic form of the artificial viscosity given by Richtmyer and Morton (1967). As a check on accuracy the total energy (kinetic, thermal, and the sum of the radiated energy) was

* Present address, Department of Physics and Astronomy, Colgate University, Hamilton, N.Y. 13346, U.S.A.

C. B. Cosmovici (ed.), Supernovae and Supernova Remnants, 251–260.

periodically compared with the value at $t=0$. Energy was conserved to better than 5% while the mass was conserved to better than 0.1%. In order to keep the calculating cost from becoming prohibitively high, the radiative heating and cooling were set to zero for $T<1000$ K. This of course limits the density maximum in the thermally unstable shell and keeps the integration time step from getting too small. It has no effect on the overall dynamics.

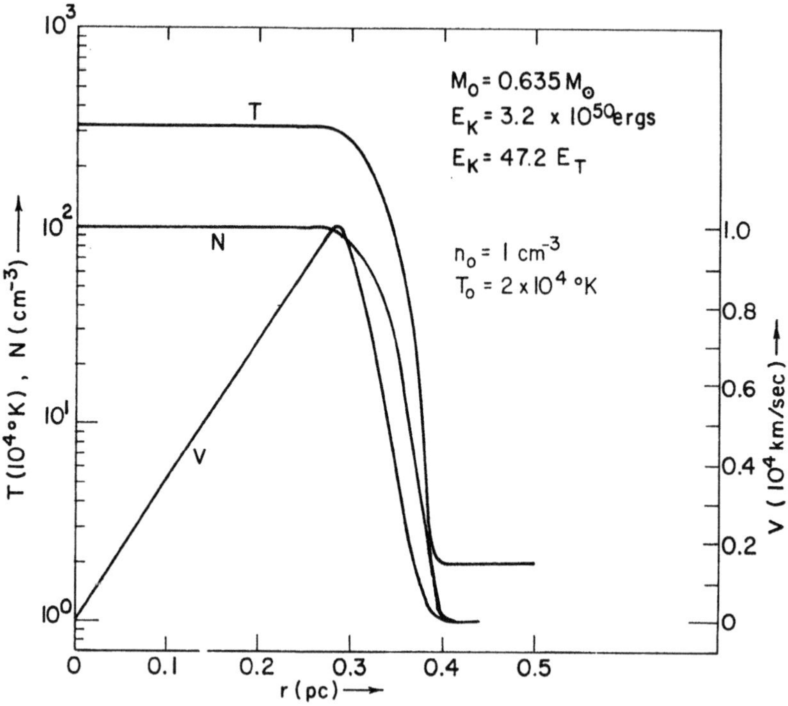

Fig. 1. The initial distributions of pensity, temperature, and velocity as a function of distance.

Figure 1 shows the initial density, temperature, and velocity distributions as a function of distance. The mass of this shell is $M_0=0.635\ M_\odot$. The kinetic energy is $E_K=3.2\times10^{50}$ ergs while the thermal energy of the shell $E_T=E_K/47.2$. The interstellar medium is assumed to have a uniform temperature $T_0=2\times10^4$K and density $n_0=1\ \text{cm}^{-3}$ of cosmic abundance and to have $\text{Ne}/N_\text{H}=1.1$. Our calculations show that for times $t>10^3$ yr the remnant dynamics are independent of the details of the initial conditions and depend only on $E_0=E_K+E_T$ and n_0.

Figure 2 shows the number density as a function of distance at various times. The time in years labels each curve and the thickness of the shell for which $N>$ $>10\ \text{cm}^{-3}$ is also shown. The density profile at 1043 yr shows the adiabatic jump of 4 at the shock front ($R\approx4$ pc) and a steady decrease as one moves inward until the remains of the original shell is encountered at $R\approx2.2$ pc. Interior to the remains of the original shell is a strong density depression due to the 'sweeping out' of material

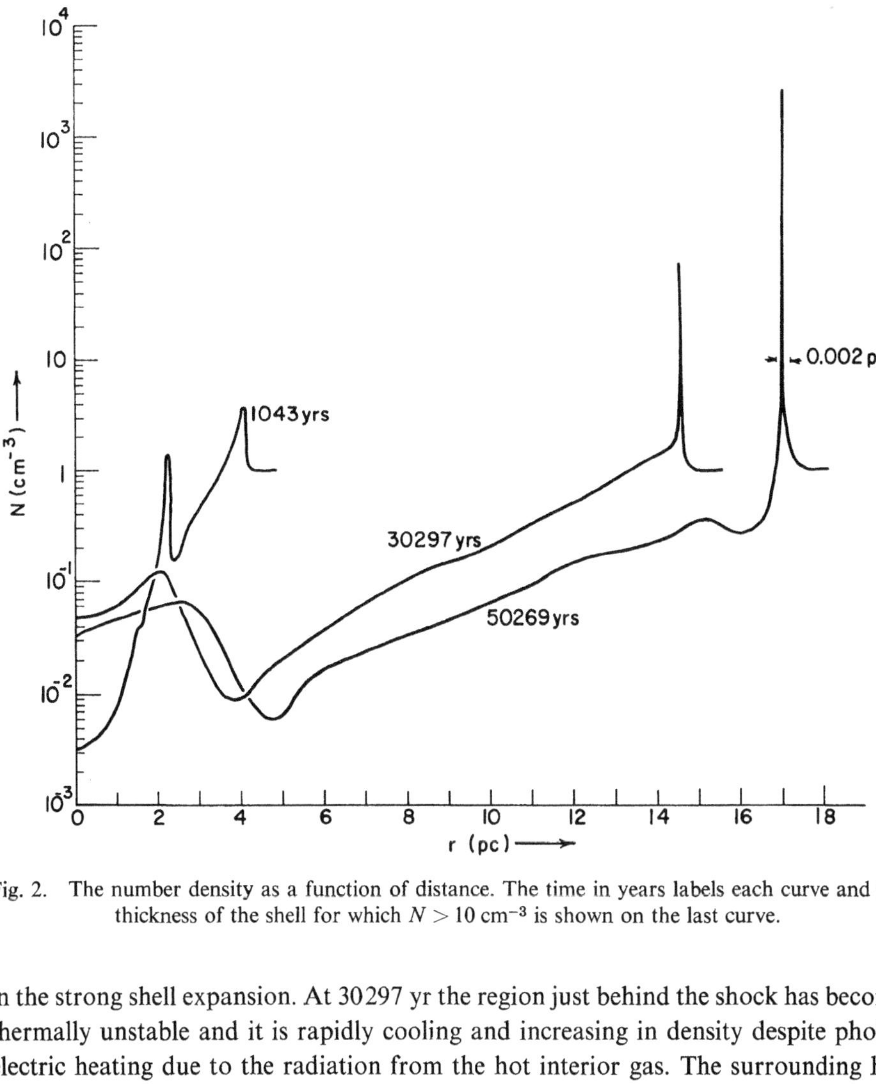

Fig. 2. The number density as a function of distance. The time in years labels each curve and the thickness of the shell for which $N > 10$ cm^{-3} is shown on the last curve.

in the strong shell expansion. At 30297 yr the region just behind the shock has become thermally unstable and it is rapidly cooling and increasing in density despite photoelectric heating due to the radiation from the hot interior gas. The surrounding hot regions push more gas into this rapidly cooling region in an attempt to maintain pressure equilibrium, thereby increasing the density. The ionization fraction in the shell for $N > 10$ cm^{-3} is $x = \mathrm{Ne}/N_H < 10^{-1}$. Henceforth gas with $x < 10^{-1}$ is defined as neutral. At 30297 yr there is 145 $M_\odot$ of neutral gas in the dense shell while the total mass effected by the blast wave at this time is 606 $M_\odot$. By now 21% of E_0 has been radiated away and the optical depth at the Lyman limit due to the neutral shell is 26. The gas interior to the dense shell is seen to be about an order of magnitude lower than $n_0 = 1$ cm^{-3} and the original shell is spread out over several parsecs. At 50269 yr the dense shell has grown in size and density. However, by this time the maximum density of the shell has been limited by setting the heating and cooling equal to zero for $T < 10^3$ K. Now the neutral gas in the shell has a mass of 506 $M_\odot$ while the total mass effected by the blast wave is 904 $M_\odot$. Still the region where $N > 10$ cm^{-3} is

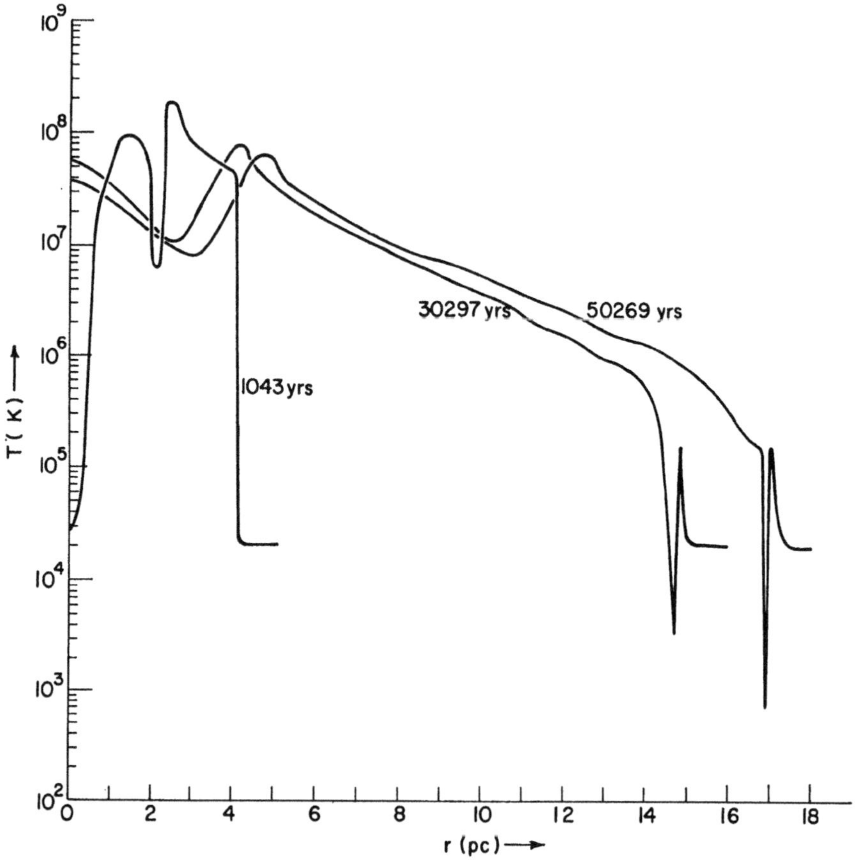

Fig. 3. The temperature as a function of distance. The time in years labels each curve.

only 0.002 pc. The density behind the dense shell continues to drop slowly while the original shell spreads and propagates outward. At this time 45% of E_0 has been radiated away by the remnant and the optical depth at the Lyman limit due to the neutral shell is 61.

Figure 3 shows the temperature distribution as a function of distance for the same times as in Figure 2. At 1043 yr there is a steep rise in the temperature at the shock and it rises steadily as one moves inward until the remains of the original shell is reached. The temperature increase toward the center is a result of the shock moving faster through these regions. The large temperature drop seen in the center is due to the strong adiabatic cooling associated with the rapid expansion of the original shell. The low pressure at the center created by the low temperature and density there is just beginning to cause the original shell to collapse towards the center. This can be seen more clearly in the next figure where the velocity profiles are given. The collapse of the original shell smooths out the temperature and density profiles near the center as can be seen in the next two curves. In these curves we see the sharp rise in temperature at the shock, the strong depression in the dense shell, and the increase in temperature towards the center. At these later times near pressure equilibrium is attained behind the dense shell. However, when the thermal instability begins to form (~28000 yr) the pressure in the dense shell is lower than the gas in front or behind it, while for later times ($t > 40000$ yr) the pressure in the shell can actually be larger than the gas just ahead and behind due to inertial effects. This is a common occurrence in thermal instabilities (see Mansfield, 1973). Steven Gull has pointed out that the effect of this pressure drop in the shell when it first forms is to make it susceptible to Rayleigh-Taylor instabilities (see the discussion following this paper).

Figure 4 shows the velocity profiles as a function of distance at various times. The curve at 1043 yr shows that very near the center there is still some expansion,

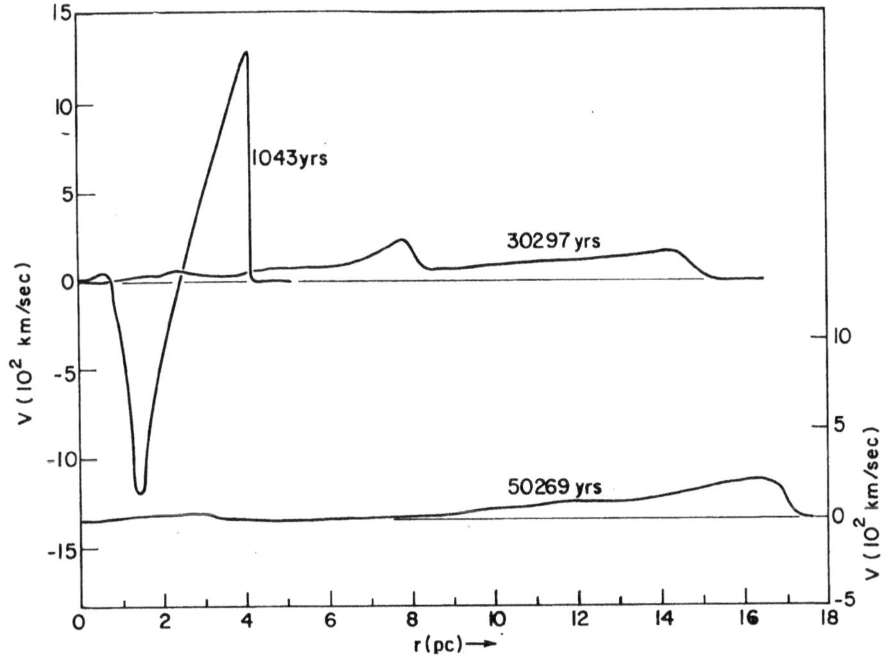

Fig. 4. The velocity as a function of distance at various times. The last curve is displaced downward for clarity.

but the large negative velocities in the range 0.5 pc $\lesssim R \lesssim$ 2.5 pc show that the collapse of the originally ejected shell is well underway. The curve at 30297 yr shows some standing wave motion interior to the dense shell which gives rise to a maximum velocity at $R \sim 8$ pc. This is a common phenomenon throughout the evolution, but it is of little physical significance since there is no appreciable mass involved. These standing waves are created whenever there are significant dynamical changes which occur on a time scale which is smaller than the sound crossing time of the blast wave.

In Figure 5 the density, ionization fraction, and temperature are plotted at 50269 yr against Lagrangian mesh points in order to illustrate more clearly the structure of the dense shell. From this Figure it can be seen that most of the shell is at high

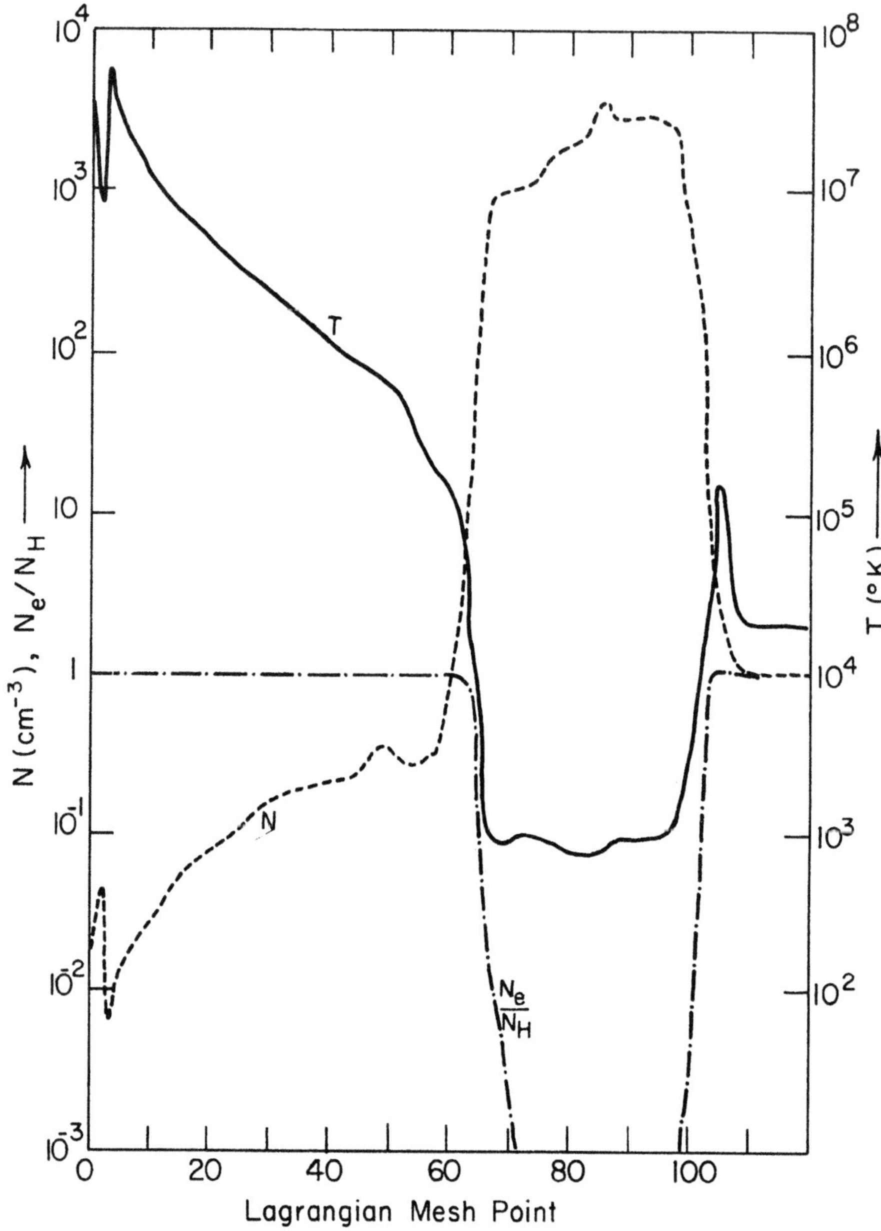

Fig. 5. The number density, temperature, and ionization fraction at 50269 yr as a function of Lagrangian mesh points.

density $N > 10^3$ cm^{-3}, low temperature $T < 10^3$ K, and low ionization fraction $x < 10^{-1}$. (Keep in mind that we have set the heating and cooling equal to zero for $T < 10^3$ K). However, in front of and behind this dense neutral region are two fairly dense, hot, and highly ionized regions ($x \gtrsim 0.8$, $T \gtrsim 10^4$ K, $N \sim 10^2$). These regions will produce optical radiation which will give rise to limb brightening. The outer edge

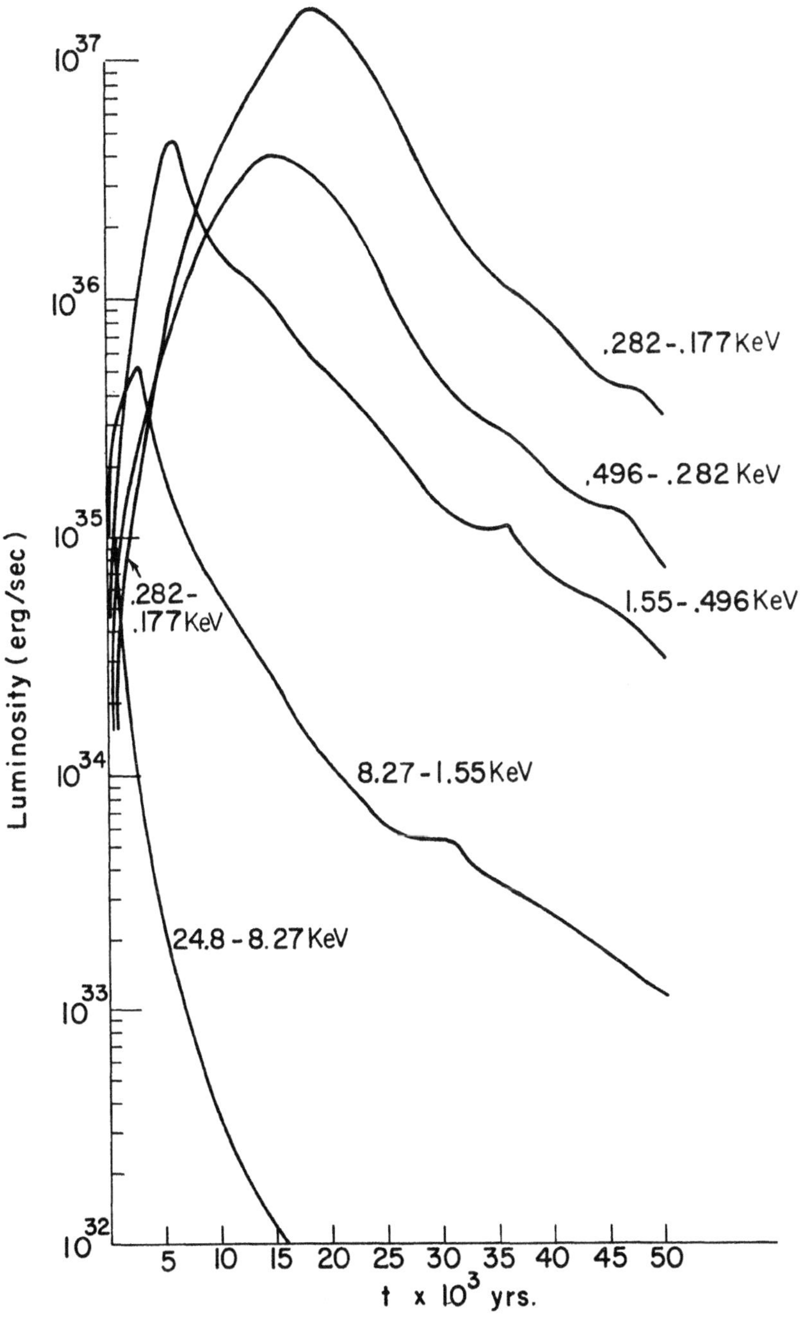

Fig. 6. The X-ray power emitted in various energy ranges which are labelled in keV as a function of time.

of the dense shell is too cool and underionized because the radiative transfer was done from the center outwards thereby neglecting the ionizing radiation from the hot ($\sim 10^5$ K) gas in the vicinity of the shock. Including this effect would give rise to a larger amount of gas in front of the neutral shell capable of radiating in the optical, although it would have no significant effect on the overall dynamics. It is also interesting to notice that there is an appreciable amount of gas with temperatures larger than T_s, the shock temperature. This is because the density does not drop off very drastically toward the center while the temperature rises strongly. This differs from the similarity solution which predicts very little mass with $T > T_s$ because of the very strong density gradients (Cox, 1972c approximates the similarity density solution by $n \propto R^9$).

In order to calculate the X-ray emission from the large mass of hot gas interior to the blast wave the calculation of the radiation from a high temperature, low density gas made by Tucker and Koren (1971) were used. The curves of power per unit volume in various energy ranges as a function of temperature presented in their Figure 2 were analytically approximated to an overall accuracy of better than 10%. Then the power emitted in each energy range from each Lagrangian shell was calculated and then the results were summed according to energy range. The results of this calculation are shown in Figure 6 where the total power emitted in each energy range is given as a function of time. The energy ranges label each curve. The results for the first $\sim 10^3$ yr are somewhat dependent on the details of the initial conditions, but for $t > 10^3$ yr the pattern is clear; the lower energy ranges attain peak power later, and the higher the energy range the sharper the peak and the steeper the decline with time. In general, it is quite clear that supernovae remnants are capable of producing a sufficient amount of thermal X-ray energy to account for many of the observations. This is in agreement with the more approximate calculations made by Tucker and quoted by Gorenstein (see the paper presented at this meeting by Dr Gorenstein).

3. Discussion

In general our detailed numerical calculations are in agreement with the predictions of the analytical approaches of Cox (1972a, b, c), Tucker (1970, 1971), and others (see the review by Woltjer, 1972). However, there are some interesting new features and differences: Cox (1972a) has calculated that there is a significant amount of gas just behind the shock front that will radiate optically. Our calculations show that this same gas ($T \gtrsim 10^4$ K, $N \sim 10^2$) is present not only just behind the shock but also just behind the dense H I shell. In other words, the cool, dense H I is sandwiched between two masses of gas that will radiate optically. Our calculations also show that the density falls off much more slowly behind the front than the variation of approximately $n \propto R^9$ predicted by the similarity solution. This means that there is actually an appreciable amount of gas with $T \gg T_s$, thus giving rise to a harder X-ray spectrum for a given shock velocity than the models of Tucker (1971), based on the similarity solutions, would predict. The present calculations also show that the hot gas created by the SNR is easily capable of producing the total fluxes and general features of the X-ray observations.

Our results are also in general agreement with the numerical model of Rosenberg and Scheuer (1973). The present calculations show more structure because of the greater accuracy of our numerical method. In addition, the more accurate treatment of the radiative cooling and the inclusion of radiative transfer and ionization balance allows us to make a much more detailed calculation of the physics of the dense shell. Perhaps the most striking outcome of this is the large mass of cool, dense, neutral hydrogen predicted by the present model. Cox (1972a) also showed that for older SNR a cool dense shell of neutral hydrogen should be formed just behind the shock front. However, the present calculations show that it should be possible to observe this H I in older SNR directly because of its large mass (hundreds to tens of thousands of solar masses depending on the age and initial parameters), well-defined morphology, and velocity structure. This of course assumes that the interstellar medium in the vicinity of the remnant is uniform enough so that the shell structure is maintained. It is encouraging to see that such an observation has been made quite recently by Assousa and Erkes (1973) for the SNR HB21. They find that their data can be consistently fitted by an expanding shell of neutral hydrogen with an expansion velocity of roughly 25 km s^{-1} and a mass of $\sim 10^3 M_\odot$ to $\sim 10^4 M_\odot$ depending on the uncertain distance to HB21. From this it is clear that HB21 is more highly evolved than the present model at 50000 yr, but nevertheless their general description is in excellent agreement with the predictions of our model.

4. Plans for Future Work

In a more detailed paper that is in preparation for publication we will present a scaling method that will allow the prediction of the behavior of all models with a given value of $E_0 N_0^2$ once one model has been numerically calculated with that value of $E_0 N_0^2$. We will then present numerical calculations for various values of $E_0 N_0^2$. In this way the relevant part of the E_0-N_0 plane for SNR will be covered. We will also show the later evolution of the present model.

Acknowledgements

Part of this work was supported by a grant from the National Science Foundation to Edwin E. Salpeter, and part by the National Astronomy and Ionosphere Center, which is operated by Cornell University under a grant from the National Science Foundation.

References

Assousa, G. E. and Erkes, J. W.: 1973, submitted for publication.
Cox, D. P.: 1972a, *Astrophys. J.* **178**, 143.
Cox, D. P.: 1972b, *Astrophys. J.* **178**, 159.
Cox, D. P.: 1972c, *Astrophys. J.* **178**, 169.

Cox, D. P. and Daltabuilt, E.: 1971, *Astrophys. J.* **167**, 113.
Cox, D. P. and Tucker, W. H.: 1969, *Astrophys. J.* **157**, 1157.
Jura, M. and Dalgarno, H.: 1972, *Astrophys. J.* **174**, 365.
Kafatos, M. C. and Morrison, P.: *Astrophys. J.* **168**, 195.
Mansfield, V. N.: 1973, *Astrophys. J.* **179**, 815.
Richtmyer, R. D. and Morton, K. W.: 1967, *Difference Methods for Initial Value Problems*, 2d ed., Interscience, New York.
Rosenberg, I. and Scheuer, P. A. G.: 1973, *Monthly Notices Roy. Astron. Soc.* **161**, 27.
Tucker, W. H.: 1970, *Astrophys. Space Sci.* **9**, 315.
Tucker, W. H.: 1971, *Astrophys. J.* **167**, L85.
Tucker, W. H. and Koren, M.: 1971, *Astrophys. J.* **168**, 283.
Woltjer, L.: 1972, *Ann. Rev. Astron. Astrophys.* **10**, 129.

DISCUSSION

M. Friedjung: It is curious that some of the gas moves back to the star. Could these be ascertained in some cases? The effects could be rather spectacular.

V. Mansfield: There would only be less than I solar mass accreted, though this might be spectacular. I have also made some calculations which indicate that deuterium formed will stay close to the star because of the piston falling back. At 80000 yr the outer radius is less than 5 pc with an average temperature of $T \sim 10^6$K and $n \sim 0.05$ cm^{-3}.

S. F. Gull: Dr Mansfield showed that there was a region of low pressure behind a radiative shock front. This means that there are regions which have a negative lapse rate and are thus unstable to convection. I have made similar calculations which included a simple model of convection and find that perhaps 1% of the kinetic energy becomes available in the form of convective motions when the thermal instability develops. This leads to the possibility of the non-thermal radio emission due to the turbulent amplification of magnetic field and acceleration of relativistic particles by convective motions (this process may also be occurring in the young shell SN remnants). This may provide an alternative to the Van der Laan model of radio emission from old SN remnants.

V. Mansfield: There is no doubt that the initial stages of the type of thermal instability that I have calculated is also unstable to Rayleigh-Taylor type instabilities as well. The question that needs study is the efficiency of this process.

SEARCH FOR CELESTIAL X-RAY BURSTS

T. L. CLINE and U. D. DESAI
NASA/Goddard Space Flight Center, Greenbelt, Md. 20771, U.S.A.

1. Introduction

Short time base astronomy is presently gaining interest because the transient phenomena involved are the most likely to be associated with high energy astrophysical processes. Just as solar flares and shock waves, rather than the quiet-time solar emissions, reveal the dynamics of high energy solar phenomena, catastrophic celestial events may produce a variety of analogous observable effects and possibly some which are entirely new. It may be that the rapidity of these celestial events, which is central to their production of great energy, will be the key that allows for their ease of observation. Pulsar glitches and X-ray flare stars have indicated a beginning to the variety of such phenomena. Studies of other fast time variations, particularly in the energetic X-ray and gamma-ray domain, as well as in the radio, infrared and other regimes, should provide new insights into such processes as the formation of pulsars, neutron stars and black holes. In particular, supernovae should ultimately be observable at the exact moment of collapse, through their production of gamma rays (Colgate, 1968). The detection of these energetic photons should reveal a considerable amount of new information about the supernova process itself, as well as calibrating the T-zero of the delayed optical emission, in order to better determine the absolute luminosity and distance of the event. Also, this information may assist in the determination of how effective supernovae are as generators of cosmic radiation. Finally, if detection techniques are sufficiently refined, it may be possible to observe the supernova gamma ray bursts from sources at vast distances, thereby providing a new tool for cosmological exploration.

In this note, we describe progress in two experimental programs, first, the search through satellite data for existing evidence for fast, >80 keV photon intensity bursts, and second, the development of very sensitive instrumentation to make possible studies of X-ray and gamma ray bursts in the 0.1 to 15 MeV region with angular resolution, energy resolution, and time resolution over the complete dynamic range of several nanoseconds to several hours.

2. Satellite X-Ray and Gamma-Ray Burst Search

Five satellites have been launched over the past ten years carrying a variety of our electron and solar X-ray detectors. Each utilized a semi or fully omnidirectional, actively shielded CsI (Tl) crystal as an energetic photon detector. Although their primary function was that of coincident annihilation spectrometers, these crystals were capable of being used independently, both in rate format and in multi-channel

C. B. Cosmovici (ed.), Supernovae and Supernova Remnants, 261–265.

energy response, in order to carry out solar flare monitoring. The time resolution of the OGO instruments varies with bit rate, but generally is better than on the IMPs, whereas the IMP energy spectra are created simultaneously rather than sequentially, as on the OGOs. In general, continuous coverage was provided from September 1966 to December 1969 and from March 1971 to the present.

We have looked for increases in our observed photon rates at a number of the published times of gravitational radiation events (Weber, 1969), at 82 of 97 times of observed galactic radio noise spikes (Hughes and Retallack, 1973), provided by V. A. Hughes (private communication), at several times of two-station coincident radio intensity increases, provided by R. Huguenin (private communication), at a number of times of rapid atmospheric fluorescence increases, provided by D. Bertsh (private communication), at the published time of an atmospheric flourescence increase suggested as a candidate event for the supernova NGC 5253 (Ögleman and Özel, 1972), and during the 4-day time interval of interest for a possible Cygnus X-3 X-ray flare. Although no X-ray flare was detected with a time history of the radio flare (Parsignault *et al.*, 1972), it was possible that there may have been one too brief to be detected by directional sensors having widely spaced sampling times. None of these searches has produced a positive result to date. Nevertheless, just as the celestial X-ray sources form such a different population of objects from the traditional astronomical sources; it may be that emitters of transient radiations of various types are distinct and separate in their immediately apparent characteristics.

3. X-Ray Increase Phenomenology

Data from the IMP and OGO satellites can be scrutinized with an alternate point of view. In addition to looking for X-ray or gamma ray bursts at the known times of other phenomena, we can instead examine the records for intensity increases that are well outside statistical variation, and occur when the satellites are unaffected by any known geophysical or solar influences. The rate of occurrence of rapid counting rate increases can then be treated as the upper limit to the number of supernova gamma ray bursts, independent of their true origin. A size spectrum of inferred upper limits can then be created as follows. Supernovae presumably occur throughout space with an approximately uniform density and frequency of one per 30 to 100 yr per unit volume which, on the average, contains a galaxy such as ours. The integral size spectrum of supernova outbursts would consequently, because of purely geometrical considerations, obey a power law of index -1.5, which can be normalized to the known rate of occurrence. This size spectrum is then converted to a spectrum of gamma ray burst upper limits, by incorporating the maximum known rate of occurrence of gamma ray intensity increases, determined to be several per year having the equivalent intensity of several hundred counts per square centimeter. This size spectrum upper limit is shown in Figure 1. Since the spectrum must obey the form $N(>S) \propto S^{-3/2}$, knowing the upper limit at one size or frequency determines an upper limit for another. Thus, a balloon experiment, requiring an average frequency of several bursts per day,

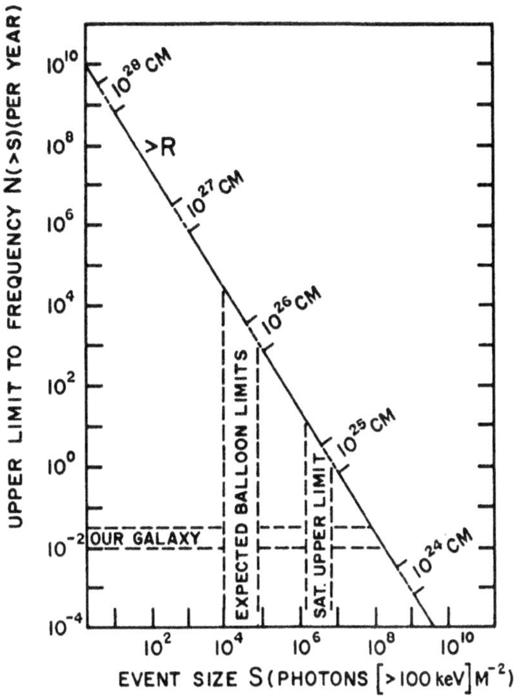

Fig. 1. An upper limit to the supernova gamma-ray burst size spectrum, inferred from the IMP-6 upper limit to the rate of observed gamma-ray intensity increases as a function of size, normalized to the rate of supernovae per unit intergalactic volume, and fitted to a distribution of $N(>S) \propto S^{-1.5}$.

must be sensitive to increases of less than 1 photon per square centimeter in intensity. This size spectrum can be tested for its compatibility with both theoretical models and observations. For example, the distances of the supernovae that would give bursts within the satellite upper limit band is 2×10^{25} cm, or 6.5 mpc. This is within a factor of two of known candidate superclusters. The intensity equivalent at that distance turns out to be 10^{48} ergs output per burst, which is also not incompatible with present ideas. Also, the integrated intensity of all events out to one Hubble radius is below the omnidirectional diffuse background at these energies, so that the upper limit spectrum does not violate known data in that sense.

4. New Detector Design

We have begun the development of new balloon detector systems with pulse burst response over the entire temporal dynamic range usable, from a few nanoseconds to a few hours. These designs were initiated so as to include μsec to msec sensitivity, suggested by Colgate (1968) as the appropriate supernova burst time, within the temporal dynamic range. Such very fast time resolution has been in the past outside the response of typical balloon or satellite systems. The limitation of transmission bit rate precludes indefinitely fast-reponse linear processing, and a triggered system is undesirable since it involves preselection of event characteristics. Our first system involves an alternative concept, namely, time interval processing. Figure 2 shows a plot of the differential probability of the occurrence of the succeeding pulse, following any given pulse, on a random time basis, $dP_0/dAT=\exp(-AT)$. This probability is a maximum for small time intervals, as is the corresponding probability per unit time interval on a linear time interval basis. However, the corresponding probability per unit time interval $\Delta\Pi_0=\exp(-AT_1)-\exp(-AT_2)$, on a logarithmic time interval scale, is a power law of index 1 for small times. Thus, a function that is a maximum

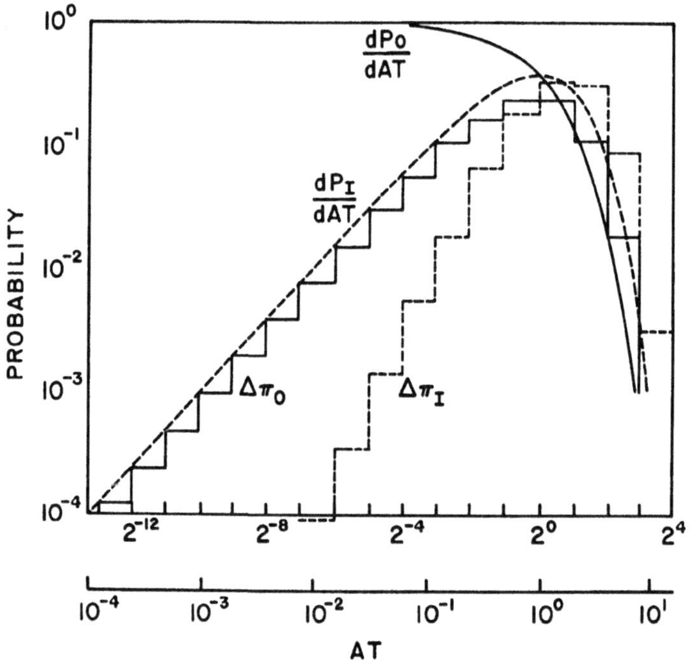

Fig. 2. A plot of differential probabilities dP/dAT, and probabilities per unit time interval $\Delta\Pi$, for succeeding events (0) and alternate events (I), on a logarithmic time basis, which comprise a Poisson distribution

at small times can be converted to one that is a minimum, in order to optimize the signal to noise ratio. Further, this improvement can be enhanced, at the expense of one-half the count rate, by converting to the probabilities for alternate successive pulses, indicated by dP_I/dAT and $\Delta\Pi_I$. The probability per unit time interval for every alternate pulse, $\Delta\Pi_I$, on a log basis, is a power law of index 2 for small times, sharpening a Poisson distribution to a narrow spike on the time interval scale. Thus, even with a very large single scintillator, and with a low data bit rate, a brief pulse burst consisting of a small number of counts can be separated from the detector background by orders of magnitude if the average time interval of event pulses is less than that of the background. Further, structure in the burst would be indicated by its departure from a random distribution. A 25-bin time interval analyzer ranges from 5 ns to 0.32 s with time interval bins of unity width, i.e., having $(T_2 - T_1)/T_1 = 1$. Such a device can be read out several times per second with the modest telemetry requirement of a few kilobits s^{-1}. This system can be incorporated into any gamma-ray detector with inherently fast response. A simple, omnidirectional detector, using about one square meter of plastic scintillator, sensitive to 100 keV to several MeV photons, is presently being prepared for use.

A second system which uses this principle and retains a semi-omnidirectional character for event sensitivity, while incorporating energy and directional resolution, is the Compton telescope. The inherent susceptibility of this system to background may be much less than in static-source astronomy because of the high instantaneous fluxes of gamma-rays contemplated. Also, the background can be reduced by the use of time-of-flight instrumentation to avoid upwards-moving secondaries. Finally, the use of mosaics provides greater sensitive area and also eliminates the azimuth uncertainty inherent in a single, two-element telescope. Thus, it should be possible to have a wide angle system with directional response, fast time resolution, and energy resolution up to about 15 MeV. A balloon-borne system incorporating all these features is also presently being developed, which gives great promise of forming the basis for a space-shuttle supernova experiment.

References

Colgate, S. A.: 1968, *Can. J. Phys.* **46**, S476.
Hughes, V. A. and Retallack, D. S.: 1973, *Nature* **242**, 105.
Ögelman, H. and Özel, M. E.: 1972, Reprint, 1972 IAU Meeting, Athens.
Parsignault, D. R., Gursky, H., Kellogg, E. M., Matilsky, T., Murray, S., Schreier, R., Tananbaum, H., Giacconi, R., and Brinkman, A. C.: 1972, *Nature Phys. Sci.* **239**, 123.
Weber, J.: 1969, *Phys. Rev. Letters* **22**, 1320.

DISCUSSION

W. P. S. Meikle: At Glasgow, we observed a phase shift in a VLF signal reflected from the ionosphere which may have been caused by an X-ray burst striking the ionosphere during the eruption of SN 1970g. The date of this event was July 28, 1970. Do you have any X-ray data for this period?

T. L. Cline: Only OGO-5 was operating then, which does not have the continuous coverage of our systems on OGO-I, -3 and IMP-6 and -7; that is, the X-ray detector is commutated on and off with a daty cycle of about 0.2. Nevertheless, the possibility of an event is certainly worth checking into, and I'll get the time of your signal from you and do that.

Note added after the Conference. Intense bursts of cosmic gamma rays of energy > 100 keV have been discovered by the Los Alamos group, occurring at the rate of several per year, using multi-satellite coincidences (R. W. Klebesdel, I. B. Strong, and R. A. Olson, *Astrophys. J. Letters* **182**, L89, 1973). A number of these events have been found to be among the most intense increases observed in the IMP-6 data discussed here. All the measurements regarding the photon identification, the time histories, and intensities of the bursts are in agreement. The differential energy spectra of these events have been measured with the IMP-6 instrument and are found to be gamma-ray in nature, with very little soft component in the X-ray domain (T. L. Cline, U. D. Desai, R. W. Klebesadel and I. B. Strong, *Astrophys. J. Letters* **185**, L1, 1973).

GAMMA RAY OBSERVATIONS FROM THE CRAB NEBULA AND NP-0532

B. PARLIER, B. AGRINIER, M. FORICHON, and J. P. LERAY
Service d'Electronique Physique, CEN Saclay, France
G. BOELLA and L. MARASCHI
Laboratorio Fisica Cosmica e Tecnologie Relative, CNR Milano, Italy
R. BUCCHERI and N. R. ROBBA
Istituto Fisica Università, Palermo, Italy
and
L. SCARSI
Laboratorio Fisica Cosmica e Tecnologie Relative CNR and Istituto Fisica Università, Palermo, Italy

1. Introduction

The gamma-ray emission from the Crab Nebula and NP-0532 has been investigated in a series of balloon flights of a multiplate Spark Chamber triggered by a Čerenkov detector-scintillator coincidence system (Parlier *et al.*, 1973; Leray *et al.*, 1972; Vasseur *et al.*, 1971), having a gamma-ray energy threshold for detection at 20 MeV. Figure 1 shows a schematic view of the gamma-ray telescope; Figure 2 the view of the telescope without its thermal protection.

The data obtained on the continuum and pulsed gamma-ray flux from the Crab from six successful flights of the gamma-ray telescope carried out in the years 1969–1971, has been recently published (Parlier, *et al.* 1973) we will give here a brief summary of the main results.

Table I gives the balloon flights relevant characteristics. The same instrument, except for minor modifications, was used in the 1969–1970 flights (Nb 5, 7, 11, 13 with an average floating altitude of 5 mb); for flights 14 and 15, carried out in 1971 (floating altitude 2 mb), a modified triggering system was used with increased efficiency on the gamma ray low energy side.

The telescope was kept pointing within a few degrees of the Crab Nebula direction by a programmed azimuthal orientation system using the Earth's magnetic field as a reference, the zenith angle of the telescope axis being constant at 26° (Figure 3).

2. Experimental Results

2.1. Total gamma-ray emission from the Crab source

Our search for gamma-ray emission from the Crab source was carried out by comparing the observed count distribution (N_0) with that expected from the atmospheric background (N_e), for circular regions centred on the Crab with radius, θ, increasing by 1° steps up to 27°.

C. B. Cosmovici (ed.) Supernovae and Supernova, Remnants, 267–281.

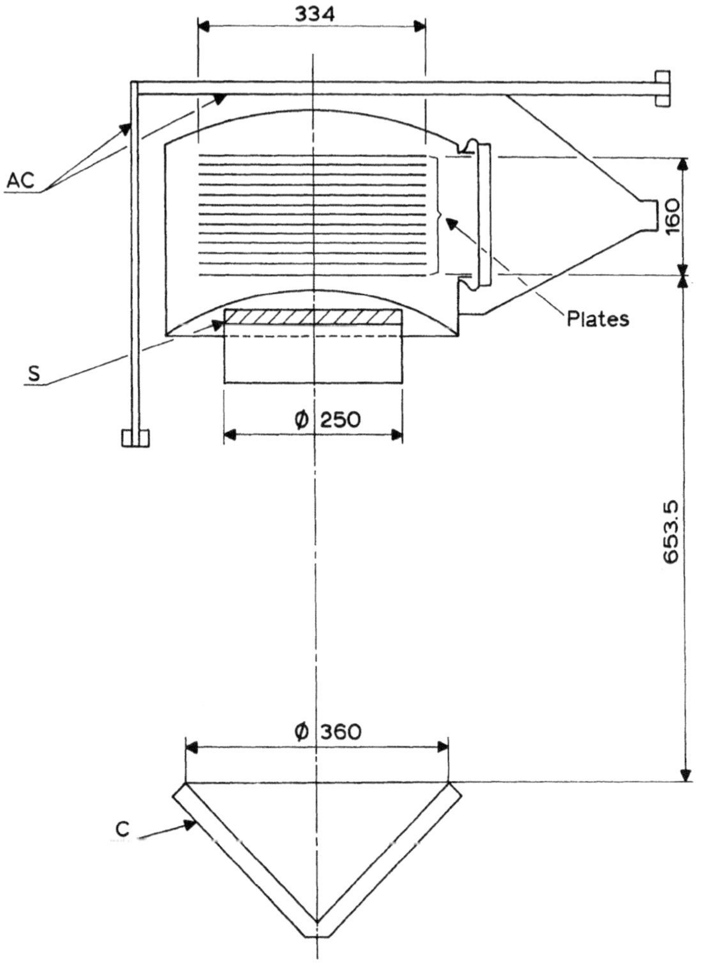

Fig. 1. Sketch of the telescope assembly. AC: anticoincidence scintillation counters; S: plastic scintillator; C: lucite directional Čerenkov counter.

For the region of sky observed, the number of counts contributed by the atmospheric gamma radiation, (N_e), was derived for each flight from the experimental data; the data from balloon flights with similar characteristics were grouped together in two sets referring to flights 5, 7, 11 and 13, and flights 14 and 15 respectively (Table I). During all balloon flights, with the exception of the last section of flight 15, the Crab Nebula was kept within a cone of $\lesssim 10°$ half angle from the telescope axis, so that the experimentally determined atmospheric background in the region neighbouring the Crab position might be overestimated, if a source were present, due to the addition of counts from the source itself. To avoid this effect the background level for the second set of flights (14 and 15) was derived from the unstabilized section of flight 15 (see Figure 3) during which the telescope axis was always pointing in direction more than 10° away from the Crab Nebula; for the first set of flights (5, 7, 11 and 13) this procedure was not possible.

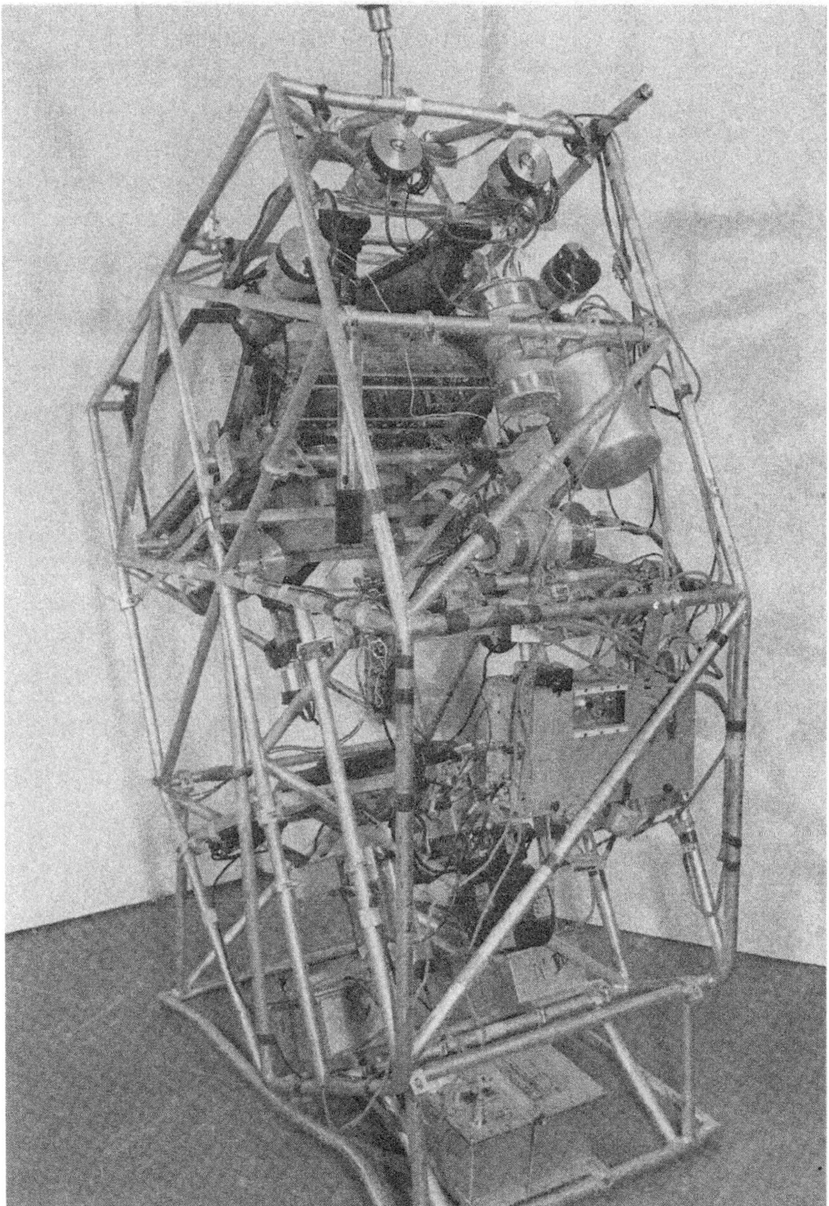

Fig. 2. View of the telescope, without its thermal protection.

TABLE I

Balloon Flights Relevant Characteristics

Flight number, date and launching site	Average ceiling altitude	Interval of time with telescope pointing toward the Crab (UT)	Sensitive time	Number of pictures at ceiling: total	D Events
(5) Sept. 8, 1969 Aire/Adour 43° 43′ N. 0° 13′ W.	5.1 mb	04^h16^m – 09^h25^m	8777 s	10776	2174
(7) Oct. 7, 1969 Aire/Adour	4.7 mb	03^h00^m – 04^h35^m	2949 s	3581	537
(11) Sept. 15, 1970 Aire/Adour	4.7 mb	03^h23^m – 06^h57^m	7050 s	9008	1476
(13) Sept. 24, 1970 Aire/Adour	5.2 mb	02^h48^m – 06^h59^m	7932 s	11133	1909
(14) Sept. 24, 1971 Aire/Adour	2.0 mb	04^h20^m – 07^h26^m	5422 s	8346	710
(15) Oct. 2, 1971 Aire/Adour	2.0 mb	03^h26^m – 08^h00^m (08^h–09^h15 non pointing)	11202 s	11060	946

The results are shown in Figure 4. In Figure 4a, which refers to the second set of flights, the source contribution shows up clearly above the background level obtained from data well outside the Crab region (curve II), the visibility of the source is reduced when the background is evaluated from the pointed section of the flights (curve III).

A check for an instrumental effect connected with the direction of the telescope axis was made on the data from the unstabilized part of flight 15. We find no excess associated with the telescope axis but an excess appears in the Crab direction (Figure 4b).

Figure 4c refers to flights 5, 7, 11 and 13; here the visibility of a point source is confined to a narrower cone, due to the combined effect of the higher level of the atmospheric background (lower balloon ceiling altitude) and its biased experimental determination in the source region.

To evaluate the Poissonian probability of the effect being due to chance fluctuation of the background, we estimate the number of events obtained in the 5° cone, corresponding to the estimated spread in arrival direction for an E^{-2} spectrum, as

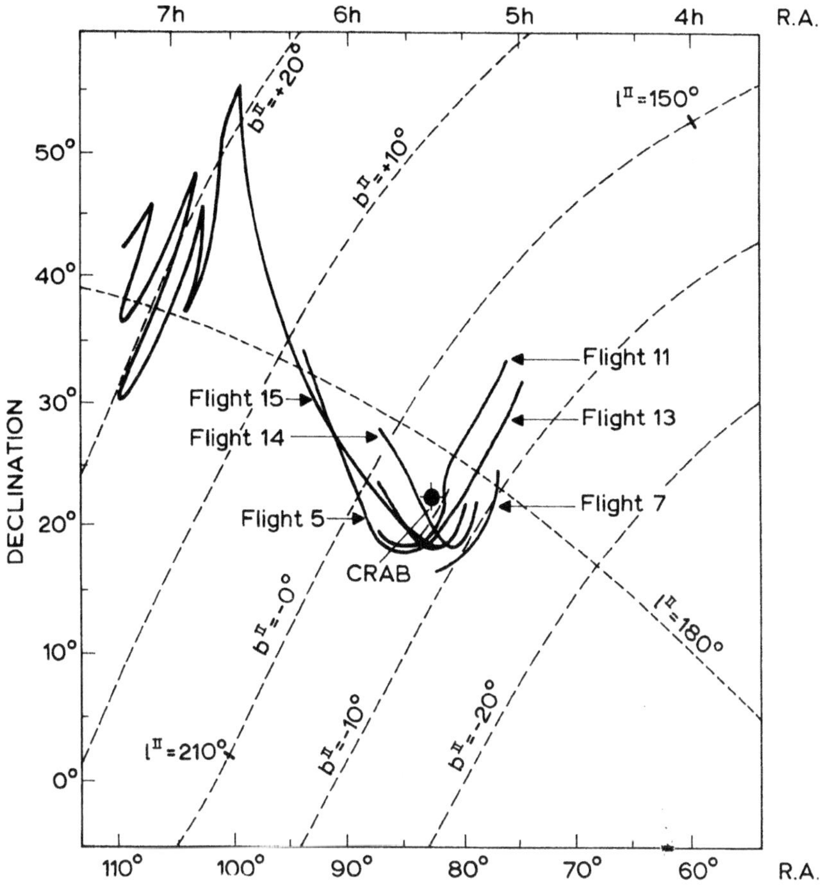

Fig. 3. Region of sky observed. The continuous lines indicate the trajectory of the gamma-ray telescope axis for the different balloon flights.

deduced from calibration at accelerators. From the sum of the two sets we obtain the probability $P=2.7\times10^{-2}$; $P=2\times10^{-3}$ is obtained for the flights 14 and 15 for which the background can be correctly evaluated.

The consistency of the data of the two sets of flights adds to the conclusion of the existence of a discrete gamma-ray source in the Crab direction.

The probability value quoted does not need to be multiplied by any number of successive trials, since the source position is fixed a priori; an incorrect statement at this subject has been by E. O'Mongain (1973) in a recently published paper.

Figure 5 shows, for flights 14 and 15, the number of counts attributed to the Crab source $N=(N_0-N_e)$ as a function of the opening angle θ of the cone centred on the source.

The continuous curve S represents N with the assumption of a gaussian distribution for the spread of the measured gamma-ray arrival directions; the curve is normalized to N_{5°, obtained experimentally at $\theta=5°$.

The asymptotic value $S=(35\pm 13)$, represents the total counts contributed by the source, the error, purely statistical, being deduced from N_0 in the 5° cone. Assuming a source of spectral shape E^{-2}, we obtain the flux value:

$$\phi_{\mathrm{Crab}}(E_\gamma \gtrsim 20 \text{ MeV}) = (9 \pm 3.5) \times 10^{-5} \text{ photon cm}^{-2}\,\text{s}^{-1}.$$

The atmospheric gamma-ray intensity evaluated from the combined sets of flights is

$$I_{\mathrm{atm}}(E\gamma \gtrsim 20 \text{ MeV}) = \\ = (1.8 \pm 0.2) \times 10^{-3} \text{ photon cm}^{-2}\,\text{s}^{-1}\,\text{sr}^{-1}\,\text{g}^{-1}\,\text{cm}^{2}$$

to be compared with the value given by Beuermann (1971) at the same vertical rigidity geomagnetic cutoff (4.5 GeV).

$$I_{\mathrm{atm}}(E\gamma \gtrsim 20 \text{ MeV}) = 2 \times 10^{-3} \text{ photon cm}^{-2}\,\text{s}^{-1}\,\text{sr}^{-1}\,\text{g}^{-1}\,\text{cm}^{2}.$$

2.2. Pulsed gamma-ray emission from NP-0532

We have carried out a temporal analysis searching for a pulsed gamma-ray emission from the Crab-Pulsar NP-0532 on the same data.

The arrival time of each gamma ray event detected is marked in UTC using as a reference a local quartz clock synchronized with MSF, Rugby and IBF, Turin stations; for the six flights of Table I, the balloon geographic position during flight

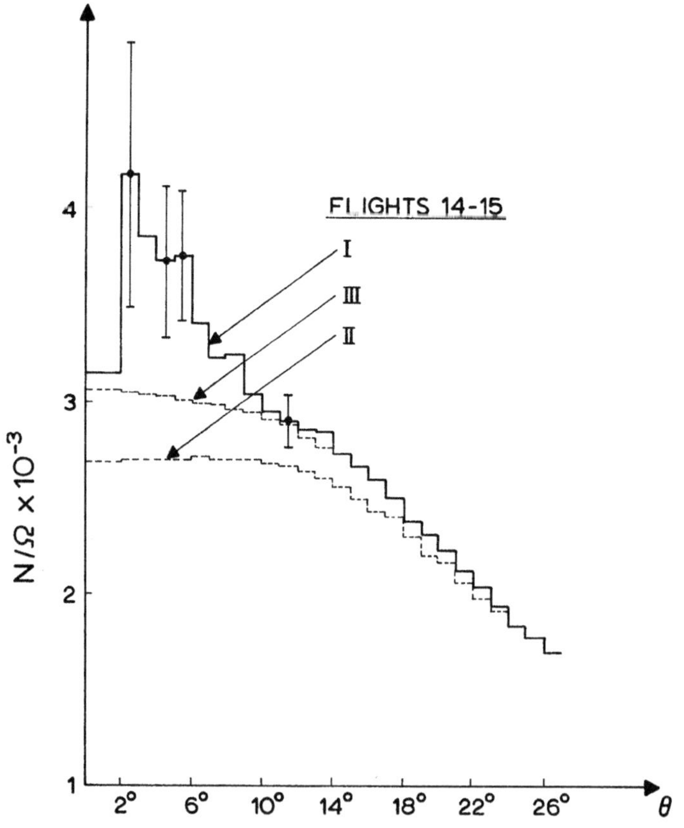

Fig. 4a.

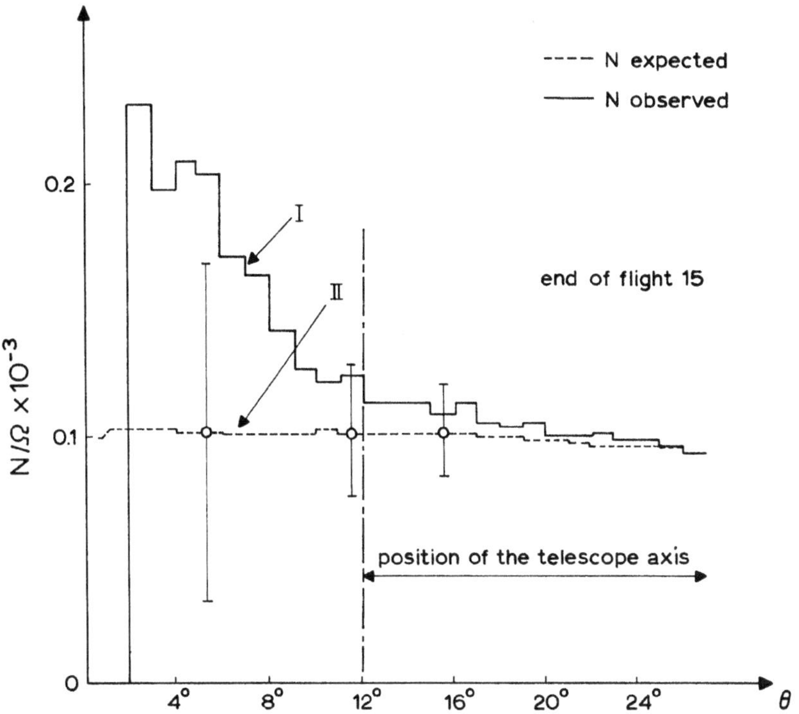

Fig. 4b.

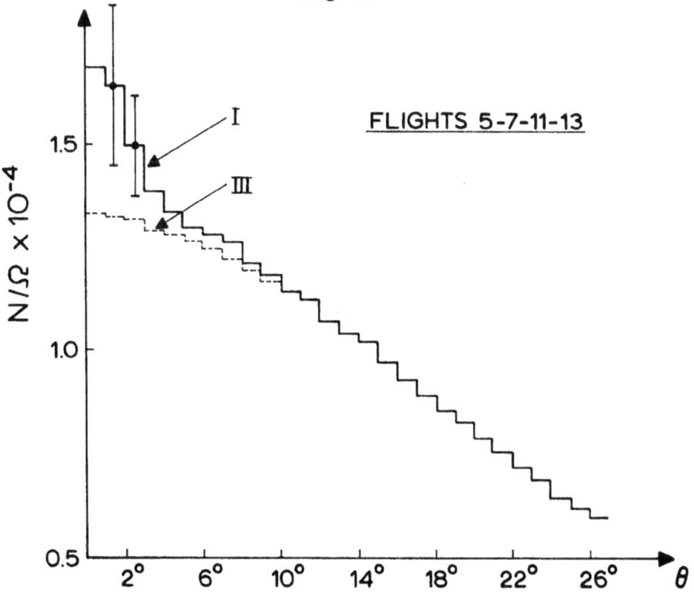

Fig. 4c.

Figs. 4a–c. Counts/steradian as a function of radius θ of the circular region centred on the Crab Nebula. Curve I: $N_0(s\tau^{-1})$; Curve II: $N_e(s\tau^{-1})$ computed from the data excluding the source; curve III: $N_e(\mathrm{sr}^{-1})$ computed from the data including the source. (a) Flights Nb 14 and 15. (b) Unstabilized section of flight Nb 15. (c) Flights Nb 5-7-11-13.

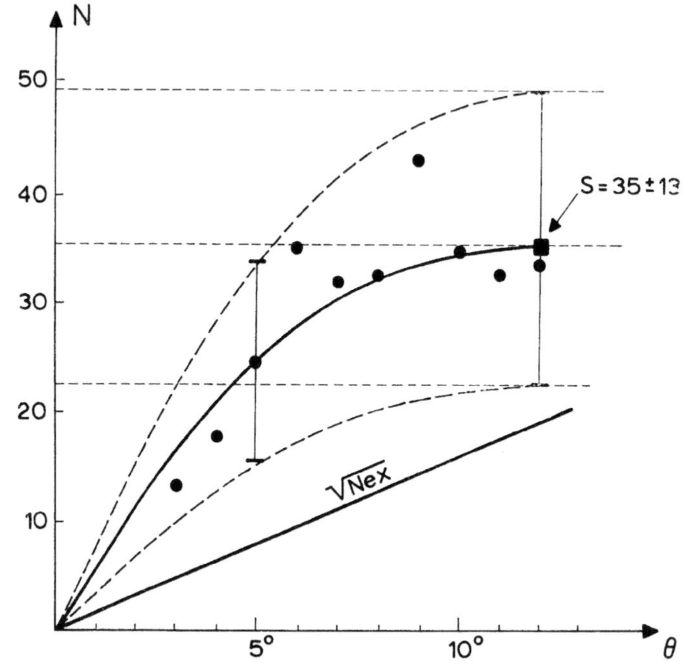

Fig. 5. Counts N, as a function of θ, for flights 14 and 15. ●, $N_S = (N_0 - N_e)$ with N_0 as indicated in Figure 4a, curve 11. The full line represents the counting rate deduced from a Gaussian distribution of the gamma-ray arrival direction normalized to N for $\theta = 5°$. $\sqrt{N_e}$ is also shown.

and the UTC marking were good enough to be consistent with a temporal analysis down to millisecond accuracy.

The UTC gamma ray event arrival times at the detector were transformed to heliobarycentric arrival times using a computer program originating from D. Richards of the Arecibo Ionospheric Observatory and the 'Solar System Ephemeris' from

TABLE II

Coefficients for temporal analysis

Flight number and data	Julian date	E	A (s^{-1})	B (s^{-2})	C (s^{-3})
(11) Sept. 15. 1970	2440844.5	0.050	30.213704173155898601	−1.929268034 E^{-10}	2.0722 E^{-21}
(13) Sept. 24. 1970	2440853.5	0.045	30.213704173155898601	−1.929268034 E^{-10}	2.0722 E^{-21}
(14) Sept. 24. 1971	2441218.5	−0.2428	30.213704198355074077844	−1.9292786572 E^{-10}	2.083258 E^{-21}
(15) Oct. 2. 1971	2441226.5	−0.2428	30.213704198355074077844	−1.9292786572 E^{-10}	2.083258 E^{-21}

Epoch (JD): 2440351.5 + 68889.8075 S in each case.

I. I. Shapiro of MIT (private communication from P. Horowitz). The pulsar phase was computed by fitting the formula.

$$N = E + At + Bt^2 + Ct^3$$

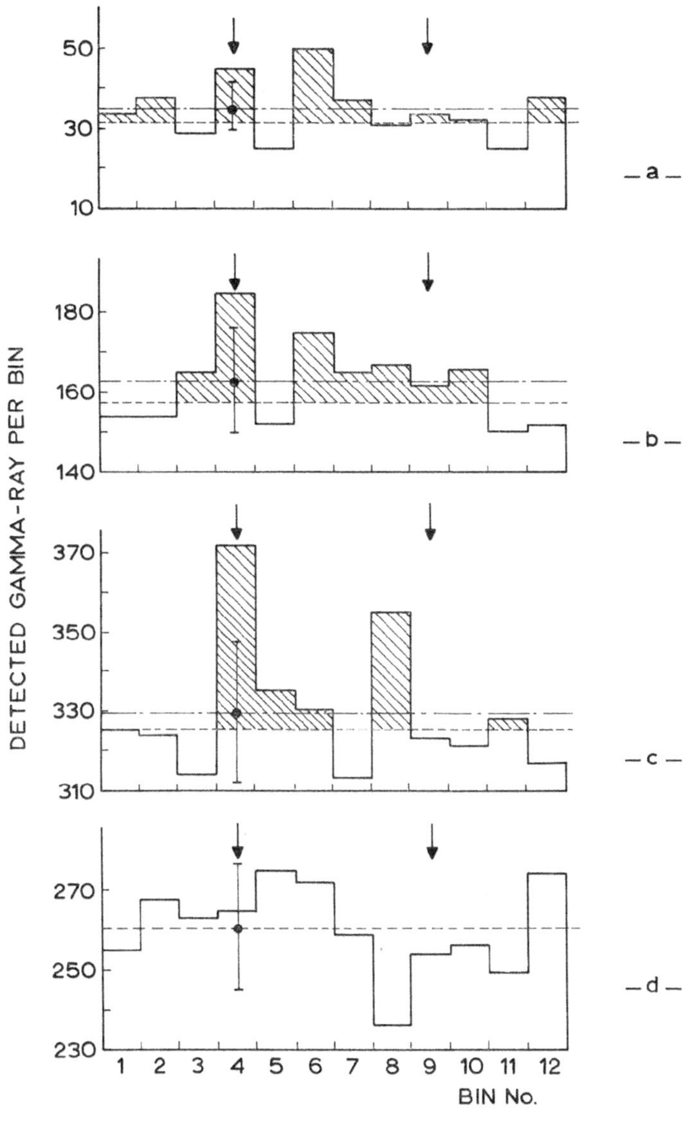

Fig. 6. Gamma-ray data as a function of the phase in the period of NP-0532 for different circular regions centred on the Crab Nebula position. The background level (dotted line) has been deduced from the value of N_e computed for the Crab Nebula source study. Mean of the phase distribution: (a) Circular region defined by $\theta \lesssim 5°$. (b) Circular region defined by $\theta \lesssim 12°$. (c) Circular region defined by $\theta \lesssim 20°$. (d) Region defined by $\theta > 20°$. The arrows indicate the phase of the NP-0532 optical pulses; center of bin 4: main pulse-center of bin 9: peak of the interpulse.

to the pulsar ephemeris where t is the time elapsed between the epoch of the adopted pulsar ephemeris and the arrival time of the event in the plane containing the barycentre of the Solar System: the coefficients are derived from radio (private communication from J. Rankin, J. Roberts and G. H. Pettengill of AIO) and optical (private communication from P. Horowitz of SAO, Harvard) observations of NP-0532.

The values used for the coefficients of the phase fitting formula are listed in Table II. The computed phase values have been directly checked for the events of the flights 11 and 13, with the NP 0532 X-ray pulsating pattern as obtained with a 20 keV–150 keV balloon borne X-ray telescope, flown simultaneously with our detector with a common event timing system (Boclet *et al.*, 1972). The observed X-ray main pulse phase values agree with the computed radio phase values extrapolated to infinite frequency at better than 1 ms.

Figures 6a–c, shows the data resulting from the superposition of all six flights of Table I and gives the phase distribution of the observed gamma rays for the circular regions of the sky centred on the Crab with radii 5°, 12° and 20° respectively; Figure 3d refers to N_0 for the region outside the 20° radius circle. In the adopted bin distribution the main NP 0532 radio-optical pulse falls at the centre of bin 4 and the peak of the interpulse at the centre of bin 9; the phase interval defined by bins 3 to 10 corresponds to the part of period perturbed by the pulsar signal as indicated by X-ray observations of NP-0532 (Boclet *et al.*, 1972; Kurfess, 1971). For the distribution of Figures 3a, b, c, the number of counts observed in the phase interval defined by bins 3 to 10 shows prominent positive fluctuations with respect to the expected atmospheric background level, in particular at the position of the primary pulsar peak. The primary pulsar peak shows up at $>2\sigma$ level in all the cone apertures up to 20° although the mean number of gamma-rays per bin has varied from 34 (Figure 6a) to 326 (Figure 6c). This effect is absent in the Figure 6d in which the contribution of gamma rays above 20 MeV from a point source in the direction of NP 0532 is negligible.

As for the discrete source analysis, the highest confidence for the primary pulsar peak is obtained for the second set of flights for which, by using the data in the 5° cone, we calculate the probability of the excess of counts in phase with the primary pulsar peak being due to chance fluctuation as $P=4\times10^{-3}$; for the sum of the two sets: $P=4\times10^{-2}$.

Here again one has not to multiply the probability value by the number of bins since the phase position of the main pulse is a priori known. The choice of the 5° cone aperture angle has been justified as corresponding to the estimated spread in arrival direction for an E^{-2} spectrum and it would be only confusing between counter and Spark Chamber experiments to compute a probability using the data referring to the whole aperture cone of the telescope as it has been done by E. O'Mongain (1973).

By taking the excess counts observed in phase with the primary radio-optical peak in the second set of flights, assuming an E^{-2} spectral shape we obtain for the pulsed flux.

$$\phi \text{NP } 0532(E_\gamma \gtrsim 20 \text{ MeV}) = (3 \pm 1.5)\, 10^{-5} \text{ photon cm}^{-2}\, \text{s}^{-1}.$$

This values, because it does not take into account the possible contribution of the secondary pulse and of the region between the two pulses (too spreaded out to be detected in this experiment) should be considered as a lower limit for the pulsar gamma-ray emission.

3. Crab Nebula and NP 0532 Energy Spectra

3.1. Total flux from the Crab (continuous plus pulsed)

Figure 7 shows the experimental situation concerning the total flux measured from the Crab source (Nebula plus NP-0532), from the X-ray up to the 10^3 MeV region. The flux value obtained in our experiment lies on the extrapolation of the energy spectrum from the X-ray energies; on the same line are in good agreement also the points obtained at 430 MeV and 1300 MeV by McBreen *et al.* (1973).

A preliminary value $\phi(E\gamma > 200\ \text{MeV}) = 3 \times 10^{-6}$ keV cm^{-2} s^{-1} keV^{-1} has been announced for the SAS-B Satellite at the May 1973 Washington 'Workshop on Gamma-ray Astronomy'.

Points 5 and 6 on the figure (Frye and Chia Ping Wang, 1969 and Clark *et al.*, 1969

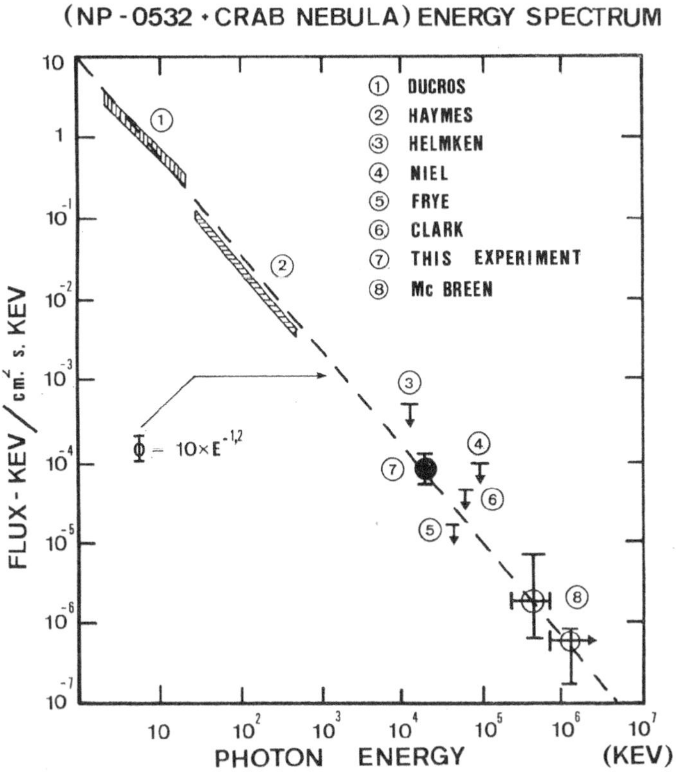

Fig. 7. X- and Gamma ray energy spectrum from the Crab Nebula. Total flux = continuous plus pulsed.

respectively) given as upper limits, and which seem to be inconsistent with the other data, can be explained why being too low. The first (Frye and Chia Ping Wang, 1969) because of an overoptimistic estimate of the accuracy in the gamma-ray arrival direction reconstitution leading therefore to an underestimate of the source strength*; the second (Clark *et al.*, 1969) due to the OSO-3 experiment, because of the masking effect of the diffuse radiation collected in the 15° acceptance cone angle of the telescope. Figure 7 clearly indicates for the Crab total flux a power law spectrum extending over all the energy range considered. It is the only example for a source (in particular for a supernova remnant) for which such a statement can be made on the basis of experimental observations.

A best fit between the available experimental points gives:

$$\phi_{(\text{Crab Nebula + NP}\cdot 0532)} = 10 \times E^{-1.2}\ \text{keV cm}^{-2}\ \text{s}^{-1}\ \text{keV}^{-1}$$

between ~5 keV and 2000 MeV.

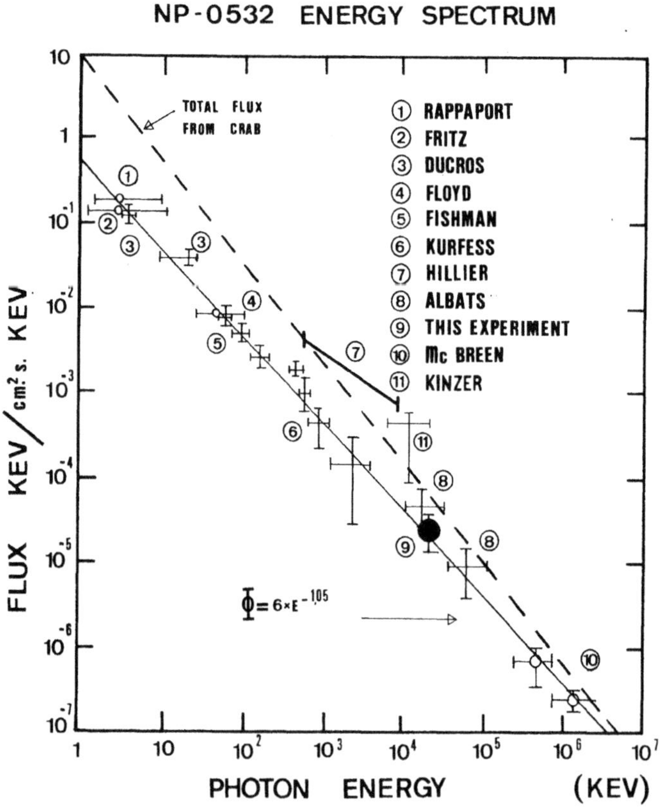

Fig. 8. X- and gamma ray energy spectrum from NP-0532.

* Moreover the upper limit quoted here contradicts the positive flux value obtained by the same authors for NP 0532 (see Albats *et al.* (1973) and Figure 8).

3.2. NP-0532 ENERGY SPECTRUM

Figure 8 shows the experimental situation concerning the pulsed flux attributed to NP-0532.

From 10 keV up to the GeV region, the experimental points seem to fit well with a power law spectrum given by

$$\phi_{\text{NP-0532}} = 0.6 \times E^{-1.05}\ \text{keV cm}^{-2}\ \text{s}^{-1}\ \text{keV}^{-1}.$$

A discrepancy exists with the data of Hillier *et al.* (1970) in the MeV region; Hillier's flux however may be too high because of two reasons (i) assumption of a hard spectrum that tends to overestimate the flux and (ii) too high an energy range assumed for the detector sensitivity.

4. Ratio ($\phi_{\text{pulsed}}/\phi_{\text{total}}$) from the Crab Source

Figure 9 gives, as a function of energy, the ratio $R = (\phi_{\text{pulsed}})/(\phi_{\text{total}})$ for the source in the Crab. From ~1 keV up to ~100 keV it exists a good number of experimental

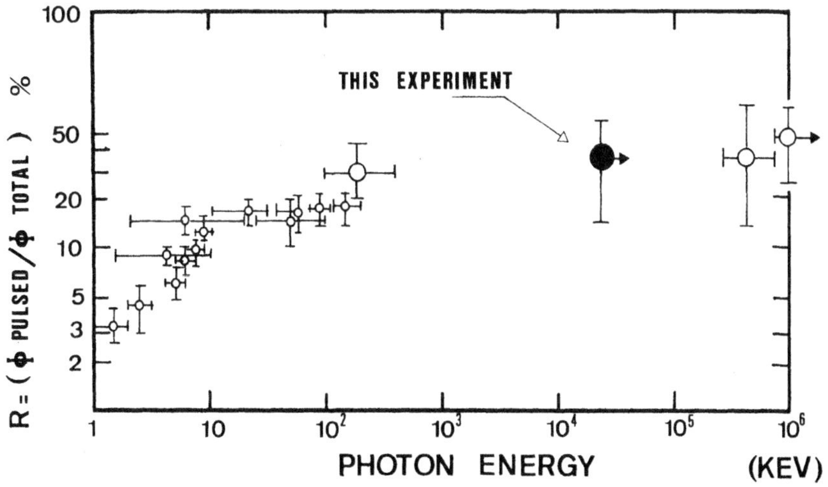

Fig. 9. ($\phi_{\text{pulsed}}/\phi_{\text{total}}$) versus photon energy for the source in the Crab.

measurements; R increases from ~2% at 1 keV to ~15% at 10 keV and thereafter it seems to have a slower, but steady increment with energy reaching ~35% at $E\gamma \sim 20$ MeV and ~50% around the Gev.

5. The Very High Energy Region ($\gtrsim 10^{11}$ eV)

Above 10^{11} eV the data concerning the celestial gamma-ray sources derive from observations with ground based instruments of the atmospheric Čerenkov radiation produced by the primary gamma-ray induced extensive showers. The SAO group, in reporting the results of observations over a 3-yr period (1969–1972) (Fazio, 1972) indicates for the Crab Nebula an average flux of $(4.4 \pm 1.4) \times 10^{-11}$ photon cm^{-2} s^{-1} at $E\gamma \gtrsim 2.5 \times 10^{-11}$ eV.

This value is more than one order of magnitude below that obtained by extrapolating the power law: $\phi = 10 \times E^{-1.2}$ keV cm^{-2} s^{-1} keV^{-1}, which fits well the data up to $\sim 10^9$ eV; a break or an equivalent change in the Crab source energy spectrum must therefore occur before 10^{11} eV.

The experimental evidence, as given by Fazio (1972) seems to indicate that the gamma-ray flux may vary with time, with the most significant flux, $(1.2 \pm 0.24) \times \times 10^{-10}$ photon cm^{-2} s^{-1} at a 5σ level, occurring 60 to 120 days after major spin-up of the pulsar NP-0532. We have checked for a similar effect in our data: among the six balloon flights (see Table I) only the flight 15 falls in an 'activity period' and apparently it shows a stronger source than the others. We wish here just to mention the coincidence: the effect, in our case, is weak and it could be accidental.

Periodic analysis of 122 hr of data collected by the SAO group, has shown no significant pulsation effect, giving an upper limit of 1.2×10^{-11} photon cm^{-2} s^{-1} for an assumed primary pulse width of 1.3 ms. On the other hand Grindlay (1972) using the same observing instrument, reports evidence of a pulsed flux from NP-0532:

$$\phi_{(E\gamma > 6.8 \times 10^{11}\ \mathrm{eV})} \simeq 1.25 \times 10^{-11}\ \text{photon cm}^{-2}\ \text{s}^{-1}.$$

Some clarification is needed.

6. NP-0532 Pulse Shape

The pulse shape of NP-0532 is experimentally rather well defined in the X-ray energy region up to 400–500 keV. The observations indicate a double structure as in the radio and optical band, with an increasing value with energy, of the secondary to primary peak energy content reaching ~ 2.5 at 1 MeV. While the primary peak remain narrow at all energies, the secondary seems to spread out; in the 100 keV–1 MeV range also the region between the two peaks is relevant and the pulse duty cycle approaches $\sim 60\%$.

At gamma-ray energies, mainly because of low statistics, it is difficult to define a precise duty cycle for the pulsed emission and more than a qualitative indication for the pulse structure.

At 20 MeV (Parlier *et al.*, 1973), the primary pulse is clearly visible, while the secondary is possibly too spreaded out to show up; at several hundred MeV and around 1 GeV the ratio primary to secondary peak seems to be about unity (McBreen *et al.*, 1973).

At still higher energies ($\gtrsim 10^{11}$ eV), the situation appear more or less the same.

Acknowledgements

The authors wish to express their gratitude to their colleagues and to the technical staff of Milan and Palermo for their fundamental contribution throughout all the experiment and to Prof. C. Dilworth for criticism and discussions. We are deeply indebted to GIFCO, the Italian Navy and Air Force and CNES launching team for having made the balloon flight possible.

Partial support of this work by the Comitato Regionale Siciliano Richerche Nucleari is gratefully acknowledged.

References

Albats, P., Frye, G. M., Zych, A. D., Mace, O. B., Hopper, V. D., and Thomas, J.A.C.W.R.U.: 1973, Preprint.

Beuermann, K. P.: 1971, *J. Geophys. Res.* **76**, 4291.

Boclet, D., Brucy, G., Claisse, J., Durouchoux, Ph., and Rocchia, R.: 1972, *Nature Phys. Sci.* **235**, 72.

Clark, G. W., Garmire, G. P., and Kraushaar, W. L.: 1968, *Astrophys. J.* **153**, 203.

Ducros, G., Ducros, R., Rocchia, R., and Tarrius, A.: 1970a, *Nature* **227**, 152.

Ducros, G., Ducros, R., Rocchia, R., and Tarrius, A.: 1970b, *Astron. Astrophys.* **7**, 163.

Fazio, G. G.: 1972, in H. Bradt and R. Giacconi (eds.), 'X- and Gamma-Ray Astronomy', *IAU Symp.* **55**, 303.

Fishman, S. J., Harnden, F. R., Johnson, W. N., and Haymes, R. C.: 1969, *Astrophys. J.* **158**, L61.

Floyd, F. W., Glass, I. S., and Schnapper, H. W.: 1969, *Nature* **224**, 50.

Fritz, G., Meckins, J. F., Chubb, T. A., Friedman, H., and Henry, R. C.: 1971, *Astrophys. J.* **164**, L55.

Frye, G. M. and Chia Ping Wang: 1969, *Astrophys. J.* **158**, 925.

Grindlay, J. E.: 1971, *Nature* **234**, 153; 1972, *Astrophys. J. Letters* **174**, L9.

Haymes, R. G., Ellis, D. V., Fishman, G. J., Kurfess, J. D., and Tucker, W. H.: 1968, *Astrophys. J.* **151**, L9.

Helmken, H. and Hoffman, J.: 1971, *Bull. Am. Phys. Soc.* **16**, 533 – DE 12.

Hillier, R. R., Jakson, W. R., Murray, A., Redfern, R. M., and Sale, R. G.: 1970, *Astrophys. J.* **162**, L177.

Kinzer, R. L., Share, G. H., and Seeman, N.: 1973, *Astrophys. J.* **180**, 547.

Kurfess, J. D.: 1971, *Astrophys. J. Letters* **168**, L39.

Leray, J. P., Vasseur, J., Paul, J., Parlier, B., Forichon, M., Agrinier, B., Boella, G., Maraschi, L., Treves, A., Buccheri, R., Cuccia, A., and Scarsi, L.: 1972, *Astron. Astrophys.* **16**, 443.

McBreen, B., Ball., Jr., S. E., Campbell, M., Greisen, K., and Koch, D.: 1973, Preprint C LNS, p. 216.

Niel, M., Vedrenne, G., and Bouigue, R.: in L. Gratton (ed.), 'New-Solar X- and Gamma-Ray Astronomy', *IAU Symp.* **37**, 325.

O'Mongain, E.: 1973, *Nature* **241**, 376.

Parlier, B., Agrinier, B., Forichon, M., Leray, J. P., Boella, G., Maraschi, L., Buccheri, R., Robba, N. R., and Scarsi, L.: 1973, *Nature Letter*.

Rappaport, S., Bradt, H., and Mayer, W.: 1970, *Nature Phys. Science* **229**, 40.

Vasseur, J., Paul, J., Parlier, B., Leray, J. P., Forichon, M., Agrinier, B., Boella, G., Maraschi, L., Treves, A., Buccheri, R., Cuccia, A., and Scarsi, L.: 1971, *Nature Phys. Science* **223**, 46.

PHYSICAL CONDITIONS IN THE FILAMENTS OF SNR FROM THEIR OPTICAL SPECTRA: IC 443

SANDRO D'ODORICO

Asiago Astrophysical Observatory, University of Padova, Italy

1. The Observations

IC 443 is, together with the Cygnus Loop, the best known old supernova remnant. A 21-cm continuum map has been obtained by Duin *et al.* (communication in this volume), where also references to earlier radio work can be found. IC 443 is very likely associated with an X-ray source (see the communication by Gorenstein in this volume). A pulsar is also probably associated with the remnant (Davies *et al.*, 1972)

In a red light photograph IC 443 shows an asymmetrical shell-like structure, the filaments on the NE side being much stronger than the others. This is interpreted

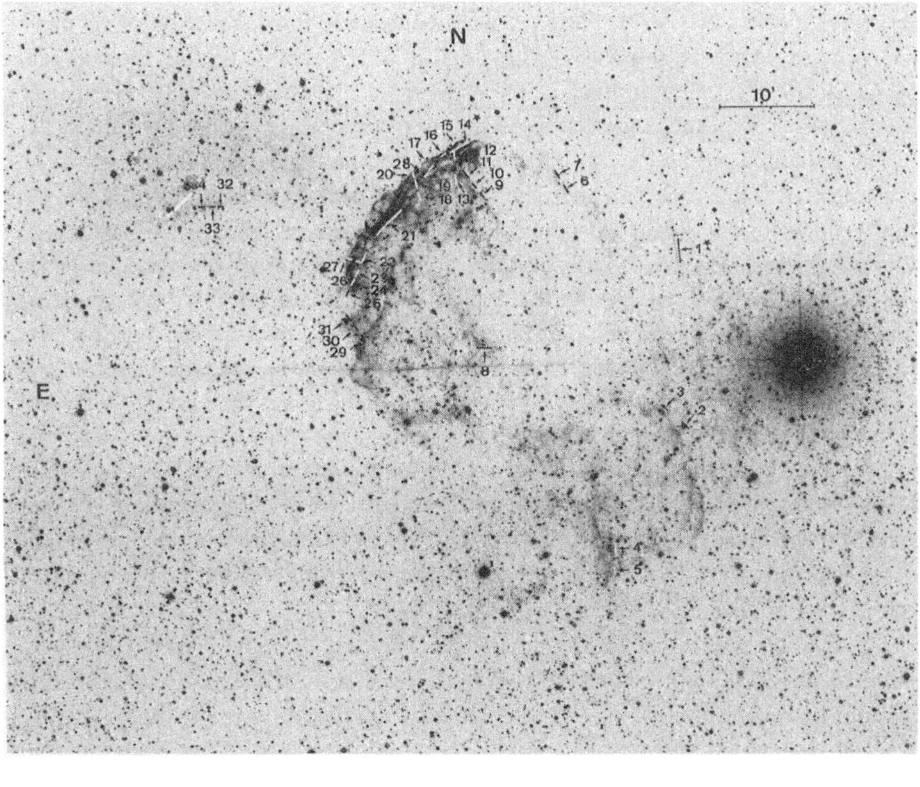

Fig. 1. An $H\alpha + [N\,II]$ photograph of IC 443 obtained with the Asiago Schmidt telescope. The positions for which we measured line intensities are indicated. Three filaments studied by Parker (1964) are marked with asterisks.

C. B. Cosmovici (ed.), Supernovae and Supernova Remnants, 283–293.

as an interaction with denser interstellar clouds, which could be associated with Sharpless H II region 259. We will come back to this point in Section 2.

Lozinskaya (1969) derived an expansion velocity of 65 km s^{-1} based on interferometric observations of the brightest part of the nebula. Parker (1964) measured relative line intensities for the brightest emission lines in the red, green and blue spectral regions at five positions in the nebula. Parker's observations being the only available at present, we thought of interest, among observations of other remnants, to obtain new spectra of IC 443, both to measure relative line intensities and to obtain radial velocities of the weaker filaments. This paper deals with the line strengths, while radial velocities will be reported in a forthcoming paper.

Figure 1 is a photograph of IC 443, obtained with the Asiago 92/67 Schmidt telescope. The combination of 103a-E plate and RG 1 filter isolates a spectral band approximately 300 Å wide at half-transmission and centered at Hα. Only Hα and λλ6583 and 6549 of [N II] give a significant contribution in this wavelength range. The photograph is light printed to allow the identification of the observed positions, but all the main features of the remnant are visible.

We used for the spectroscopic observations the Asiago image-tube spectrograph described by Bertola (1972). The scale perpendicular to the dispersion is 127″ mm^{-1}. The spectra cover the interval λλ5800–7500 at 125 Å mm^{-1}. In this spectral range IC 443 shows the emission lines λ6300 of [O I], λλ6549, 6584 of [N II], Hα, λλ6717, 6731 of [S II]. Exposure times up to 45 min were needed for the weaker filaments.

The reduction method is the following. Calibration plates were taken with a spot sensitometer at the same time than the sky plates. The light source of the sensitometer has a similar wavelength dependence than the phosphor light of the image tube. Calibration and sky plates were developed together and traced with the microphotometer of the Asiago Observatory. Most of the 17 spectra used in this work are of high quality and the measured photographic densities are all on the linear portion of the calibration curve. The heights of the exploring slit of the microphotometer were chosen to separate non homogeneous emission features on the spectra. They were usually kept between 20 and 40″

The relative intensities had to be corrected for the sensitivity of the spectrograph-image tube system. The corrections were computed from observations of standard stars and their energy distributions given by Hayes (1970); however the correction is significant only for the λ6300 line intensity. The reddening was not taken in account, but again it should be significant only for λ6300.

The line intensity ratios are summarized in Table I. Column 1 gives the position numbers as in Figure 1, column two gives a three steps classification of the observed features according to their brightness, the last five columns give the intensity ratios. Data for the λ6300 line are rather scanty because in many spectra the line is contaminated by the night sky emission at the same wavelength. In a few cases where the contribution of the sky was subtracted, the intensity given in Table I is a lower limit.

It's important to note that the filaments in IC 443 appear less sharp than those in the Cygnus Loop and a diffuse emission component is present. We have indicated

TABLE I

Line intensities

Position	Bright. class[b]	$I(H\alpha)/I(6584)$	$I(6717)/I(6731)$	$I(H\alpha)/I(6300)$	$I(S\ II)^a/I(N\ II)$	$I(H\alpha)/I(S\ II)$
1[c]	I	1.5–1.67	1.48:	–	1.54:	0.76:
2	Id	2.11:	1.29:	–	2.20:	0.72
3	IId	1.5	1.19	3.96	1.66	0.68
4	III	1.5	1.31	–	2	0.56
5	Id	1.5	1.17	–	1.44	0.78
6	I	1.49	1.21	–	1.37	0.82
7	I	1.34	1.09	–	1.31	0.76
8	I	1.84:	1.25	–	2.1	0.66
9	I	1.23	1.32	–	1.47	0.62
10	I	1.27	1.02	–	1.3	0.73
11	II	1.48	1.32	5	1.69	0.66
12	III	1.2	0.95	–	1.47	0.61
13	Id	1.36:	1.22:	–	1.65:	0.62:
14[c]	III	1.81–1.75	1.02–1.2	–	1.65–1.75	0.72–0.83
15	III	1.51	1.23	–	1.18	0.95
16	III	1.81	1.28	2.16	1.68	0.81
17	III	1.88	1.23	–	1.40	1
18	Id	2	1.1	–	2.37	0.63
19	III	2.1:	1.41	–	2.09	0.75
20	IId	1.69:	1.21	–	1.27	1:
21[c]	III	2–1.85	1.42–1.50	–	2.1–2.7	0.51–0.71
22	III	1.79	1.20	3.24	1.6	0.84
23	Id	1.36	1.29	3.65	1.68	0.61
24	III	1.95	1.28	2.89	2.45	0.60
25	Id	1.64	1.28	–	1.68	0.73
26	III	2.15	1.29	3.29	2.36	0.68
27	III	1.99	1.24	–	2.27	0.66
28	IId	1.64	1.23	–	1.37	0.72
29	III	2.26:	1.38	–	1.62:	0.65
30	Id	1.62	1.02	–	1.78	0.68
31	III	2.17	1.06	2.25	2.66	0.61
32	I	2.18	–	–	1.14	1.43
33	I	2.1:	1.44	–	1.64:	0.96
34	I	2.16	1.40	–	1.44	1.13

[a] Assumed $I(6549) = \frac{1}{3} I(6583)$.
[b] I: weak; II: average; III: strong emission; d: diffuse.
[c] Parker (1964).

in column 2 when the intensity ratios are relative to a diffuse region; however, this classification should be kept as indicative when the region is weak.

From the scatter of the results for positions where several observations were available, we estimate the uncertainty of a given intensity smaller than 15%. We have used a column in Table I for those cases in which the quality of the spectra or the microphotometer tracings were poor.

In Table I the intensity ratios for three filaments measured by Parker (1964) are also given. Since the intensities are measured at different positions within a filament, we have indicated in each case the range of the observed values.

2. Discussion of the Emission Line Intensities

Physical models for the filaments in SNR have been proposed by several authors. Cox's (1972) work is a valuable example of a model which predicts strengths of observable emission lines. The comparison of the predicted intensities with those observed by Parker (1967) in the best studied filament of the Cygnus Loop provides a good test of the model and of the assumptions on the pre-supernova interstellar density and magnetic field. This approach is not useful with our data because the number of observed emission lines is small. Since our observations are obtained at several positions in the object, we are able to investigate through these few lines how physical conditions and abundances are likely to change within the remnant. We will discuss particularly the electron density, the temperature and the nitrogen and sulphur abundances.

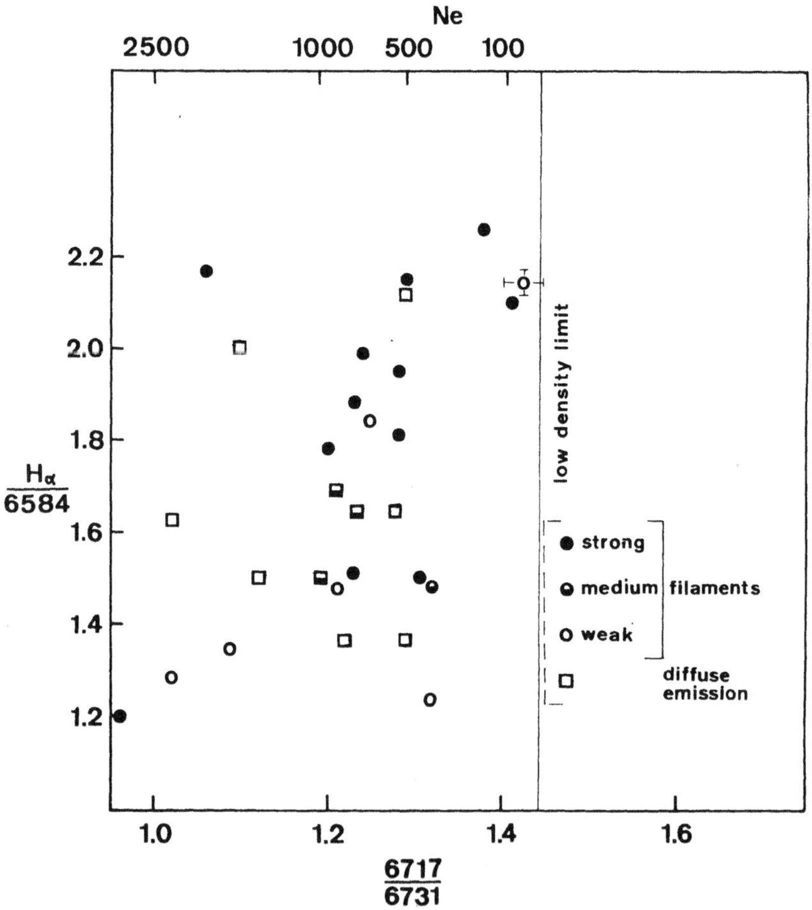

Fig. 2. $I(6717)/I(6731)$ plotted as a function of $I(\mathrm{H}\alpha)/I(6584)$ for all the observed filaments. The density scale is taken from Saraph and Seaton (1970). The point with the bars represents the range of intensities in the most east ward filament (see Section 2C).

A. ELECTRON DENSITY

In Figure 2 we have plotted $I(6717)/I(6731)$ vz $I(\mathrm{H}\alpha)/I(6584)$ for all the observed positions. The $I(\mathrm{H}\alpha)/I(6584)$ ratio is either a temperature or a nitrogen abundance indicator. The ratio $I(6717)/I(6731)$ of [S II] is a density indicator (Saraph and Seaton, 1970; Krueger *et al.*, 1970).

Most filaments have densities between 500 and 1000 electrons cm^{-3}. Observational errors can be responsible for the values observed outside this range, but some higher density points are probably real. Parker's (1964) values of the [O II] doublet at $\lambda\lambda$3727–29 Å give densities lower than 400 cm^{-3}. The difference with the values from the [S II] is most likely an effect of the different ionization potentials. After being compressed and ionized by the shock wave, the interstellar material gets progressively cooler and denser: the [S II] lines originate from a region of lower ionization and higher density than those of [O II]. The same effect has been observed by Peimbert (1971) in the fast moving knots of Cas A. It is also present in the Crab, where the [O II] ratio gives densities $\simeq 1000\ \mathrm{cm}^{-3}$ (Osterbrock, 1957), while [S II] gives $N > 2500$ (D'Odorico, unpublished). Also from Figure 2 we infer that there is no correlation of density with brightness, appearance or the $I(\mathrm{H}\alpha)/I(6584)$ intensity ratio of the filaments.

B. IONIZATION, TEMPERATURES AND CHEMICAL ABUNDANCES

Theory of stellar evolution and observations of young SNR, such as the Crab and Cas A (Woltjer, 1958; Peimbert and Van den Bergh, 1971) suggest the presence of peculiar abundances in the material ejected in the interstellar space after the supernova explosion. We do not have, however, a good estimate of the mass ejected as a function of supernova type, and we do not know whether and how a complete mixing of supernova and interstellar material takes place.

For IC 443 we will assume that overabundances can still be detected in its present stage of evolution. At a distance of 1.5 kpc, with a homogeneous pre-supernova interstellar density of 2 cm^{-3} and the angular diameter of 33′, the swept up mass would be 70 $M_\odot$. If the mass ejected from the supernova is about 10 $M_\odot$ we should indeed still observe some overabundance effects, if present.

Figure 3 is a plot of $I(\mathrm{H}\alpha)/I(6584)$ vz $I[\mathrm{S\ II}]/I[\mathrm{N\ II}]$; the main result is that H^+ and S^+ are roughly proportional to each other. The variation of $I(\mathrm{H}\alpha)/I(6584)$ cannot be ascribed to a double ionization of nitrogen because sulphur would also be strongly doubly ionized and the $I[\mathrm{S\ II}]/I[\mathrm{N\ II}]$ ratio would not increase.

The observed trend can be explained either with a temperature or a nitrogen abundance variation in individual filaments. The $I(\mathrm{H}\alpha)/I(6584)$ ratio is related to $N(\mathrm{N}^+)/N(\mathrm{H}^+)$ and since N^{++} is negligible and $N(\mathrm{N}^0)/N(\mathrm{N}^+) \approx N(\mathrm{H}^0)/N(\mathrm{H}^+)$ we can assume $N(\mathrm{N})/N(\mathrm{H}) \approx N(\mathrm{N}^+)/N(\mathrm{H}^+)$. Thus the observed range of the $I(\mathrm{H}\alpha)/I(6584)$ values gives either a variation of the temperature between 8000 and 10000 K with Orion abundances for nitrogen (as given by Peimbert and Costero, 1969) or a variation of the N abundance by a factor 2 with a temperature of 8000 K for all filaments.

Computations were carried with Pengelly's (1964), case B, Hα emission coefficient, and with the collision strengths by Czyzak *et al.* (1968) and the atomic transition probabilities by Garstang (1968) for N^+.

Further informations can be derived from Figure 4, where we have plotted I (Hα)/I (6584) vz I (Hα)/I [S II]. Although this latter ratio changes little, as previously noted, the filaments are not randomly distributed in the plane. Those with the higher and the smaller I (Hα)/I(6584) values have both low values of I (Hα)/I [S II]. Among the filaments in the lower half of the curve (small I (Hα)/I (6584)) many are classified as diffuse and the majority is located outside the bright, NE front of the remnant. This strongly suggests that the trend has a physical meaning. Two interpretations are possible: according to the first we find a higher sulphur abundance where the interaction with the interstellar medium is smaller; in the second the weaker filaments have a region where S is ionized and H and N neutral more extended than expected from their Hα/6584 temperature.

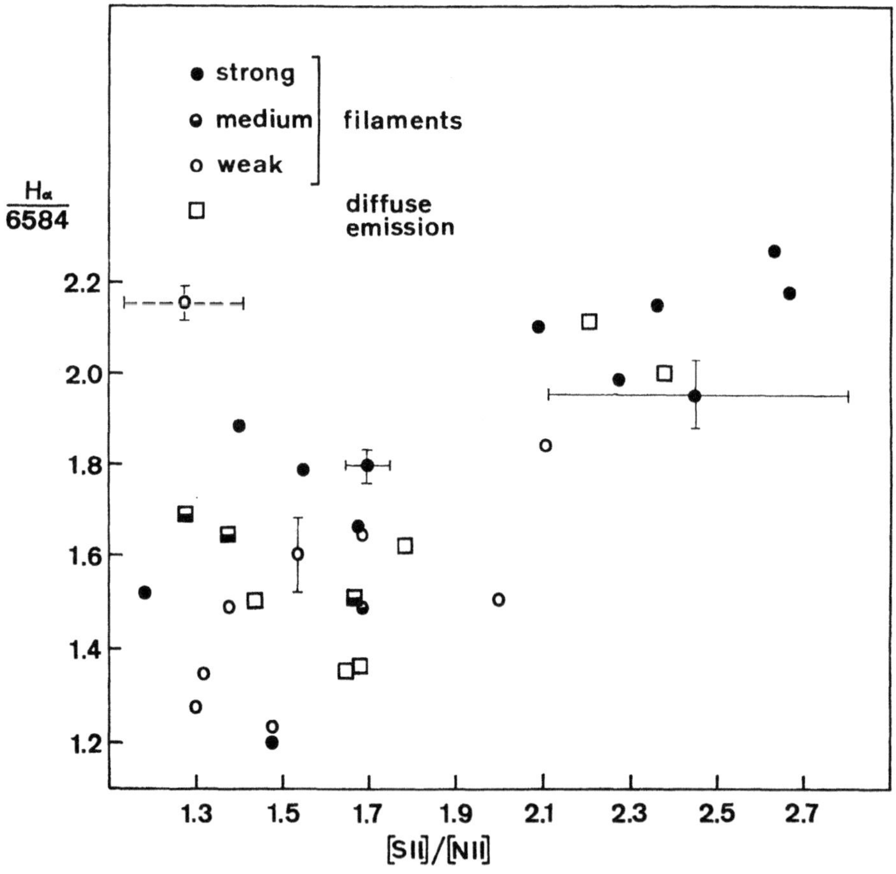

Fig. 3. A plot of I(S II)/I(N II) as a function of I(Hα)/I(S II) for all the observed filaments. The continuous bars indicate the spread of values observed in Parker's filaments, the broken bar represents our observations of the most eastward filament (see Section 2C).

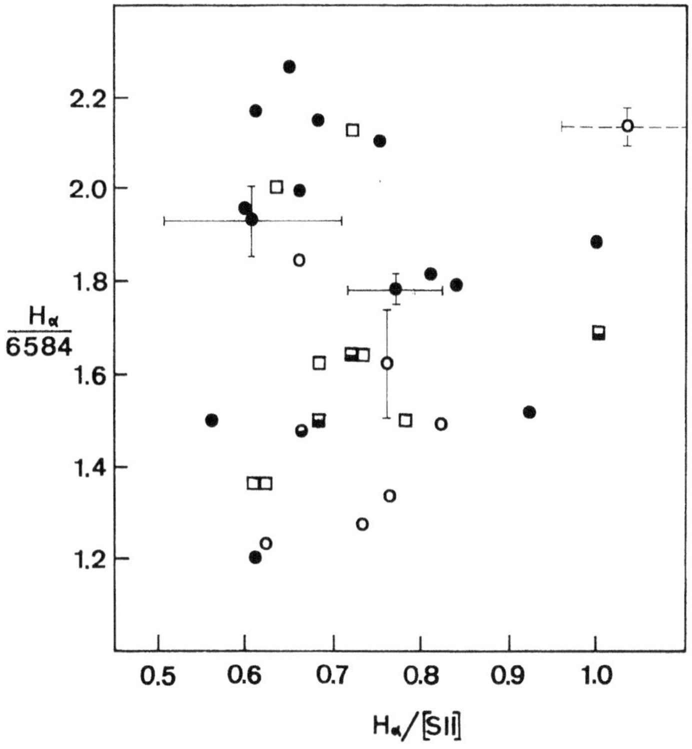

Fig. 4. A plot of $I(H\alpha)/I(6584)$ as a function of $I(\mathrm{H}\alpha)/I(\mathrm{S\ II})$. All the symbols as in Figure 3.

The few data on the [O I], λ6300 line indicate a stronger emission from those filaments which have a higher I (Hα)/I (6584) ratio; this result confirms that the variation is a temperature effect if in the filaments, as it is likely, higher ionization regions are also at higher temperature.

To summarize, the interpretation of the observations is not unique. There are three possibilities which are not mutually exclusive and could contribute together to the observed intensities:

(1) The N and S abundances are the same for all the filaments, the temperatures vary in a range of about 2000 K, the filaments of lower brightness and at higher temperature are characterized by a higher degree of stratification.

(2) The temperature varies as in case 1, the N abundance is constant, the S abundance is higher in filaments of lower brightness and at higher temperature.

(3) The temperature is about the same in all the filaments, the N abundance varies within a factor of 2, the filaments with a higher N abundance are also more abundant in sulphur. This third possibility is rather improbable, because it seems unlikely that the N abundance changes gradually from filament to filament, without any correlation with brightness or location in the remnant.

C. THE FILAMENT NORTH-EAST OF IC 443

Positions 32, 33 and 34 belong to the filament at the boundary of an H II region, north-east of the brighter part of IC 443. The physical conditions result clearly different from those observed in the filaments of the remnant. The value of $I(6717)/I(6731)$ is close to the low density limit, that is the density is about 10 electrons cm^{-3}. The intensity ratios given in Table I and plotted in Figures 2, 3 and 4 can be interpreted twofolds: with a nitrogen abundance as in the remnant and no significant amount of N^{++} we must invoke a lower S^{+} content and most likely a sulphur underabundance; with the remnant abundances for sulphur and nitrogen, both these elements have to be significantly doubly ionized. Again the two possibilities are not mutually exclusive, but the second one seems more probable because of the low density observed in the filament. A search for a possible ionizing star is much needed; if such star is not found, we shall have an observational evidence that the interstellar gas is pre-ionized by the radiation flux from the hot gas in the filaments.

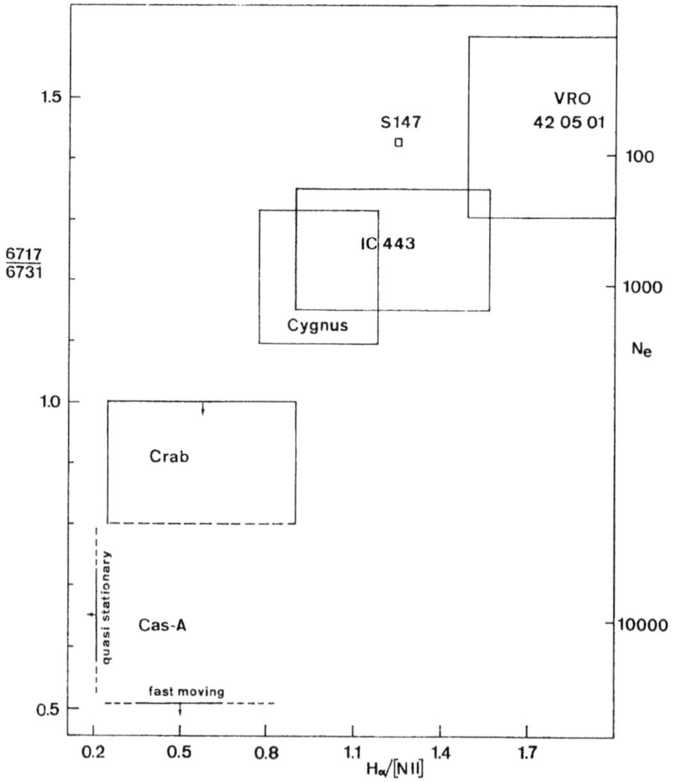

Fig. 5. A plot of $I(6717)/I(6731)$ versus $I(H\alpha)/I(N\,II)$ for all the remnants with observations in the spectral range λλ 6000–7000 Å. Sources for the line intensities in the text. When not given, $I(6549)$ was assumed $\frac{1}{3}$ of $I(6584)$. The rectangles indicate the spread of the intensity ratios within a remnant.

3. A General View on SNR

In Figures 5 and 6 IC 443 is compared with other remnants for which there are observations in the spectral range $\lambda\lambda$6300–6800 Å. Sources for the line intensities are as follows. S 147: Parker (1964); Cygnus Loop: Parker (1967); IC 443: this paper; VRO 42 05 01: D'Odorico (unpublished); Crab nebula: Trimble (1970) for [S II]/[N II] and Hα/[N II], D'Odorico (unpublished) for the [S II] ratio; Cas A: Peimbert and Van den Bergh (1971). For S 147 only one filament was observed, the same is true for each type of filaments in Cas A, where in addition Hα and [N II] were not seen in the fast moving filament. For all other remnants the variation of the intensities within the remnant is indicated by the dimensions of a rectangle. These dimensions are related to the number of filaments observed and therefore give only a lower limit of the true spread of the intensities in the object.

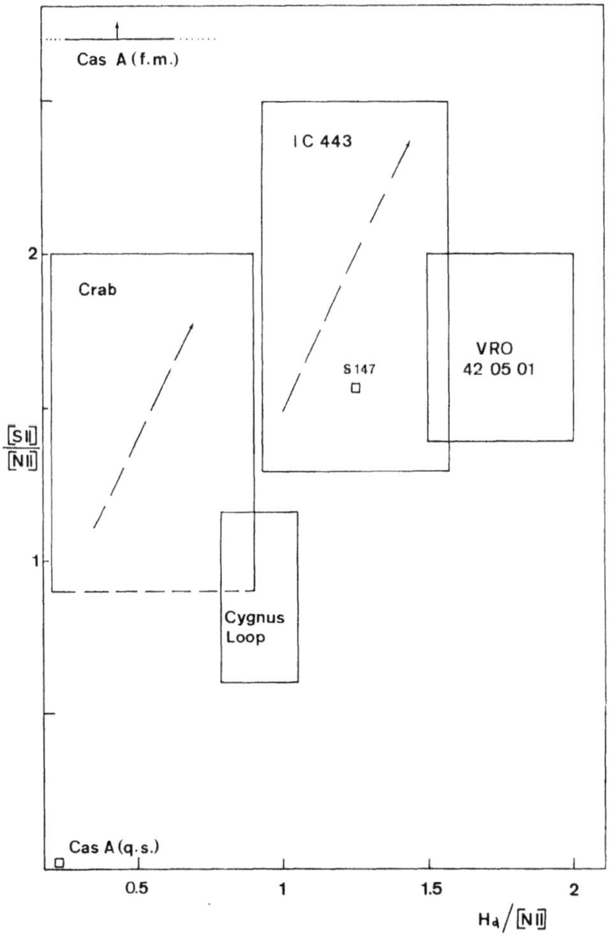

Fig. 6. A plot of I(S II)/I(N II) versus I(Hα)/I(N II) for all remnants with observations in the spectral range $\lambda\lambda$ 6000–7000 Å. Sources for the line intensities in the text. The rectangles and the arrows indicate respectively the spread of the values and the trend of the values within the remnant.

In Figure 5 the remnants are clearly distributed along an evolutionary path. The filaments in the high velocity younger remnants have higher [S II] densities. The variation of the I (Hα)/I [N II], that we have discussed in Section 2B, has to be mainly a temperature effect because we expect higher temperatures in the more energetic filaments.

From Figure 5 it is clear that the physical conditions within a remnant change little in comparison with the difference among remnants in different evolutionary stages.

The presence of quite different remnants on the observed sequence is also an experimental evidence that physical conditions in the filaments are quite independent from the supernova type.

The outcome of Figure 6, where we have plotted I [S II]/I [N II] vz I (Hα)/I [N II], is less straightforward. For the Crab nebula Trimble (1970) gives only an upper limit for I [S II]/I [N II] and indicates the observed trend, which turns out to be the same than that observed in IC 443 (I [S II]$\propto I$ (Hα)). The high value of [S II]/[N II] found for Cas A fast moving filament and also for the Crab relatively to the Cygnus, could be due to a higher sulphur abundance, that later on is diluted in the interaction with the interstellar medium. But there is an alternative explanation that we have already discussed in Section 2B the presence of a higher stratification in the high velocity filaments. The [S II] lines in Cas A originate from a region about 10 times denser than in IC 443. If in this layer sulphur is fully ionized, while N is in part neutral, the observed high value of I [S II]/I [N II] can be explained.

Finally, we just point out that an helpfull tool in understanding the abundance problem is the comparison of SNR spectra with those of other filamentary nebulae, that is nebulae where a smaller energy shock is present. Lozinskaya and Esipov (1971) have discussed some of these objects.

So far we have obtained spectroscopic observations of A 21, NGC 2359 and NGC 6888. The results will be published in a forthcoming paper.

Acknowledgements

I want to thank Dr J. W. Erkes for introducing me to supernovae remnants, and Mr M. Franceschi who patiently made many versions of the drawings.

References

Bertola, F.: 1972, *Proc. ESO/CERN Conference on Auxiliary Instrumentation for Large Telescopes* (ed. by Laustsen and Reiz).
Cox, D. P.: 1972, *Astrophys. J.* **168**, 159.
Czyzak, S. J., Krueger, T. K., Martins, P. de A. P., Saraph, H. E., Seaton, M. J., and Shemming, J.: 1968, in D. E. Osterbrock and C. R. O'Dell (eds.), 'Planetary Nebulae', *IAU Symp.* **34**, 138.
Davies, J. G., Lyne, A. G., and Seiradakis, J. H.: 1972, *Nature* **240**, 229.
Garstang, R. H.: 1968, in D. E. Osterbrock and C. R. O'Dell (eds.), 'Planetary Nebulae', *IAU Symp.* **34**, 138.
Krueger, T. K., Aller, L. H., and Czyzak, S. J.: 1970, *Astrophys. J.* **160**, 921.
Lozinskaya, T. A.: 1969, *Sov. Astron.-A.J.* **13**, 192.
Lozinskaya, T. A. and Esipov, V. F.: 1971, *Sov. Astron.-A.J.* **15**, 353.

Osterbrock, D. E.: 1957, Publ. Astron. Soc. Pacific **69**, 227.
Parker, R. A.: 1964, *Astrophys. J.* **139**, 493.
Parker, R. A.: 1967, *Astrophys. J.* **149**, 363.
Peimbert, M. and Costero, R.: 1969, *Bol. Obs. Tonantz. Tac.* **5**, 3.
Peimbert, M. and Van den Bergh, S.: 1971, *Astrophys. J.* **167**, 223.
Peimbert, M.: 1971, *Astrophys. J.* **170**, 261.
Pengelly, R. M.: 1964, *Monthly Notices Roy. Astron. Soc.* **127**, 145.
Saraph, H. E. and Seaton, M. J.: 1970, *Monthly Notices Roy. Astron. Soc.* **148**, 367.
Trimble, V.: 1970, *Astron. J.* **75**, 926.
Woltjer, L.: 1958, *Bull. Astron. Inst. Neth.* **14**, 39.

DISCUSSION

A. G. De Bruyn: Recently Mathewson has detected a new technique for discriminating between SNR and H II regions based on the different ratio of Hα [S II]. Can this effect be fully explained by a T_e or n_e difference? Or is also an abundance effect present?

S. D'Odorico: As I have mentioned in my talk, there is some evidence of sulphur overabundance in Cas A (Peimbert, 1971) and possibly in IC 443. My personal feeling is, however, that the relative strength of the [S II] lines with respect to Hα is mainly due to the presence in the filaments of a cool, very dense region where only S is ionized.

M. F. McCarthy: Have you made any comparison of SNR spectra with Campbells Hydrogen Envelope Star (HD 184738) which is a WR star plus a Planetary Nebula located in Cygnus. A study of relative intensities in this source might prove informative.

S. D'Odorico: No, I have no spectra of this object.

HIGH RESOLUTION 21 CM CONTINUUM OBSERVATIONS OF SOME GALACTIC SUPERNOVA REMNANTS

R. M. DUIN, R. G. STROM, and H. VAN DER LAAN
Sterrewacht Leiden, Leiden, Holland

1. Introduction

The Westerbork Synthesis Radio Telescope has made observations of some galactic supernova remnants at 21 cm continuum. The observations of Tycho's remnant were carried out by Strom and Duin. Further details about the instrument and the measurements can be found in Strom and Duin (1973).

The observations of IC 443 were made by Duin and Van der Laan. A detailed discussion is postponed until data at other wavelengths will be available.

2. Tycho's Remnant

With the Westerbork Synthesis Radio Telescope 21 cm continuum observations were made of the young and very regularly shaped remnant of Tycho. Figure 1 shows the total intensity map, which has a resolution of 24″ × 27″ in RA and Dec. This

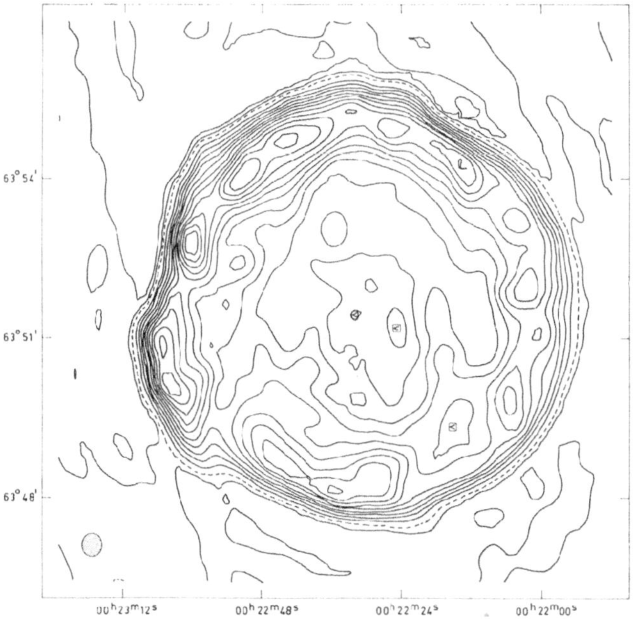

Fig. 1. Contours of total intensity for Tycho's Remnant. Above the map zero level, which is dashed, the contour interval is 13.75 K. The symbol '<' enclosed in a square indicates a local minimum. The shaded ellipse represents the synthesised half-power beam.

C. B. Cosmovici (ed.), Supernovae and Supernova Remnants, 295–301.

map looks very similar to previous results by Baldwin (1967) at the same wavelength. But in addition to the total intensity also the other Stokes parameters were mapped, and Figure 2 shows a vector representation of the polarised intensity and polarisation angle distribution of the 21 cm radiation. At higher frequencies (Weiler and Seielstad, 1971; Kundu and Velusamy, 1971) the vectors lie in a circular pattern, implying a radial magnetic field, but here we see a typical 'cell' structure: regions where the position angle remains constant, while large changes define adjacent regions, the transitions being smaller than our beam.

Of course beam depolarisation must occur in these narrow transition regions, where the polarisation angle changes rapidly. However, inside the cells very little depolarisation occurs since the polarisation at higher frequencies has the same percentage. Because we only see rotation and virtually no intrinsic depolarisation, we interpret

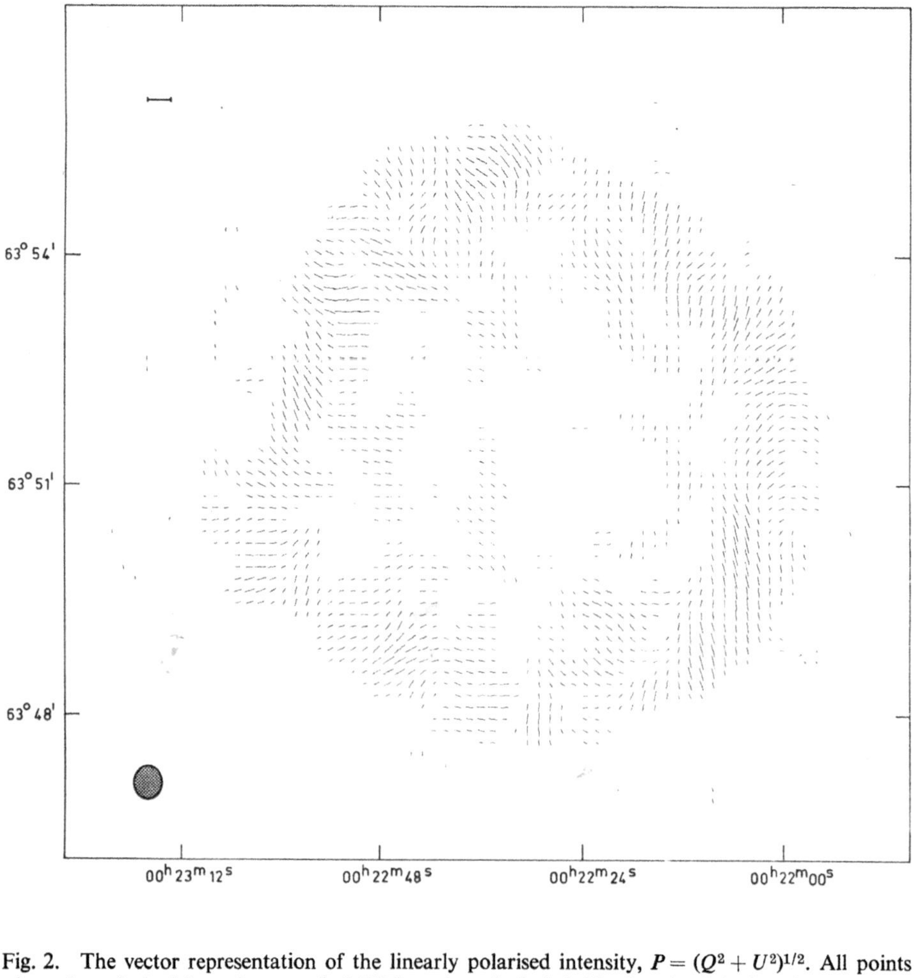

Fig. 2. The vector representation of the linearly polarised intensity, $P = (Q^2 + U^2)^{1/2}$. All points where the polarised intensity exceeds 5.5 K have been included. The length of the bar shown on top corresponds to a polarised intensity of 33 K.

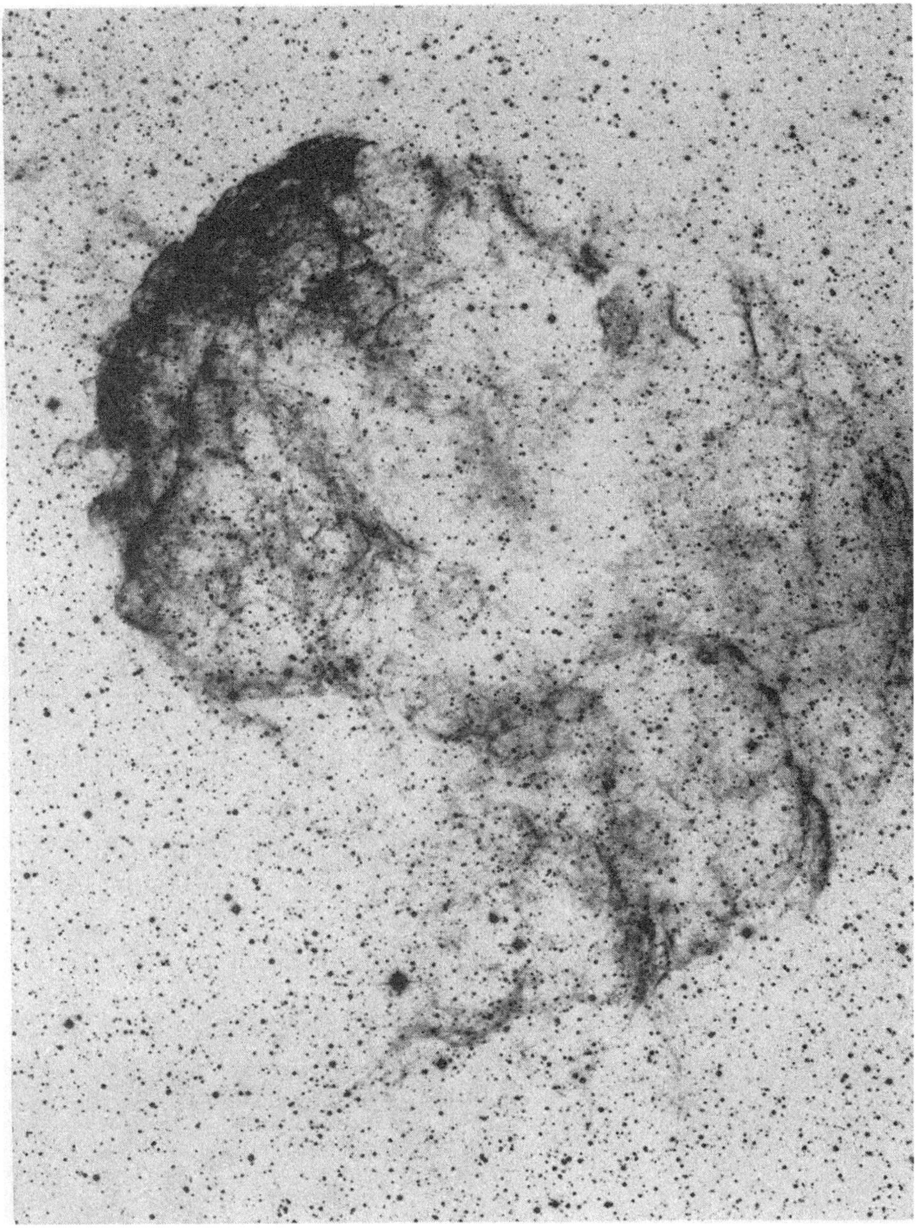

Fig. 3. A copy of the red Palomar Survey plate of IC 443. The main contribution to the optical picture comes from line radiation (Hα).

this as an indication of a thin Faraday screen with scale sizes on the order of the cells, lying outside the radio emitting region.

As we know that Tycho's remnant is decelerated considerably (van den Bergh, 1971) and has swept up a fair amount of interstellar material, we identify this Faraday screen with the matter and magnetic fields swept up by the remnant. Our conclusion therefore is that the shell of relativistic particles is surrounded by a thin relatively dense screen of hot gas which produces the weak optical emission and is responsible for the structure in the polarisation map at 21 cm wavelength. For further details the reader is referred to Strom and Duin (1973).

3. IC 443

The much older supernova remnant IC 443 was also observed with the Westerbork array at 21 cm continuum. More information about IC 443 is presented elsewhere in this volume, therefore only a few characteristics are recalled. Figure 3 shows a

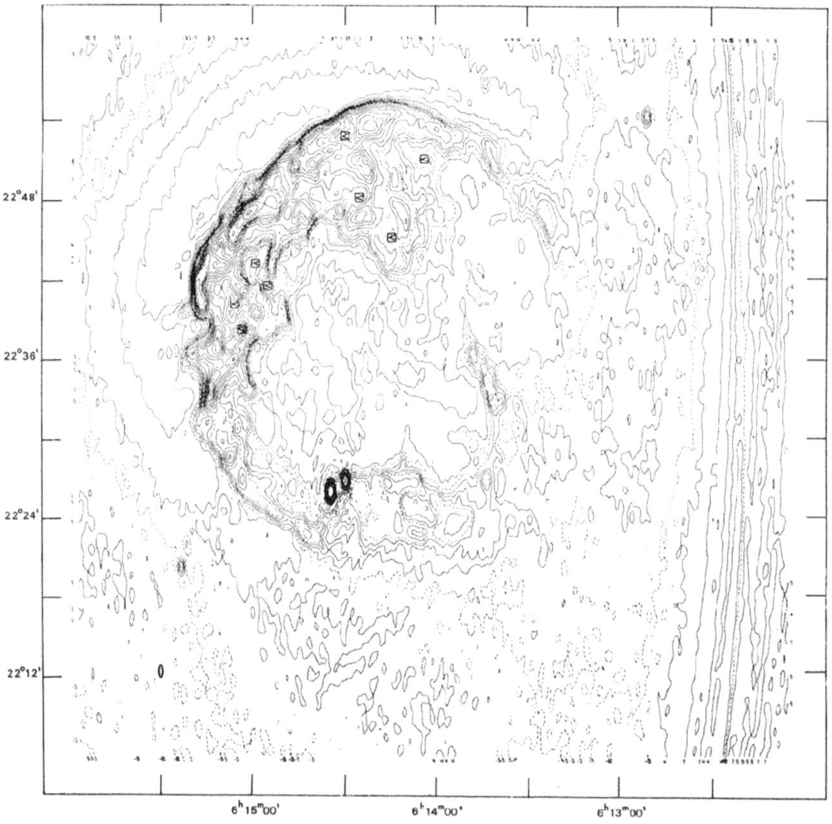

Fig. 4. Contour map of the total intensity of IC 443 at 21 cm continuum. The outermost dashed contour is the apparent zero level (see text). From that level the contour intervals between pairs of dashed contours are 1.1, 2.2, and 4.4 K respectively. The symbol '<' enclosed in a square indicates a local minimum. The shaded ellipse in the SE corner represents the half power beam.

reproduction of the red Palomar Sky Survey plate. There is a pulsar associated with the remnant, located outside this photograph to the West. From the slowing down rate of the pulsar we deduce a lifetime of about 70000 yr. To the north-east lies an H II region, which is possibly spatially correlated to IC 443. It is likely that the expanding shell encountered a larger density in this direction than in the south-west direction, and a value of $n_H = 10\ \text{cm}^{-3}$ (Locke *et al.*, 1964) for the interstellar medium to the north-east seems reasonable. The expansion velocity of this part of the nebula is about 65 km s^{-1} (Lozinskaya, 1969) and from measurements of Parker (1964) the filaments in this part have densities on the order of $n_e = 400\ \text{cm}^{-3}$ and temperatures of $T_e = 10^4$ K as is confirmed by the results of D'Odorico.

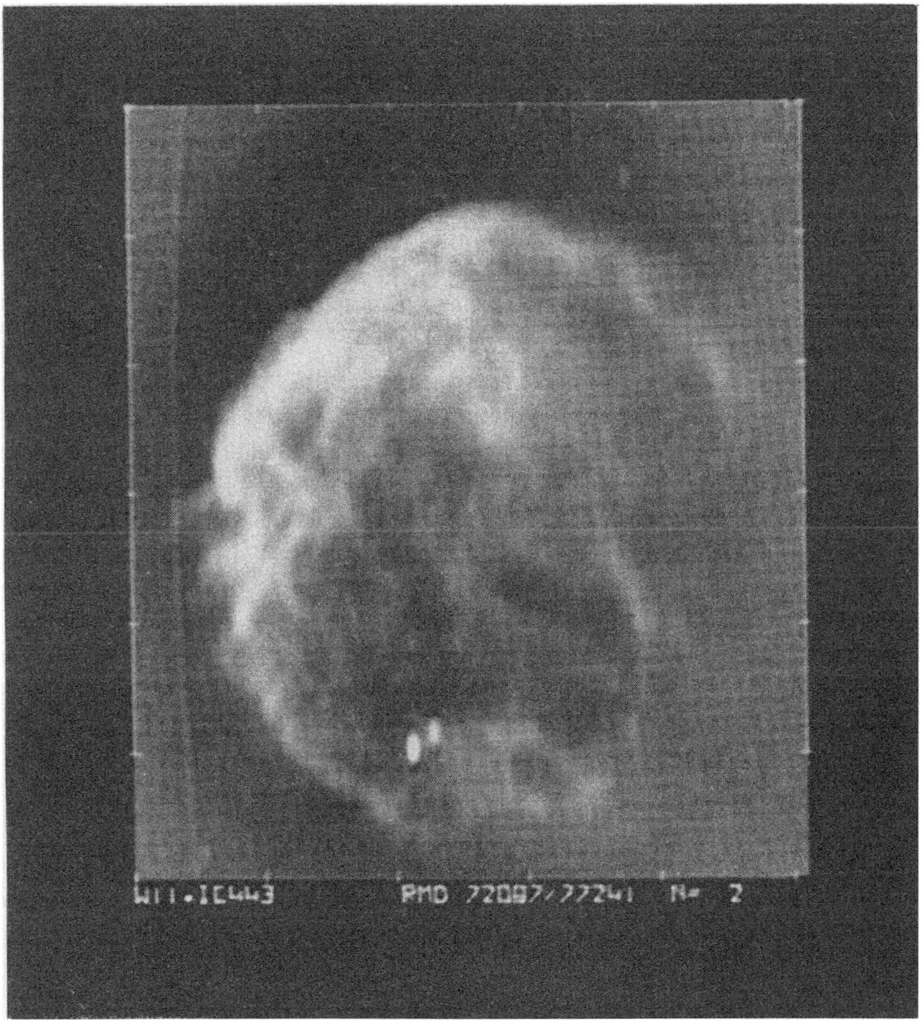

Fig. 5. The 'radio-photograph' of IC 443. The same information as in the contour map (Figure 4) is present, but converted to a photograph using a computer-controlled cathode-ray tube.

Figure 4 shows the contour map of the 21 cm continuum radiation. We see the same shell as in the optical picture in the north-east, but in the south-west fewer details are seen as a result of the primary beam attenuation of a single element of the array. This decreases the signal to noise ratio away from the center of the observed field, which is at RA $=6^h14^m37^s$ and Dec. $=22°42'6''$. Attention must be paid to the fact that since the shortest baseline used is 36 m any structure larger than 9′ is more than a factor 2 attenuated and so a large fraction of the total flux is not represented in this map.

The same data as in the contour map are present in Figure 5, but transformed to a photograph. When we compare this 'radio' photograph to the photograph of Figure 3, which is mainly Hα line radiation, we note the remarkable similarity between the two. Every radio ridge corresponds to a filament and regions free of filaments appear as gaps in the radio photograph. This spatial coincidence was already noted by Hill (1971), but our smaller beam enables us to confirm this in much more detail.

For an H II region we expect a spatial coincidence between the optical line emission and the radio free-free emission. At this moment we can not rule out completely the possibility that the peaks in our map are of thermal origin. If however the radiation at 21 cm is non-thermal, which seems to be the case on the basis of the available spectral index information (Hill, 1971) and the absence of any H109α recombination line radiation (Dickel and Milne, 1972), we need a mechanism to enhance the synchrotron volume emissivity in the filaments.

Our suggestion is that the gas behind the shock, propagating at 65 km s^{-1}, which therefore has a temperature of $T_e = 10^5$ K and which has four times the density in front of the shock undergoes unstable cooling under pressure equilibrium. The pressure of the hot gas is proportional to 4×10 cm$^{-3} \times 10^5$ K, while the gas in the filaments has a pressure proportional to 400 cm$^{-3} \times 10^4$ K. If the magnetic field is also compressed with the matter, then the synchrotron volume-emissivity in the filaments is enhanced sufficiently to account for the correspondence of Figures 3 and 5. In that case the absence of polarisation at 21 cm wavelength can easily be explained by the high Faraday dispersion in the filaments. For a magnetic field of $H = 10^{-4}$ G. and an electron density of $n_e = 400$ cm^{-3} we find, taking 0.02 pc. as a typical pathlength in the filaments a rotation measure of RM $= 650$ rad m^{-2}.

Acknowledgements

The Westerbork Radio Observatory is operated by the Netherlands Foundation for Radio Astronomy with the financial support of the Netherlands Organisation for the Advancement of Pure Research (ZWO). R.M.D. also thanks ZWO for a grant.

References

Baldwin, J. E.: 1967, 'Radio Astronomy and the Galactic System', *IAU Symp.* **31**, 337.
Bergh, S. van den: 1971, *Astrophys. J.* **168**, 37.
Dickel, J. R. and Milne, D. K.: 1972, *Austr. J. Phys.* **25**, 539.

Hill, I. E.: 1972, *Monthly Notices Roy. Astron. Soc.* **157**, 419.
Kundu, M. R. and Velusamy, T.: 1971, *Astrophys. J.* **163**, 231.
Locke, J. L., Galt, J. A., and Costain, C. H.: 1964, *Astrophys. J.* **139**, 1071.
Lozinskaya, T. A.: 1969, *Sov. Astron.* **13**, 192.
Parker, R. A. R.: 1964, *Astrophys. J.* **139**, 493.
Strom, R. G. and Duin, R. M.: 1973, *Astron. Astrophys.* **25**, 351.
Weiler, K. W. and Seielstad, G. A.: 1971, *Astrophys. J.* **163**, 455.

DISCUSSION

P. Gorenstein: Do you have measurements of 21 cm absorption from interstellar gas along the line of sight? When this is compared to X-ray absorption along line of sight we obtain information on the chemical composition of the interstellar gas.

R. M. Duin: For Tycho only absorption measurements with very bad velocity resolution have been made in Westerbork.

J. R. Dickel: We have some 21 cm data on IC 443 and more is being reduced by De Jager. There appears to be a decrease in the 21 cm emission at all velocities right in the direction of IC 443 itself but we cannot separate out the effect of absorption by Hydrogen in front and a possible hole in the hydrogen cloud caused by the SNR. In any event, trying to match the profile with models of the hydrogen distributions in that direction ($l = 189°$, $b = +3°$) we find that there are no components of emission in the direction of IC 443 which do not show a decrease at the SNR's positions so we must conclude that the SNR should lie at the distance of the Perseus feature at 2.2 to 3 kpc, considerably more distant than previously suggested.

A. S. Wilson: I seem to remember that Moffat (paper in *Monthly Notices*, 1971) found that for the Cygnus loop the radio radiation was also closely associated with the optical filaments but showed some tendency to be systematically behind (or in front of, I don't remember) the expanding filaments. Is there then any evidence for a similar effect in IC 443 or do the filaments and radio emission agree to within the limits of the resolution?

R. M. Duin: The filaments and the radio emission agree very well. Of course because of the larger distance of IC 443 our linear resolution is worse than Moffats linear resolution.

S. F. Gull: To determine the extent of Faraday rotation across your map of 3C10 you compare your results with the high frequency observations by Kundu and Velusamy and Weiler and Seielstad. Since these observations have a much poorer resolution than yours, can you be sure that the high frequency polarization is every where radial, and does not show similar cellular structure to that found at 1400 MHz?

R. M. Duin: The high frequency polarisation measurements had a resolution comparable to the size of cells observed in the Westerbork 1400 MHz map, and they should have been able to detect any similar effect.

J. R. Dickel: The 11 cm cells are similar but less significant, in qualitative agreement with Duin's conclusion.

W. M. Goss: I disagree with Duin's statement that the Westerbork observations of 3C10 (Tycho's SNR) do not give a reliable measure of the total H I optical depth. Our work has a frequency resolution of 27 km s^{-1} and does in fact give the total integrated H I along the line of sight. This frequency resolution is sufficient to resolve the individual spiral arms.

OBSERVATIONS OF THE STRUCTURES OF TYCHO'S AND KEPLER'S SUPERNOVA REMNANTS AT 2.7 AND 5.0 GHz

S. N. HENBEST
Dept. of Physics, University of Cambridge, Cambridge, England

1. Introduction

The fluid dynamical model of Gull (1973a, b) predicts that the radio emission of young supernova remnants occurs in instabilities at the boundary between the ejected mass and the 'swept-up' interstellar medium, and that no central energy source is necessary. High resolution radio observations of the thickness, polarization properties and structure of the shells of such objects provide a test for these theories, and also enable an upper limit to be set on the flux from point sources within the shell which could be supplying it with energy.

The remnants of Tycho's and Kepler's supernovae are of particular interest in this respect as they are both of known age, both are believed to have originated in type I

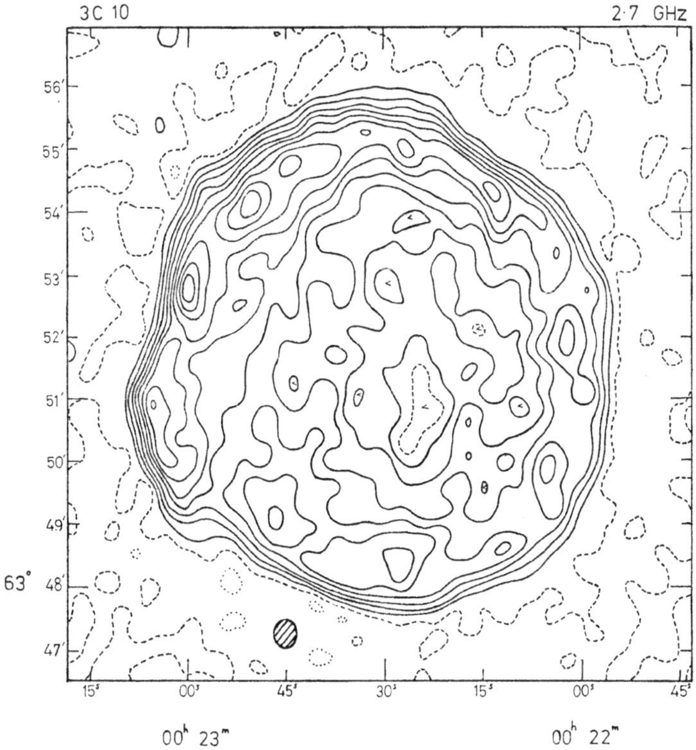

Fig. 1. Tycho's supernova remnant at 2.7 GHz. The contour interval is 7.6 K; the zero contour is dashed, negative contours are dotted and isolated minima are indicated by the symbol $<$. The shaded ellipse represents the half-power beamwidth ($24'' \times 27''$).

C. B. Cosmovici (ed.), Supernovae and Supernova Remnants, 303–305.

supernovae (Minkowski, 1968), and in neither has a pulsar been detected (Reifenstein *et al.*, 1969). They have therefore been mapped at 2.7 and 5.0 GHz with the One Mile telescope at Cambridge (Elsmore *et al.*, 1966), which uses the principle of aperture synthesis.

2. Tycho's Supernova Remnant (3C 10)

Figure 1 shows the total intensity map of 3C 10 at 2.7 GHz, with a half-power beamwidth of $24'' \times 27''$. It is very similar to the 1.4 GHz maps of Baldwin (1967) and Strom and Duin (1973) at the same resolution, implying that there is no significant change in spectral index over the remnant. This is in contrast with the results of Weiler and Seielstad (1971) who find a flattening of the spectrum in the north and south of the source at 2.88 GHz. These authors may not have made sufficient allowance for the effect of their first grating response, which is 7′ from the main beam in declination, the diameter of 3C 10 being 7.7′. The shell is very thin in places, particularly in the north-west, and contains seventeen peaks of emission. The outer edge is unresolved in many places even at twice this resolution.

The 5.0 GHz map of Tycho's supernova remnant, also with a half-power beamwidth

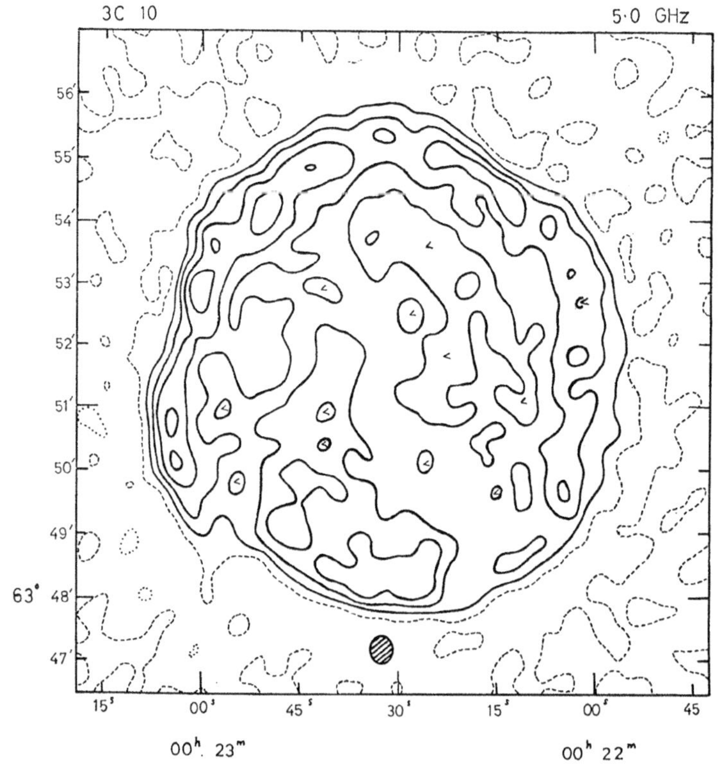

Fig. 2. Tycho's supernova remnant at 5.0 GHz. The contour interval is 3.0K; zero and negative contours, and isolated minima are represented as in Figure 1. The half-power beamwidth is again $24'' \times 27''$.

of $24'' \times 27''$, is shown in Figure 2. There is no significant difference between this and the 2.7 GHz map, again implying a constant spectral index over the source.

The absence of unresolved point sources on either map implies a 3σ upper limit on any such sources within the shell of 19 mfu at 2.7 GHz and 30 mfu at 5.0 GHz. (1 mfu $= 10^{-29}$ W m^{-2} Hz^{-1}).

The signal-to-noise ratio is too low to allow useful polarization data to be obtained at this resolution for either frequency, but 2.7 GHz results at lower resolution agree with the polarization measurements of Weiler and Seielstad (1971) and also Dickel and Hermann (private communication).

3. Kepler's Supernova Remnant (3C 358)

Our preliminary results are in agreement with the work of Dickel (private communication) at 2.7 GHz, who finds this remnant to be somewhat elliptical with a rather thick shell (thickness/outer radius $\sim$0.6) of diameter about 3.5′.

References

Baldwin, J. E.: 1967, 'Radio Astronomy and the Galactic System', *IAU Symp.* **31**, 337.
Elsmore, B., Kenderdine, S., and Ryle, M.: 1966, *Monthly Notices Roy. Astron. Soc.* **134**, 87.
Gull, S. F.: 1973a, *Monthly Notices Roy. Astron. Soc.* **161**, 47.
Gull, S. F.: 1973b, *Monthly Notices Roy. Astron. Soc.* **162**, 135.
Minskowski. R.: 1968, *Nebulae and Interstellar Matter*, Chapter 11, University of Chicago Press, Chicago.
Reifenstein, E. C., III, Brundage, W. D., and Staelin, D. H.: 1969, *Astrophys. J.* **156**, L125.
Rosenberg, I.: 1970, *Monthly Notices Roy. Astron. Soc.* **151**, 109.
Strom, R. G. and Duin, R. M.: 1973, *Astron. Astrophys.* **25**, 351.
Weiler, K. W. and Seielstad, G. A.: 1971, *Astrophys. J.* **163**, 455.

ON THE STRUCTURE OF THE PULSE OF NP 0532 AT γ-RAY ENERGIES

L. MARASCHI and A. TREVES

Istituto di Fisica dell'Università and Laboratorio di Fisica Cosmica e Tecnologie Relative del CNR, Milano, Italy

1. Introduction

Recent observations have extended the γ-ray spectrum of the Crab Nebula pulsar (NP 0532) to $\sim 10^{12}$ eV. A review of the experiments was given at this conference by Parlier *et al.* (1973).

The relevance of various γ-ray emission mechanisms has been discussed by Apparao and Hoffman (1970), by Zheleznyakov and Shaposhnikov (1972) and by Maraschi and Treves (1973). Synchrotron radiation may explain the pulsed flux of NP 0532 from the optical band to the highest energies. The contribution of Compton scattering may be significant only above 100 MeV and would arise at the speed of light cylinder.

In the light of these results the experimental data on the double structure of the pulsed emission at γ-ray energies are considered. On the basis of a simple model an explanation of the variation of the ratio of the two pulses with energy is proposed.

2. Experimental Data

The pulse of NP 0532 shows a double structure at optical and X-ray frequencies. The ratio of the energy contents in the two components varies regularly from the optical to the hard X-ray band (1 eV–100 keV). Above 100 keV, owing to the low statistics, the shape of the pulse is rather uncertain. Most observations indicate a double structure as at lower frequencies. The observed ratios of the secondary to the primary pulse component at various energies are reported in Figure 1. It is apparent that above 1 MeV the experimental situation is confused. The uncertainties may be even larger than indicated in the figure, since the width of the secondary pulse seems to vary with energy, and different authors do not use homogeneous criteria for estimating the pulse ratio. If one neglects the point of Grindlay (1972) at $\sim 10^{12}$ eV, which contrasts with the measurements of other observers at about the same energy (Fazio, 1972), one can deduce from the figure that above 1 MeV the ratio of the pulses does not increase with energy and possibly decreases.

3. Discussion

The origin of the multiple structure of pulses of pulsating sources at present is not well understood. The most natural interpretation is that subpulses correspond to different emission regions. The sites of emission considered in most pulsar models

C. B. Cosmovici (ed.), Supernovae and Supernova Remnants, 307–309.

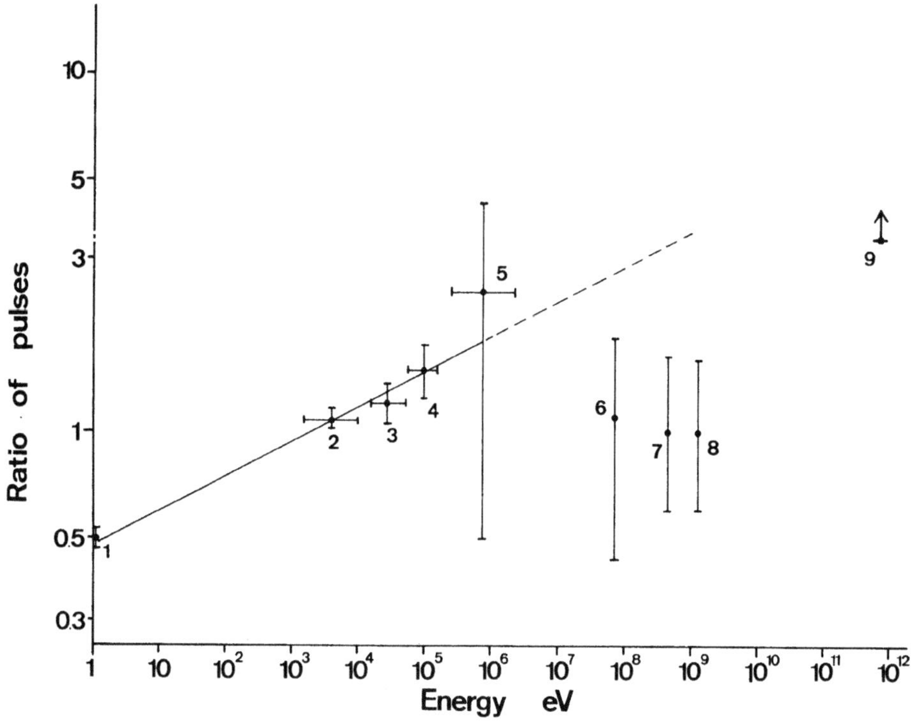

Fig. 1. Ratio of energy in the secondary pulse to energy in the main pulse of NP 0532. No secondary pulse was observed by Albats *et al.* (1972) between 10 and 100 MeV and by Parlier *et al.* (1973) above 20 MeV. For the observations see, (1) Papaliolios and Horowitz (reported in Rappaport *et al.*, 1971); (2) Rappaport *et al.* (1971); (3), (4) Boclet *et al.* (1972); (5) Orwig *et al.* (1971); (6) Browning *et al.* (1971); (7), (8) McBreen *et al.* (1973); (9) Grindlay (1972).

(e.g. Ruderman, 1972) are the polar caps, at the surface of the star, and the speed of light cylinder. We tentatively associate the two subpulses of NP 0532 with the two regions. It seems natural that the collimation of the beam of radiation should be narrower where the magnetic field is higher. This suggests that the sharper main pulse is produced at one polar cap, the other cap being not observable for geometrical reasons. The secondary pulse which is broader, would arise at the speed of light cylinder. With this scheme we try to interpret the observations.

The regularity of the pulse ratio curve between 1 eV and 1 MeV seems to indicate that the mechanism producing each peak is the same over the whole energy interval, and from the polarization of the optical radiation it is natural to consider synchrotron emission. The dependence of the pulse ratio with energy may be explained by a slightly different index γ of the electron spectra in the two regions of emission, $\Delta\gamma \simeq 0.1$. Above ~ 1 MeV the pulse ratio curve has a change which suggests a saturation of synchrotron emission in the secondary pulse. As mentioned in the introduction above 100 MeV the dominant mechanism of production of γ-rays at the speed of light cylinder may be Compton scattering. In that case the ratio of the energy content in the two pulses is not related to the values observed at lower energies. In the intermediate region (1–100 MeV) the pulse ratio depends on the detailed shape of the synchrotron and Compton spectra.

If a sharp break in the electron spectrum occurs at energies between 4.8×10^{10} eV and 4.8×10^{11} eV, corresponding to synchrotron photons of 1–100 MeV for a magnetic field $H_{\perp} \simeq 2.5 \times 10^4$ G, one should expect a minimum of the pulse ratio between the energy of the break and 100 MeV. In fact the Case-Melbourne measurements between 10 and 100 MeV (Albats *et al.*, 1972), and the French-Italian observations above 20 MeV (Parlier *et al.*, 1973) give no evidence of a secondary pulse. At 70 MeV Browning *et al.* (1971) find a ratio ~ 1, but the peaks are shifted of 3 ms with respect to the optical ones.

A clarification of the experimental situation between 1 and 100 MeV is required to allow a detailed discussion of the proposed model for the origin of the two peaks.

Acknowledgement

We are grateful to C. Dilworth for a critical reading of the paper.

References

Albats, P., Frye, Jr., G. M., Zych, A. D., Mace, O. B., Hopper, V. D., and Thomas, J. A.: 1972 *Nature* **240**, 221.
Apparao, K. M. V. and Hoffman, J.: 1970, *Astrophys. Letters* **5**, 25.
Boclet, D., Brucy, G., Claisse, J., Durouchoux, Ph., and Rocchia, R., 1972, *Nature* **235**, 69.
Browning, R., Ramsden, D., and Wright, P. J.: 1971, *Nature* **235**, 128.
Fazio, G. C.: 1972, in H. Bradt and R. Giacconi (eds.), 'X- and Gamma Ray Astronomy', *IAU Symp.* **55**, 303 (contains a review of the results up to May 1972).
Fazio, G. C., Helmken, H. F., O'Mogain, E., and Weekes, T. C.: 1972, *Astrophys. J. Letters* **175**, L117.
Grindlay, J. E.: 1972, *Astrophys. J. Letters* **124**, L9.
McBreen, B., Ball, Jr., S. E., Campbell, M., Greisen, K., and Koch, D.: 1973, submitted to *Astrophys. J.*
Maraschi, L. and Treves, A.: 1973, *XII Int. Cosmic Ray Conf., Denver*, Paper No. 541.
Orwig, L. E., Chupp, E. L., and Forrest, D. J.: 1971, *Nature Phys. Sci.* **231**, 172.
Parlier, B., Vasseur, J., Leray, J. P., Forichon, M., Agrinier, B., Boella, G., Maraschi, L., Buccheri, R., Sacco, B., and Scarsi, L.: 1973, *Nature Phys. Sci.* **242**, 117.
Rappaport, S., Bradt, H., and Mayer, W.: 1971, *Nature Phys. Sci* **229**, 40.
Ruderman, M.: 1972, *Ann. Rev. Astron. Astrophys.* **10**, 427.
Zheleznyakov, V. V. and Shaposhnikov, V. E.: 1972, *Astrophys. Space Sci.* **18**, 166.

DISCUSSION

P. Gorenstein: Did you do the same calculation for the Vela pulsar? Can you tell how the strength of the pulse varies with frequency?

A. Treves: No, not yet. I'm very much interested in the problem of the Vela pulsar and hope to get to it later.

SEARCH FOR GAMMA RAY LINES FROM SUPERNOVAE AND SUPERNOVA REMNANTS*

E. L. CHUPP**, D. J. FORREST, A. N. SURI,
RON ADAMS and C. TSAI

University of New Hampshire, Durham, N.H. 03824, U.S.A.

1. Introduction

An important astrophysical problem on which gamma ray astronomy has particular bearing is the study of gamma ray lines from supernovae (SN) in the process of development or from their remnants (SNR). Since the existence of many SNR is an experimental fact, most estimates of possible gamma ray emission concern the end result of the SN process. Exactly what occurs in the initial phase and how it is triggered is subject to considerably more speculation. Only a brief summary of some of the more pertinent predictions and observations will be given here.

The field of gamma ray astronomy received its initial impetus several years ago from the predicition by Morrison (1958) that nuclear emission lines from various radioactive nuclei could be observable from balloon or satellite instruments. Later Clayton and Craddock (1965) predicted gamma ray line emissions from the Crab Nebula based on the californium light decay curve hypothesis. Searches for these nuclear lines below 1.5 MeV have been made with balloons by Haymes *et al.* (1968) using a large area NaI spectrometer in an active collimator. The experiment was capable of setting a limiting flux of 4×10^{-4} photons cm^{-2} s^{-1} at 0.5 MeV. Also Jacobson (1966) has used a high resolution Ge(Li) detector to improve the measurements, but both experiments gave null results.

More recently Clayton *et al.* (1968) have calculated the gamma ray line intensities from type I Supernova assuming that Si burning takes place in the supernova shell, giving rise to Ni^{56}. Electron capture decay of Ni^{56} leads to several gamma ray lines which may be observable for as long as a year following the SN event. The lines which are strongest are at 0.812 MeV, 0.847 MeV, 0.511 MeV, and 1.31 MeV and could be discernible as lines from a SN at a distance of 1 mpc. Very recently Brown (1973) has made Monte Carlo calculations of the transport of gamma ray lines out of the expanding shell of a type I SN produced by Si burning. The calculations show that Compton scattering could produce a strong continuum with no lines resolvable below 0.8 MeV.

Clayton (1971) has extended calculations on explosive nucleosynthesis to include the production of Fe^{60}. The Fe^{60} undergoes β^- decay to Co^{60} which in turn undergoes

* This paper could not be presented at the Conference.
** At present Alexander von Humboldt Senior Fellow and Honorary Fulbright Fellow at the Max-Planck-Institut für Physik und Astrophysik, Institute für Extraterrestrische Physik, 8046 Garching, Federal Republic of Germany.

C. B. Cosmovici (ed.), Supernovae and Supernova Remnants, 311–316.

β^- decay to Ni^{60} giving the characteristic cascade pair of gamma rays at 1.17 MeV and 1.33 MeV. He estimates that supernovae remnants toward the galactic center may be sufficiently intense in gamma ray lines of Co^{60} (1.17 MeV, 1.33 MeV) and Sc^{44} (1.156 MeV) to give fluxes at the Earth of

$$F_{60}\,(\text{galactic}) \approx 1.7 \times 10^{-5} \rightarrow 1.7 \times 10^{-3}\ \gamma\text{'s cm}^{-2}\,\text{s}^{-1}$$
$$F_{44}\,(\text{galactic}) \approx 4.5 \times 10^{-4}\ \gamma\text{'s cm}^{-2}\,\text{s}^{-1}.$$

A recent GSFC conference on the Gum Nebula (Maran, 1971) has brought to light a possible additional source for gamma ray lines. If the Gum Nebula is a result of the hypothetical Vela X supernova (PSR 0833-45) and cosmic rays are produced in the process (Ramaty *et al.*, 1971), then gamma ray lines may be emitted as a result of energetic particle interactions in the nebular gas. The authors predict fluxes of several gamma ray lines from different processes. For the reaction Si^{28} (p, p'γ) giving a line at 1.78 MeV a flux of $<8\times10^{-3}\ \gamma$'s cm^{-2} s^{-1} is predicted on the basis that $\frac{2}{3}$ of the mass of the SN ejecta is silicon. Similarly, estimates are made for line emission fluxes at 4.43 MeV and 6.13 MeV from C^{12} (p, p'γ) and O^{16} (p, p' γ) respectively. The results are:

$$F\,(4.43\text{ MeV}) \simeq 4 \times 10^{-2}\ M_C/M_\odot\ \gamma\text{'s cm}^{-2}\,\text{s}^{-1}$$

and

$$F\,(7.14\text{ MeV}) \simeq 1.5 \times 10^{-2}\ M_O/M_\odot\ \gamma\text{'s cm}^{-2}\,\text{s}^{-1}.$$

Burbidge (1966) has suggested that the energy released in burning 10^{-2} solar masses of hydrogen could be emitted in MeV gamma rays in 10^3 s during a SN outburst. In this case specific gamma ray lines at 0.51 MeV, 2.2 MeV and 5.5 MeV would be produced but the flux for a SN at 4 mpc might only be $\sim10^{-4}$ photons cm^{-2} s^{-1}.

The above examples suggest which gamma lines may be expected; however, flux estimates may be uncertain by several orders of magnitude either way.

2. Description of Experiment

Figure 1 shows a schematic of the detector and its arrangement in the rotating wheel compartment of the OSO-7 spacecraft. The gamma ray monitor (Higbie *et al.*, 1972), designed to measure the gamma ray spectrum from 0.3 to 10 MeV, consists of a 3″ × 3″ NaI(Tl) cyrstal surrounded by a cup-shaped anticoincidence shield of CsI(Na). This arrangement also enhances the relative sensitivity for lines by supressing the Compton continuum. The instrument is automatically calibrated regularly each orbit at day/night and night/day transitions with a Co^{60} source giving photopeaks at 1.17, 1.33 and 2.50 MeV. The calibration source (Forrest *et al.*, 1972) consists of a neutron-activated cobalt compound in a plastic scintillator button, which is viewed by a photomultiplier. The beta decay of the Co^{60} provides a gating pulse which is used to prevent the coincident gamma rays from Ni^{60} from being recorded during a normal gamma ray spectrum accumulation. At each day/night or night/day transition of the spacecraft, a signal automatically puts the Co^{60} beta pulse in coincidence with

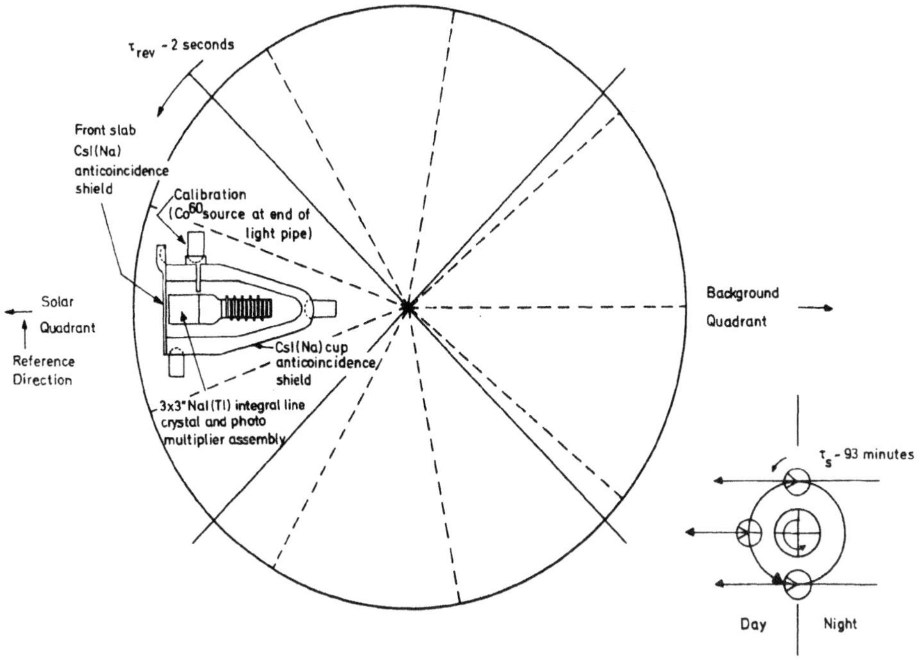

Fig. 1.

a signal in the main gamma ray detector, which is due in this case primarily to an absorption of one of the gamma rays at 1.17 MeV or 1.33 MeV. Occasionally, both gamma rays are simultaneously absorbed, giving a sum peak at 2.50 MeV. During normal operation, when the calibration lines are gated out of the spectrum, there is about 5% leakage of Co^{60} lines into their respective channels. This scheme allows the energy of a well defined peak to be determined to about 2% and verifies that the energy resolution is maintained at ~8% (referred to 662 keV). The energy resolution improves with energy approximately as $E^{-1/2}$. The gamma ray monitor has an auxiliary X-ray detector (not shown in Figure 1), consisting of a NaI(Tl) crystal of diameter $1\frac{1}{4}''$ and thickness $\frac{1}{4}''$, covered by a 10 mil thick Be foil. The X-ray energy spectrum from 7.5 to 120 keV is analyzed by a four channel analyzer. A complete X-ray spectra is taken every 30 s.

During the normal mode of operation (see Figure 1) the gamma ray pulse height spectra are accumulated over 90 wheel rotations (taking approximately 3 min) in both solar and background quadrants. The gamma ray spectrum is analyzed with a 377-channel quadratic pulse height analyzer and stored in a buffer until readout. The solar and background directions are the same during the nighttime part of the orbit so the background quadrant looks at the sky during the night. A command allows the viewing quadrants to be shifted ninety degrees (90°) counter clockwise as shown in Figure 1. Both the normal and alternate modes were used in acquiring the data discussed here. All data discussed were from satellite nighttime.

A special point should be made though; since the satellite spends about $\frac{1}{3}$ of its time at night and $\frac{1}{4}$ of this time viewing in the direction of any one quadrant, there is only about a 10% chance of observing a burst which lasts less than ~2000 s.

3. Analysis of Data

The OSO-7 experiment has allowed a partial search to be made for gamma rays associated with a supernova in NGC 5253 (1972) and a remnant associated with the pulsar 0833-45. In Table I we list the gamma ray emissions that have been studied corresponding to each source.

TABLE I

Gamma ray search technique

Source	Gamma emissions studied
Gum Nebula (possible remnant)	From low energy cosmic ray interactions: $Si^{28}(p, p' \gamma)$; $E = 1.78$ MeV C^{12} $(p, p' \gamma)$; $E = 4.43$ MeV O^{16} $(p, p' \gamma)$; $E = 6.13$ MeV cf: Ramaty and Boldt: 1971, GSFC Report-X-683-71-375.
SN in NGC 5253 Apparent type I Observed near maximum on May 13, 1973 cf: Kowal, 1972, *IAU Circular*, No. 2405.	*X-rays* (10–120 keV) April 10–May 23, 1972 *Gamma ray lines* Co^{56} (0.847 MeV) cf: Clayton *et al.*: 1969, *Astrophys. J.* **155**, 75. also 0.51 and 2.2 MeV *Continuum* 350 keV–1 MeV cf: Brown: 1973, *Astrophys J.* **179**, 607.

In the case of the Gum Nebula the scan plane of the gamma ray monitor passed near the Gum Nebula for several weeks while gamma ray spectra were being accumulated in the normal solar and background quadrants. During the time period analyzed, the alternate quadrant 4 (see Figure 1) could receive gamma rays from at least 50% of the Gum Nebula region as given by Brandt *et al.* (1971). We therefore adopted a plan of operation that switches data taking from our normal quadrants 1 (solar) and 3 (background) for a few orbits a day to alternate quadrants 2 and 4. Thus several spectra are available which should indicate any detectable fluxes from the Gum Nebula. In this case a search was made for nuclear lines from Si^{28} (1.78 MeV), C^{12} (4.43 MeV) and O^{16} (6.13 MeV) by comparing spectra taken from the Gum direction with equivalent non-Gum spectra. Under optimum viewing conditions the detector receives radiation from ~80% of the Gum Nebula, if the full nebula is a source of gamma rays. No evidence for line emission exists at a flux level of 5×10^{-2} cm^{-2} s^{-1} for either the lines at 1.78, 4.43 or 6.13 MeV. These flux limits allow limits to be placed on the relative amounts of Si$(M_{Si}/M_\odot)$ and C$(M_C/M_\odot$ and O$(M_O/M_\odot)$ present in the nebula. The current results are $M_{Si}/M_\odot < 3$; $M_C/M_\odot < 1$;

$M_0/M_\odot < 3$ if the cosmic ray flux currently existing in the nebula is that predicted by Ramaty and Boldt (1971). By summing spectra we hope to be able to reduce the mass limits by at least an order of magnitude.

The type I supernovae (Kowal, 1972) in NGC 5253 was discovered on May 13, 1972 with a magnitude of 8.5. However, the time of its maximum magnitude may have been May 6, or earlier (IAU circular 2421). During the time from April through May, NGC 5253 was near the direction of view of the background quadrant during satellite night. By scanning through the X-ray data from April 10 to May 23, we have found no indication of enhanced X-ray radiation from this direction over background.

The average background rate of X-ray channel 1 is about 200 cts s^{-1}. The statistical fluctuation of the background rate in a normal scan at three standard deviations is 21 cts s^{-1}, therefore taking 8 cm^2 as the sensitivity for the energy range 7.5 keV $\rightarrow$ $\rightarrow$ 15 keV, the limiting flux would be about $\frac{21}{8} \times 7.5 \times 1.6 \times 10^{-9} = 3 \times 10^{-8}$ erg cm^{-2} s^{-1}. The distance of the SN from the Earth is roughly about 4 mpc; therefore, estimation of the maximum average power emitted initially from the SN or until several weeks after the outburst gives about 4×10^{44} erg s^{-1} at about 10 keV. At $\sim$50 keV the maximum average power that could be emitted is $\sim 3 \times 10^{43}$ ergs s^{-1}.

According to Sklovskii (1973) the power of X-ray radiation from a type I SN 2 to 3 months after the outburst could be 3×10^{42} ergs s^{-1}. Hence the current upper limit for the average power several weeks after the outburst is a factor of 10^2 larger than that of the SN 2–3 months after the outburst. On the other hand Sklovskii (1973) also points out that prior to 2 or 3 months after the outburst the supernovae's envelope absorbs the X-ray radiation from within the shell.

A search was also made for specific unshifted gamma ray line emissions before and after the optical maximum at 0.51 MeV, 0.847 MeV and 2.2 MeV. In all cases no evidence exists for enhanced emission with a limiting line flux at $\sim 10^{-2}$ cm^{-2} s^{-1} for a transit burst. In the case of the Co^{56} gamma ray line at 0.847 MeV predicted by Clayton *et al.* (1969), spectra were summed for 1300 s or ten full data accumulations on May 3, 1972, which should be near the time of the SN light curve maximum. Since no evidence existed for a line at this energy the limiting flux is $\leqslant 6 \times 10^{-3}$ (0.847 MeV photons cm^{-2} s^{-1}). Finally, in view of the calculations mentioned above which suggest significant Compton broadening of the spectral lines predicted by Clayton *et al.* (1969), we have searched for an enhanced continuum over the range 350 keV to 1 MeV. Specific searches were made on May 1, 3 and 23 all with null results, giving a limiting flux of 3×10^{-2} photons cm^{-2} s^{-1} in the 650 keV energy band specified above.

4. Conclusions

The brief analysis described here has not given any positive indication of X-rays or gamma ray lines or continuum from a recent relatively nearby supernova. However, since the live time available for observing a transient burst (<2000 s) was quite small, any such event might have been missed even though the instrument is capable of detecting a burst of 10 keV photons with an energy flux of $\sim 10^{-7}$ erg cm^{-2} s^{-1}

or 1 MeV photons with an energy flux of $\sim 10^{-8}$ erg cm^{-2} s^{-1}. A supernova of the apparent magnitude (8.5) of SN-1972-Kowal is relatively rare, only four with a magnitude less than 10 having been observed between 1885 and 1971 (Kowal and Sargent, 1971). Therefore in order to significantly increase the chances of detecting X-ray or gamma ray SN events, the sensitivity of the instrument should be increased by at least two orders of magnitude and a balloon or satellite system developed with a 100% observational live time for any part of the sky. Such a system would greatly increase the prospects of detecting a transient burst or continuous line or continuum spectral emissions.

All the possibilities of detecting gamma ray lines from these objects are not yet exhausted however; since summing of selected spectra can reduce line flux limits to $\sim 10^{-3}$ photons cm^{-2} s.

Acknowledgements

This work was supported by NASA under Contract Nas 5-11054 and NGR 30-002-021. We are thankful for the contributions of several persons in the fabrication and data analysis phase of the experiment, particularly Dr P. Higbie, A. A. Sarkady, I. U. Gleske, S. Foss, Sue Croteau, Margaret Simmons, Kishore and Dipika Patel, P. Ferguson, P. Dunphy and Dr C. Reppin. We also express our appreciation to Mary Miklos Chupp for editing the manuscript.

References

Brandt, J. C.: 1971, Goddard Space Flight Center Report X-683-71-375, p. 5.
Brown, R. T.: 1973, *Astrophys. J.* **179**, 607.
Burbidge, G.: 1966, in *High Energy Astrophysics* (ed. by L. Gratton), Academic Press, New York, p. 145.
Clayton, D. D. and Craddock, W. L.: 1965, *Astrophys. J.* **142**, 189.
Clayton, D. D., Colgate, S. A., and Fishman, G. J.: 1969, *Astrophys. J.* **155**, 75.
Clayton, D. D.: 1971, Nature **234**, 291.
Forrest, D. J., Higbie, P. R., Orwig, K. E., and Chupp, E. L.: 1972, *Nucl. Inst. MTDS* **101**, 567.
Haymes, R. C., Ellis, D. V., Fishman, G. J., Kurfess, J. D., and Tucker, W. H.: 1968, *Astrophys. J.* **151**, L9.
Higbie, P. R., Chupp, E. L., Forrest, D. J., and Gleske, I. U.: 1972, *IEEE Trans. Nuc. Sci.* **NS-19**, 606.
Jacobson, A. S.: 1968, Doctoral Thesis, University of California, San Diego.
Kowal, C. T.: 1972, *IAU Circular*, No. 2405.
Kowal, C. T. and Sargent, W. L. W.: 1971, *Astron. J.* **76**, 756.
Maran, S. P. (Ed.): 1971, Goddard Space Flight Center Report X-683-71-375.
Morrison, P.: 1958, *Nuovo Cimento* **7**, 858.
Ramaty, R. and Boldt, E. A.: 1971, Goddard Space Flight Center Report X-683-71-375, p. 97.
Sklovskii, I. S.: 1973, *Sov. Astron.* **16**. 749.

OPTICAL OBSERVATIONS OF SUPERNOVA REMNANTS SIMEIZ 59 AND W28*

T. A. LOZINSKAYA
Sternberg Astronomical Institute, Moscow, U.S.S.R.

As a part of the program of investigation of kinematics and physical conditions in supernova remnants, detailed interferometric observations of two new objects were carried out. The first is the nebula Simeiz 59 (Sharpless 104); its size (diameter) is 8′, identified with radiosource G 74.8+06 (see Figure 1). The second is a very faint filamentary nebula in the central region of the non-thermal radio source W28; its size (diameter) is about 30′ (see Figure 2 from Milne and Wilson (1971)). In both cases an expansion of the supernova optical remnants was found. The velocity of expansion and radial velocity were evaluated and some conclusions about their nature were drawn.

The observations in the lines 6563 Å and 6584 Å were carried out. A high contrast Fabry-Pérot etalon and contact image converter mounted in Cassegrain focus of the Sternberg Astronomical Institute 125-cm reflector were used. Fabry-Pérot etalon free spectral range was 700 km s^{-1} in radial velocity scale. Linear dispersion in the

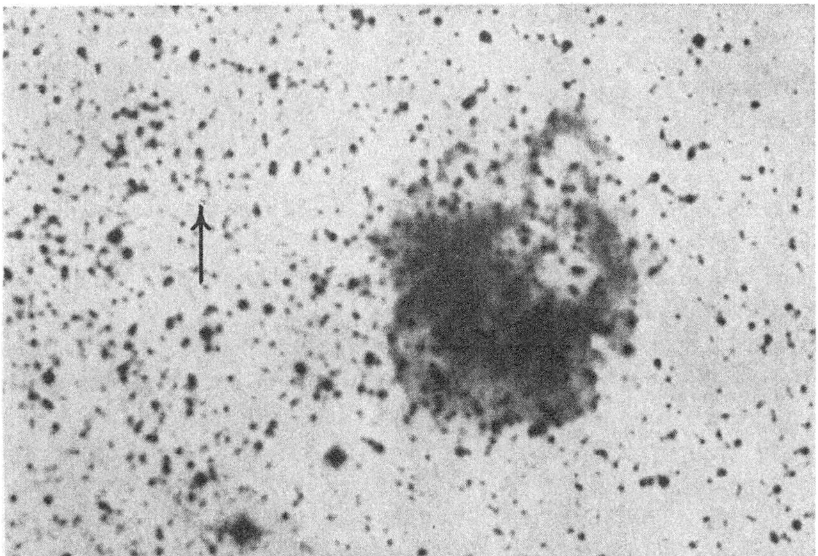

Fig. 1.

* This paper could not be presented at the Conference.

C. B. Cosmovici (ed.), Supernovae and Supernova Remnants, 317–320.

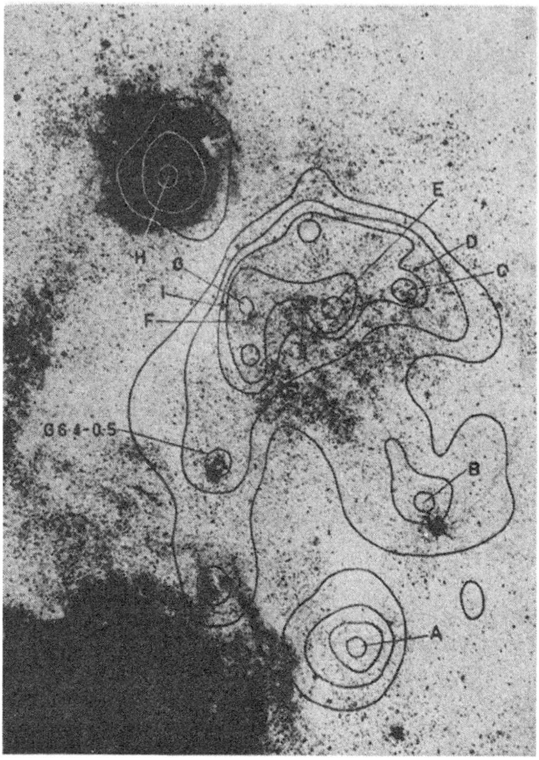

Fig. 2.

central interferometric ring was 5–7 Å mm^{-1}, actual spectral resolution corresponded to 20 km s^{-1}.

15 interferometer plates of Simeiz 59 in the line Hα were obtained. 20 plates of the filamentary nebula W28 in Hα line and 12 plates in 6584 Å were obtained. The mean exposure was 30–60 min. The radial velocity and the half-width of the line in 50–80 points of each nebula were evaluated.

Statistical reduction of the relation between radial velocity and line halfwidth on one hand and distance from the Simeiz 59 centre on the other hand was carried out. An expansion of the supernova remnant was found with the velocity 12 ± 5 km s^{-1}. The radial velocity of the object was found to be $+18 \pm 3$ km s^{-1} relative to the Local Standard of Rest.

The contrast of the faint filaments of W28 above the galactic background appeared to be higher in the line 6584 Å than in Hα line. The observations of W28 in the 6584 Å line resulted in the discovery of expansion effect with the mean velocity of 40–50 km s^{-1}. The W28 mean radial velocity according to the observations in Hα line is $+18 \pm 5$ km s^{-1} relative to the Local Standard of Rest.

The results obtained allow us to reach new conclusions concerning the investigated supernova remnants.

W28

The radio source W28 consists of three separate extended sources: two H II regions M20 and W28-A, and a non-thermal shell – the supernova remnant. The W28 source was thoroughly investigated in the radiofrequency range, in $H_{109\alpha}$ and OH lines (Milne and Wilson, 1971). The above mentioned radial velocity of the filamentary shell shows that the supernova remnant is not projected on a dense complex of dust and gas, consisting of M20 and W28-A, but is submerged in it. (The radial velocities of three objects are the same). The kinematic distance in the direction of W28 ($l=6°$) is not reliable. We shall take the radio distance to the supernova remnant, according to Milne (1970), $R=1.5$ kpc, linear radius $r=6.5$ pc.

The undisturbed density of the interstellar gas can be calculated based on the observation in 21 cm in absorption (Radhakrishnan *et al.*, 1972). If the size of the absorbing cloud is about 10^{20} cm, the density of the neutral atomic hydrogen is $\varrho=3\times10^{-23}$ g cm^{-3}. But a considerable part of the hydrogen in a dense complex of dust and gas may be in the molecular state. Indeed, the thermal radio flux density and $H_{109\alpha}$ brightness temperature of W28-A is two times as high as that of M20. Nevertheless, Hα-radiation of W28-A was not observed neither on our plates nor on the red Palomar Sky Survey print. Thus the lowest possible local absorption in the region W28-A calculated according to the Hα brightness of M20 is $4\overset{m}{.}5$. It corresponds to the dust density 10^{-4} g cm^{-2} along the line of sight. If the size of the absorbing cloud is about 10 pc (half the distance between M20 and W28-A), the dust density $\varrho_{dust}=3\times10^{-24}$ g cm^{-3}. Consequently, $\varrho_{gas}=3\times10^{-22}$ g cm^{-3}. This proves that most of the gas near W28 is in the molecular state. Taking this approximate value of gas density, expansion velocity 40–50 km s^{-1}, radius 6.5 pc, the kinetic energy of the ejected shell was calculated as $E=3\times10^{50}$ erg and the age of the remnant $t=60000$ yr. The identical calculation for the supernova remnants IC 443 and Cygnus Loop gave the kinetic energy of the ejected shell 5×10^{49} erg and 2×10^{50} erg and age 6×10^4 yr and 9×10^4 yr correspondingly. Thus W28 is a typical supernova remnant of type II.

Simeiz 59 (G 74.8+0.6)

Milne (1970) and Downes (1971) identified the radio source G 74.8+0.6 as a supernova remnant of type II and calculated by radio method the distance to be $R=9$–12 kpc and the linear radius to be $r=6$–8 pc.

But the results of our optical observation contradict this conclusion, namely:

(1) According to the Schmidt model of galactic rotation the distance of 9 kpc corresponds to the radial velocity of about 50 km s^{-1}. Our result is $Vr=+18\pm3$ km s^{-1}. Such difference cannot be explained either by errors of observation or by a peculiar radial velocity of the object.

(2) A knowledge of the expansion velocity, $V=12$ km s^{-1}, and radius, $r=6$–8 pc, can be used to determine the undisturbed density of the interstellar gas. With the usual value of the type II supernova kinetic energy of ejected shell $E=10^{50}$–10^{51}

erg, the undisturbed density of the gas will be too high, namely: $n_0 = 10^3$–10^4 cm^{-3}.

Thus Simeiz 59 cannot be a supernova remnant of type II. The nebula probably was formed after a type I supernova explosion in a dense interstellar gas cloud. If it is really a supernova remnant of type I the radio method does not let us determine the distance reliably enough as there are no type I supernovae with known distance. The kinematic distance to Simeiz 59 according to the radial velocity obtained is $R = 3$ kpc; the corresponding radius is $r = 3$ pc. In this case $n_0 = 10$–100 cm^{-3} if the kinetic energy of ejected shell is usual for type I supernova $E = 10^{47}$–10^{48} erg. Such density is possible if the supernova exploded in a dense interstellar gas cloud. The values of r and V may be used to determine the age of the object $t = 10^5$ yr.

In 1969, investigating the filamentary nebula Simeiz 22, the author came to the conclusion that this nebula is also a supernova remnant of type I. The distance to Simeiz 22 was determined 3.5 kpc, radius 4 pc, expansion velocity about 45 km s^{-1}, age 3.5×10^4 yr (Lozinskaya, 1969). Thus, already two supernova remnants of the late stage of evolution are known. The question arises how two supernova remnants of type I may exist during 3×10^4–10^5 yr, while Kepler's supernova which exploded about 500 yr ago was so quickly dissolved in the interstellar gas. Shklovsky suggested that if a supernova event occurs in a dense cloud, then the conservation of the shell takes place. The expansion of the shell is rapidly slowed down by the interstellar gas but relatively high optical and radio brightness of the shell are retained.

The fact, that two supernova remnants of type I occurred in dense clouds at the distance of 3 kpc from the Sun during 10^5 yr gives the mean frequency of about 1 supernova explosion in 250 yr. It corresponds to the usually adopted frequency of type I Supernova explosion in spiral galaxies.

References

Downes, D.: 1971, *Astron. J.* **76**, 305.
Lozinskaya, T. A.: 1969, *Sov. Astron. J.* **46**, 730.
Milne, D. K.: 1970, *Austr. J. Phys.* **23**, 425.
Milne, D. K. and Wilson, T. L.: 1971, *Astron. Astrophys.* **10**, 220.
Radhakrishnan, V. *et al.*: 1972, *Astrophys. J. Suppl. Ser.* **24**, 203.

PART VI

THEORIES ON SUPERNOVAE AND SUPERNOVA REMNANTS

SUPERNOVAE AND THEIR REMNANTS

L. WOLTJER
Department of Astronomy, Columbia University, N.Y., U.S.A.

Supernovae of type II must be related to rather massive stars. They appear to occur rather exclusively in the spiral arms of galaxies and this fact alone sets a lower limit of about 5 $M_{\odot}$ to their main sequence mass. If we believe that all massive stars above some critical value become supernovae (and none below this value) the supernova rate would indicate masses in excess of 10 or 20 $M_{\odot}$, the precise value depending on the particular choice of the luminosity function. Arguments of this type, however, do not take into account the possible dependence of stellar evolution on the presence of a binary system and other factors like rotation.

It should be stressed that these mass values refer to the main sequence. There is some indication of mass loss prior to the supernova outburst – to which we shall return later – and little is known about the actual presupernova mass, except that it probably still is several times that of the Sun.

As such a rather massive star evolves the interior will pass through successive phases of nuclear burning at increasing temperatures. Ultimately some instability follows, resulting for example from neutrino emission or the iron-helium phase change. This leads to collapse of the core, possibly to form a neutron star. At the same time neutrinos generated during the collapse, explosive nuclear reactions or possibly electromagnetic wave pressure from the collapsing object generates an outward moving shock in the envelope which ultimately is ejected. The details of how this happens depend on the density profile in the supernova. For example if the density decreases as fast as in a polytrope of order 3, the shock may further out become relativistic and cosmic rays may be accelerated, as proposed by Colgate; if the envelope is more comparable to that of a red giant interesting fragmentation effects may occur which ultimately result in the formation of some deuterium. However in all cases the bulk of the matter should come out non relativistically.

According to the results of Dr Searle an 'atmosphere' of about 0.3 $M_{\odot}$ is ejected at an average velocity of 10000 km s^{-1} for a total energy of 3×10^{50} erg. A larger amount of gas – perhaps as much as 5 $M_{\odot}$ from Dr Gordon's estimates – follows at a lower velocity of 2000 km s^{-1} for a kinetic energy of 2×10^{50} erg, making the total for the whole shell 5×10^{50} erg. Of course these numbers are all uncertain but they show that probably of the order of 10^{51} erg or less of energy is ejected in a supernova event, a figure that agrees well with independent estimates from the energetics of supernova remnants. The total binding energy of the neutron star is of the order of 10^{53} erg and thus 99% of the total energy is 'lost', probably in the form of neutrinos.

As matter and energy are injected into the surrounding interstellar medium, a supernova remnant is formed. In the earliest stages the matter will in essence expand freely but soon a significant amount of interstellar matter is swept up. A strong shock

C. B. Cosmovici (ed.), Supernovae and Supernova Remnants, 323–327.

will separate the undisturbed interstellar medium from the matter of the remnant and behind the shock the density will be up by a factor of four, while the temperature will be high. The temperature will be such that approximately $\frac{3}{2}kT=\frac{1}{2}mV^2$, with m the mean mass per particle and V the shock velocity. Putting in all the proper constants it is found that for $V=100$ km s^{-1}, T is somewhat less than 200000 K. The thermal energy per unit mass is proportional to T, but the rate of cooling at high T only to $T^{1/2}$ (Bremmsstrahlung). Consequently the cooling is unimportant as long as the shock velocity is high. The free expansion phase is therefore followed by a phase in which a hot cloud of gas of increasing mass expands with essentially constant energy into the surrounding matter. As the expansion slows down further the cooling time becomes smaller and ultimately the situation is one where the shock heated gas cools almost immediately. Pressure effects are now unimportant in the equation of motion of the gas and therefore linear momentum is preserved, i.e., MV=constant. The kinetic energy is $\frac{1}{2}MV^2$ and we therefore find that the energy of the remnant goes down proportionally with the velocity. The change over from the adiabatic to the momentum conserving phase occurs at about 200 km s^{-1} under typical conditions.

From observations of the optical emission lines in supernova remnants we may estimate the local interstellar density and hence the mass of a supernova remnant. From the observed velocity of expansion we may then obtain the present total kinetic energy and extrapolating back to the end of the adiabatic phase also the initial energy. On the basis of the observed velocities of the filaments in the Cygnus Loop and IC443 initial energies of the order of 10^{50} erg are obtained. If as we shall discuss later – the actual expansion velocities are larger than those in the filaments, energies closer to 10^{51} erg would follow. We note that in order of magnitude these estimates are in satisfactory agreement with those made independently from the spectra of the supernova. So far we have dealt with the supernova event in terms of a spherically symmetrical expansion into a uniform medium. We next consider the effect of rotation and magnetic fields. If the supernova has the angular velocity expected for a giant in a state of solid body rotation the resulting neutron star will rotate with a period of the order of 100 ms or somewhat more, just right to give typical pulsar periods. If the interior magnetic field on the main sequence is of the order of 100 G (compared to magnetic fields at the surface of the Sun of 1 G and in the atmospheres of peculiar A stars of up to 10^5 G) the field on the neutron star will be a typical pulsar field of 10^{12} G.

Somewhat larger fields and rotation rates deeper down in the star could have very interesting effects. As shown by Wilson and Leblanc the combination of a poloidal magnetic field and differential rotation may cause the ejection of jets during the moment of maximum compression. What happens is that the rotation winds the field into a mainly toroidal configuration. The repulsion of the toroidal loops then forces the jets out parallel to the rotation axis. In a case calculated by Wilson and Leblanc the energy in the jets was several times 10^{51} erg. In real magnetic stars the field is in general not at all symmetric between the two poles. As a result one would expect the jet ejection to be a quite asymmetrical process. In this case the neutron star would have to absorb the recoil momentum from the asymmetric ejection and neutron star velocities of the order of a few hundred km s^{-1} would be possible. It is interesting to note that the pulsar in the Crab has a transverse velocity of about 100 km s^{-1}, the one near IC443 of 250 km s^{-1}, while for several other pulsars velocities of the order of a few hundred km s^{-1} have been inferred from interstellar scintillations.

If the magnetic and rotational axes of the pulsar are not aligned strong electromagnetic waves are generated. These waves and also the electric fields near the pulsar accelerate particles. The positive and negative charges accelerated near the pulsar follow different trajectories and the resultant current systems generate a more static magnetic field. The relativistic electrons moving in the static field emit synchrotron radiation; a very similar type of radiation results from the interaction of the electrons with the electromagnetic wave field (synchro-compton radiation). The main practical difference between the two types of radiation is that the synchro-compton radiation should have a few percent of circular polarization in addition to the linear polarization of ordinary synchrotron radiation.

In the optical continuum emission of the Crab Nebula no intrinsic circular polarization is found at a level of 0.1% and apparently over the main body of the nebula the field is of the static variety. It has been suggested by many investigators, however, that near to the pulsar – in the wisp region – the electromagnetic wave field dominates. This could explain the local emission minimum near the pulsar: the electromagnetic waves sweep the relativistic electrons out and the emission in that region would be low. If so the shell structure of objects like Cas A perhaps be understood as due to a relatively stronger wave field. Since stellar magnetic fields seem to have a rather random orientation with respect to the stellar rotation axes and since the wave field would be strong when the angle between the two axes is large, the relative rarity of non-shell sources like the Crab could then also be understood. Of course in later phases the interstellar magnetic field swept up by the shell becomes also important. In general one would expect the internal or external fields squeezed against the shell to be tangential to the shell. However the radio polarization pictures show that frequently radial fields predominate. This need not indicate really radial fields; more probably this is the result of Rayleigh-Taylor type instabilities in the shell which make the field lines wavy. Also the shearing between stationary density concentrations in the interstellar or circumstellar medium and the moving shell (like in Cas A) may have a similar effect. Inhomogeneities in the interstellar or circumstellar medium may have important effects in the evolution of supernova remnants. If a shock meets a denser cloud, the shock transmitted into the cloud will have a lower velocity. As a result in the denser regions the temperature will be lower than would be estimated from the over all shock velocity. This may be explanation for the comparitively low X-ray temperature for objects like Cas A and Tycho, because the denser regions contribute most strongly to the emission (proportional to n^2). In objects like the Cygnus Loop the same effect may be occuring; as first suggested by Tucker the filaments would then move more slowly than the main shock and this could allow us to reconcile the rather high X-ray temperature (3×10^6 K) with the low velocity of the filaments

($100\ \mathrm{km\ s^{-1}}$ corresponding to 2×10^5 K). A similar situation might prevail in Vela X, where it is perhaps confirmed by the low age of the pulsar (12000 yr), which suggests an expansion velocity several times larger than that of the filaments. However the region is quite complex and the precise dynamics of the Vela shell remain to be elucidated. In any case we should not generalize the picture for the Cygnus Loop too much: in the case of IC443 it is already quite difficult to fit the age evaluated on the basis of the expansion velocity of the filaments ($\sim$35000 yr) with that of the pulsar (95000 yr) and it would be difficult to believe that the basic shock moves faster than the filaments. However in the case of IC443 the identification of an X-ray source with the supernova remnant still seems rather uncertain in view of the large error box (6 deg^2) of the Uhuru position for the X-ray source.

The presence of inhomogeneities is seen most clearly in Cas A, where the fast filaments of the supernova shell are intermixed with slow, dense filaments which must have been present before the supernova outburst. This is especially interesting because as Van den Bergh and Peimbert have shown the composition of both kinds of filaments is anomalous. It is difficult to determine the composition unambiguously, but the most likely interpretation of the spectra is that the moving filaments have a very large overabundance in oxygen (with respect to hydrogen), and that the stationary filaments are overabundant in nitrogen. This suggests that the stationary filaments are made of matter in which CNO cycling has taken place and which was lost by the pre-supernova well before te outburst. At the time of the supernova explosion only the layers of the star in which nuclear burning had proceeded to oxygen were still left, and it therefore seems probable that the pre-supernova had a mass much smaller than its main sequence mass.

The detailed electromagnetic evolution of the supernova remnants is very much interwoven with the evolution of the pulsars – assuming that all remnants have an associated pulsar. A pulsar which loses its rotational energy through purely electromagnetic torques will have an energy output of the form

$$L = \frac{L_0}{(1 + t/\tau)^2}$$

with L_0 the initial energy output at time $t=0$, and τ a characteristic time, which depends on the initial pulsar parameters. Setti and I have studied the evolution of the nebular magnetic field assuming that some fixed fraction of L is added to the magnetic energy content of the remnant. As the remnant expands magnetic energy is lost, however, by the expansion effects. As a result it appears that the field energy first increases – reaching half its maximum value at about $t=\frac{1}{2}\tau$. Subsequently it reaches a very flat maximum until at about $t=20\tau$ it declines again to half its maximum value. Pacini and Salvati recently have investigated the evolution of the particle content of the remnant assuming that a fixed fraction of L goes into relativistic electrons with a power law energy spectrum. Both expansion losses and synchrotron radiation losses are taken into account. The latter are so severe in the earlier phases that the whole present relativistic electron population of an object like the Crab

Nebula must have originated well after the initial outburst. In the later phases of evolution the energy input from the pulsar becomes comparatively negligible and the evolution proceeds along the lines of the expansion models discussed first by Shklovsky and subsequently by many other authors.

A more detailed discussion of most topics discussed here, as well as a full set of references may be found in our article in *Ann. Rev. Astron. Astrophys.* **10** (1972).

DISCUSSION

R. M. Duin: You can remove the discrepancy of the lifetime of the I 443 pulsar and the lifetime calculated from the expansion velocity by putting IC 443 at 3 kpc. The large dispersion measures also points to a larger distance.

L. Woltjer: Perhaps so, but it still would be impossible to have the basic shock much faster than the filaments as appears to be needed in the case of the Cygnus Loop and Vela X in order to explain the X-ray emission.

F. Zwicky: Using composite analytical photography of polarisation one obtains a basket weave pattern. As predicted in 1952 this pattern has now moved. Nobody seems to have given any explanation.

L. Woltjer: If one looks at the detailed polarisation maps, I'm not sure that this pattern is still so clearly discernible. Part of it, I suppose is related to the moving wisps which, as Scargle showed, seem to propagate all through the nebula.

F. Zwicky: Because of the radiation separating electrons from protons and high radiation pressure on ions with strong resonance lines large electric potential differences are built up which only collapse at a late stage and produce cosmic rays. The generation magnitude and final collapse do not seem to have been investigated further after the first papers of Zwicky 35 yr ago.

L. Woltjer: I would have expected the charge separation potential to be self limiting in that it would cause the electrons and protons not to separate too much.

W. M. Goss: Do you think it is significant that the Crab has an internal pulsar and no shell structure while Cas A, Tycho etc. have no detected pulsar and a shell structure in the remnant?

L. Woltjer: It is difficult to say on the basis of one object. If the difference in structure is related to the predominance of the electromagnetic wave field in Cas A, then this would be related to the angle between Ω and H in the pulsar and perhaps this would affect the chance of seeing the pulsar.

L. Biermann: I seem to remember models of a star's explosion, in which nothing was really there. (No neutron star or black hole.)

L. Woltjer: There certainly are such models. However I think that the evidence for a comparable formation rate of supernovae and pulsars suggests that at least in a good fraction of the cases something is left.

D. Branch: If the Le Blanc-Wilson jet ejection mechanism really occurs for many supernovae, do you think that we should see a wider variety of optical characteristics near maximum light, depending on the angle between ejection and the line of sight? Both type I and type II spectra always show blue-shifted absorption lines.

L. Woltjer: I would not expect that the jet would emit much optical radiation, and in any case it would be accompanied by a more spherically symmetric shell, as in the standard supernova models.

A. Treves: How would your picture of the e.m. evolution be modified if the remnant was a black hole?

L. Woltjer: Certainly every thing would be quite different and I'm not sure that the effective generation of magnetic fields and relativistic electrons would occur.

CARBON DETONATION SUPERNOVAE

JEAN-ROBERT BUCHLER
Belfer Graduate School of Science, Yeshiva University, New York, N.Y., U.S.A.

Strong circumstantial evidence exists that stars in the mass range 4–8 $M_\odot$ lead to supernova explosions and to the formation of pulsars. On the one hand is the empirical study of Gunn and Ostriker (1970) which puts the masses of the parent stars of pulsars in roughly the above range. Sofia (1972) observes that no pulsars are found in binary systems, whereas 75% of early-type stars are; he then correlates the pulsar birthrate with the occurrence of O and B runaway stars and concludes that for such a binary system to be disrupted during the supernova explosion, the progenitors of pulsars have to have a mass $\gtrsim 5\ M_\odot$. Iben (1973b) notes that the above stars go through the cepheid stage during their evolution and that the birthrate of cepheids and pulsars is comparable. On the other hand, evolutionary calculations by Rose (1969), Arnett (1969), and especially Paczynski (1970) and Barkat (1971) show that the above stars gradually develop degenerate carbon cores of $\sim 1.4\ M_\odot$ surrounded by a loosely bound envelope. The remarkable convergence of the structure of these cores leads to an almost unique presupernova model. The possible effects of rotation, which are ignored in the preceding work, have been discussed by Weidemann (1971).

A major obstacle to the evolutionary link between the above stars and pulsars is the fact that if ignition occurs in the degenerate carbon cores at a central density of 3×10^9 g cc^{-1} as predicted by Paczynski, an ensuing detonation wave will disrupt the star leaving no remnant and ejecting too many iron peak elements into space to square with observed abundances (Arnett, 1969; Wheeler *et al.*, 1970; Bruenn, 1971). It was subsequently pointed out by Colgate (1970) that electron capture in the detonated material could cause a reimplosion of part of the core. This problem was studied in detail by Barkat *et al.* (1970, 1971, 1972) who concluded that for electron captures to operate on a dynamic time scale, central densities in excess of 6×10^9 g cc^{-1} were required. A subsequent hydrodynamic calculation by Bruenn (1972) and Buchler *et al.* (1973) indicated that central densities in excess of 2.5×10^{10} g cc^{-1} were needed for reimplosion.

A parametrization of the evolution of the growing carbon cores prior to ignition was made by Barkat *et al.* (1972). This allowed a systematic study of the effects of the various uncertainties in the carbon-oxygen content of the cores, of the rate of mass addition at the He burning shell, and of the nuclear screening factors. The conclusion was reached that ignition could only be pushed as high as 6×10^9 g cc^{-1} without doing violence to physics. A more recent evaluation of nuclear screening factors by deWitt *et al.* (1973) would even reduce this density to 4×10^9 g cm^{-3}.

These studies left the carbon detonation model in a desperate shape. As a possible way out of the impasse, Iben (1973a) suggested that perhaps a sufficient temperature

C. B. Cosmovici (ed.), Supernovae and Supernova Remnants, 329–332. *All Rights Reserved*

inversion builds up during the evolution so that ignition occurs off-center. His hope was that an inwards moving detonation wave could impart sufficient momentum to the core to cause a collapse to neutron star densities while a concomitant outwards moving detonation wave would eject the outer part of the core. Subsequent hydrodynamic calculations of off-center detonations (Buchler *et al.*, 1973) have shattered this hope. The inwards moving detonation wave causes only a relatively small compression at the center followed by a bounce. Subsequently the whole core gets disrupted just as in the central detonation case. The reason for the complete disruption is the inability of electron captures at low density $\lesssim 10^{10}$ g cc^{-1} to remove enough pressure to have a gravitationally bound remnant.

The deadlock was lifted by Paczynski (1972) who pointed out that Urca shells in the presence of convection could be extremely efficient in regulating the carbon burning. In view of the complexity of the behavior of convectively driven Urca losses, there still exists at present a considerable amount of uncertainty in the subsequent evolution.

We are nevertheless able to estimate what the most likely evolution is under the influence of convective Urca losses. The degenerate carbon cores are growing in mass and are very close to the Chandrasekhar limit. They may become dynamically unstable as carbon burning occurs non-violently. This can occur either through an increase in the mean molecular weight per electron (μ_e) due to electron captures on the products of carbon burning (Couch and Arnett, 1973), or through mass additions at the He burning shell, or finally through general relativistic effects. A maximum central density before collapse of $\sim 10^{10}$ g cm^{-1} has been estimated by Wheeler *et al.* (1968).

We note that such unstable cores could also be formed in a binary system in which one of the stars has evolved to a carbon dwarf stage and the companion, in its red giant phase, is shedding mass on it. The smaller mass ejected and the longer lifetime of the system have let Wheeler (1973) to analyze their possible connection with type I supernovae.

In consequence the hydrodynamic evolution of collapsing carbon core models has been computed (Wheeler *et al.*, 1973a). Since the exact nature and development of the dynamic instability are still uncertain, the initial models have been parametrised according to their temperature with an initial central density of 10^{10} g cm^{-3}. The models are isothermal, spherically symmetric, ignore the effects of rotation and magnetic fields, and are constructed to be in hydrostatic equilibrium. Collapse is initiated by increasing the mass in each zone by one percent. We do not believe that this particular way of inducing instability has an appreciable effect on the subsequent dynamic evolution.

The evolution of the cores is followed by a standard numerical hydrodynamic code which includes an equation of state for an arbitrarily degenerate and relativistic electron gas, perfect ion gas and radiation. The ^{12}C–^{12}C nuclear reaction rates are based on the work of Patterson *et al.* (1969) and the screening rates are those of Salpeter and Van Horn (1969). The very small neutrino losses have been neglected during the fast dynamic collapse prior to ignition. The code incorporates the effects of nuclear statistical equilibrium in the detonated matter as well as electron captures (Barkat *et al.*, 1972).

Models have been studied corresponding to three different initial isothermal temperature profiles, $T=2.5\times10^8$, 4×10^8 and 6×10^8 K. In view of the high degeneracy, the temperature has a negligible effect on the initial structure of the models. However, it defines the collapse adiabat of the cores and determines the density at which the collapsing models ignite and form a detonation wave. We know from the work previously described that, in view of the high sensitivity of electron captures on density, this ignition density determines the ultimate fate of the core, namely implosion or disruption.

The ignition densities for the 2.5×10^8, 4×10^8 and 6×10^8 K models have been found to be 2.7×10^{10}, 2×10^{10} and 1.4×10^{10} g cc^{-1} respectively.

The ultimate fate of the 6×10^8 K model was complete disruption. The two cooler models behaved differently. The inner layers first started to expand following the passage of the detonation wave but their motion quickly reversed because of the overwhelming electron capture rates. The outer few zones ($\sim0.1\ M_\odot$) obtained large outwards velocity ($\gtrsim10^9$ cm s^{-1}) and were ejected.)

The numerical code used did not include the effects of general relativity, nor did it take into account neutrino transport. We also had to assume for numerical reasons that matter streams uniformly into the neutron star. The neglect of an expected shock wave near the neutron star surface may not be too serious since matter achieves supersonic infall velocities. The interaction of neutrini with the infalling matter is also unlikely to drastically alter the infall dynamics as Wilson's (1971) calculations show. A more accurate and complete calculation of the physics of the formation of the actual neutron star is in progress (Wheeler *et al.*, 1973b). Preliminary results indicate that the mass ejected, in particular, is not sensitive to the dynamics near the center where the present treatment is least accurate.

We can thus state in conclusion that the present results provide some hope to link main sequence stars in the mass range 4–8 $M_\odot$ with type II supernovae and with pulsars.

We wish to acknowledge the close collaboration with Drs Z. Barkat and J. C. Wheeler. We also acknowledge the support of the National Aeronautics and Space Administration and of the National Science Foundation, the hospitality of the NASA Goddard Institute for Space Studies in New York, and a research grant by the Ministère des Affaires Culturelles of Luxembourg.

References

Arnett, W. D.: 1969, *Astrophys. Space Sci.* **5**, 180.
Barkat, Z.: 1971, *Astrophys. J.* **163**, 433.
Barkat, Z., Buchler, J. R., and Wheeler, J. C.: 1970, *Astrophys. Letters* **6**, 117; 1971, *ibid.*, **8**, 21.
Barkat, Z., Wheeler, J. C., and Buchler, J. R.: 1972, *Astrophys. J.* **171**, 651.
Buchler, J. R., Wheeler, J. C., and Barkat, Z.: 1971, *Astrophys. J.* **167**, 465; 1972, *ibid.*, **172**, 713; 1973, submitted to *Astrophys. Space Sci.*
Bruenn, S. W.: 1971, *Astrophys. J.* **168**, 203; 1972, *Astrophys. J. Suppl.* **24**, 283.

Colgate, S. A.: 1971, *Astrophys. J.* **163**, 221.
Couch, R. G. and Arnett, D. W.: 1973, *Astrophys. J. Letters* **180**, L 101.
Gunn, J. E. and Ostriker, J. P.: 1970, *Astrophys. J.* **169**, 972.
Iben, I.: 1973a, *Astrophys. J.* **178**, 433; 1973b, Conference on Explosive Nucleosynthesis, Austin, Texas.
Mazurek, T. J.: 1972, thesis, Yeshiva University and private communication.
Ostriker, J. P. and Gunn, J. E.: 1971, *Astrophys. J. Letters* **164**, L 95.
Paczynski, B. E.: 1970, *Acta Astron.* **20**, 47; 1972, *Astrophys. Letters* **11**, 53.
Rose, W. K.: 1969, *Astrophys. J.* **155**, 491.
Sofia, S.: 1972, *Astrophys. J.* **172**, 53.
Weidemann, V.: 1971, *Astrophys. Letters* **9**, 155.
Wheeler, J. C.: 1973, to be published.
Wheeler, J. C., Barkat, Z., and Buchler, J. R.: 1970, *Astrophys. J. Letters* **162**, L 129.
Wheeler, J. C., Buchler, J. R., and Barkat, Z.: 1973a, submitted to *Astrophys. J.*
Wheeler, J. C., Wilson, J. R., and Buchler, J. R.: 1973b, in preparation.
Wheeler, J. C., Hansen, C. J., and Cox, J. P.: *Astrophys. Letters* **2**, 253.
Wilson, J. R.: 1971, *Astrophys. J.* **163**, 209.
de Witt, H. E., Graboske, H. C., and Cooper, M. S.: 1973, to be published.

DISCUSSION

J. R. Dickel: Would the He burning shell that gets thrown off by a $<4\ M_\odot$ star look like an SN but just not form a neutron star? Also, is it at all possible to form 35 $M_\odot$ SN's in case Moore's analysis is correct with more complete data?

J. R. Buchler: It is believed that stars of mass smaller than 4 $M_\odot$ do not give rise to supernovae. The envelope probably gets driven off by the radiation pressure generated in the He and H burning shells as Paczynski finds. Rose and Smith find that even prior to that the He flash could be responsible for pushing off the envelope. In either case the result would be a white dwarf and a planetary nebula. Stars in the approximate mass range 20–50 $M_\odot$ form an iron core surrounded by Si, O, C, ... shells and become dynamically unstable due to the iron-helium phase transition. Wilson's calculations show that they form a black hole; however, the explosive nature of the overlying layers of fuels were not taken into account and it could be that a detonation wave gets generated which might eject these layers giving rise to a supernova and the formation of a neutron star. This problem is currently under study. As concerns the stars larger than 50 $M_\odot$, if they exist, Barkat finds that they run into the pair-creation instability and get completely disrupted upon ignition of oxygen.

THE CONTINUUM RADIO SPECTRA OF SUPERNOVA REMNANTS

JOHN R. DICKEL

University of Illinois Observatory, Urbana, Ill. 61801, U.S.A.

There are several things we can learn about supernova remnants (SNR) by studying their continuum radio spectra. The first is that nearly all have power-law spectra showing that they must certainly emit by the synchrotron process. Second, turnovers in the low frequency spectra of several SNR can probably be related to absorption by the surrounding interstellar medium perhaps giving information on the interaction of the supernova and its remnant with its surroundings. A third characteristic which can be investigated is changes in the spectrum caused by effects intrinsic to the SNR themselves. We describe below the various observational effects and some of the implications which can be obtained from them.

1. Power-Law Spectrum

The mean spectral index of SNR is about -0.5, somewhat flatter than typical extra galactic sources. The spectral index α is defined by the equation $S_f = \text{const.} \times f^{\alpha}$ where S_f is the observed flux density at a given frequency f. From the theory of synchrotron radiation it can be shown that α can be related to the energy spectral index γ of the relativistic particles by $\alpha = -(\gamma - 1)/2$ where γ is defined by $N(E) = \text{const } E^{-\gamma}$ in which $N(E)$ is the number density of particles with energy E. The observed values of α range from near 0.0 (e.g. 3C58) to about -0.8 (e.g. Cas A) implying that different remnants have very different energy distributions for their relativistic particles. Figure 1 shows representative spectra of several SNR. That of G21.9$-$0.5 is typical of sources with strict power law spectra and the other four sources will be discussed below. Unfortunately, to date it has not been possible to correlate the spectral indices of the remnants with any other observed properties. Thus the cause of the different distributions of the radiating particles and their relation to the dynamics of the expanding shell remains uncertain. The differences, however, must be attributed to the intrinsic mechanism producing the relativistic electrons as losses due to the expansion of the remnant will not change the shape of the spectrum (Scheuer and Williams, 1968), and radiation losses, affecting the high energies first, will cause a curvature in the spectrum rather than change the slope of the straight-line portion (Kardashev, 1962). Furthermore, most SNR are too young to have lost appreciable energy by radiation anyway. Although different remnants show very different spectra, most claims of a variation in spectral index across an individual SNR have not withstood the test of accurate data and careful calibration.

C. B. Cosmovici (ed.), Supernovae and Supernova Remnants, 333–336.

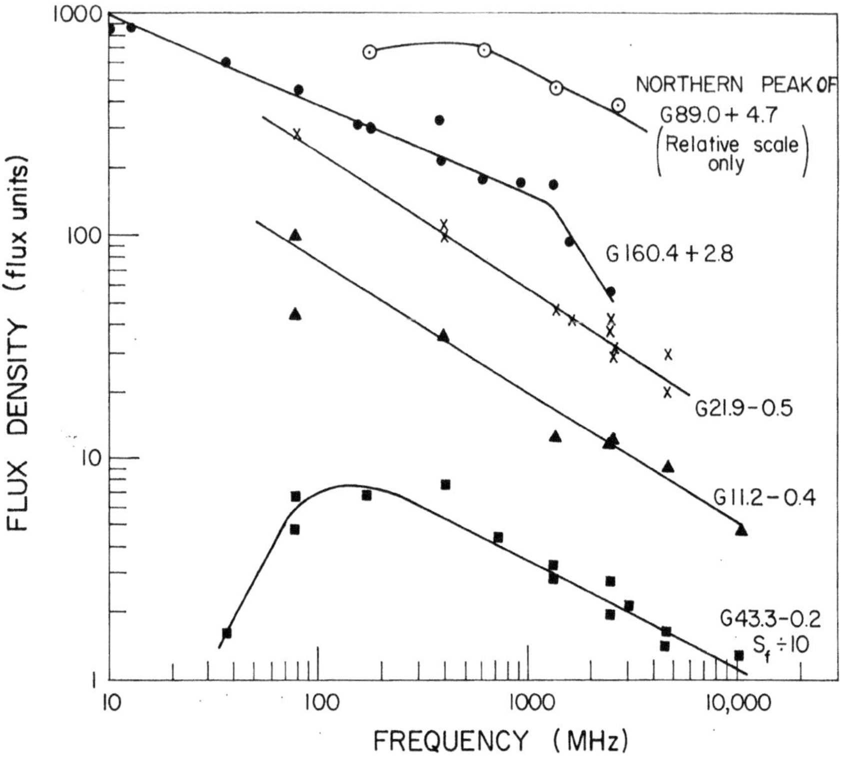

Fig. 1. Representative continuum radio spectra of several supernova remnants illustrating various features described in the text. The scale of the ordinate is in flux units where 1 flux unit = 10^{-26} Wm^{-2} Hz^{-1}. The references for the data on the northern peak of G89.0 + 4.7 (HB21) are given in Erkes and Dickel (1969); those for G160.4 + 2.8 (HB9) are in Willis (1973); and those for G21.9 − 0.5 (Kes 69), GI1.2 − 0.4, and G43.3 − 0.2 (W49B) are Dickel (1973).

2. Low Frequency Absorption of the SNR Emission

While the spectral index across the old remnant HB21 (G89.0+4.7) appears to be nearly constant as seen on spectral index maps between various frequencies by Willis (1973) and by Erkes and Dickel (1969) there may be a tendency for the spectrum to be flatter in the northern part of the source. The plot of the spectrum of this region in Figure 1 indicates that the effect is actually a curvature of the spectrum in that area. It may be possible to explain this curvature by free-free absorption of the remnant's emission by ionized hydrogen in front of the source. The optical depth τ at a given frequency f (in MHz) can be found from the formula

$$\tau = 3.7 \times 10^5 f^{-2} T_e^{-3/2} \int_0^L N_e^2 \, \mathrm{d}l ,$$

where T_e is the excitation temperature of the medium in Kelvin's, N_e is the electron density in cm^{-3} and l is the path length in parsecs (adapted from Oster, 1961). To get a turnover at a frequency as high as about 250 MHz would require rather extreme conditions but these might have been provided by a fossil Stromgren sphere initiated when the supernova erupted or by X-ray emission as the remnant cools. A large neutral hydrogen shell around HB21 has a significant enhancement near the position of the spectral anomaly which is also an area where the expansion of the remnant's shell appears retarded (Erkes and Turner, 1970; Assousa and Erkes, 1973). The present neutral hydrogen density in that region is the order of 10 cm^{-3} (Assousa and Erkes, 1973) and presumably the effects of the supernova could have ionized much of this material. The observed parameters require a present emission measure, $\int N_e^2\, dl$, of about 10 cm^{-6} pc if we assume a temperature of 100 K for the cloud. Such a picture would require a current fractional ionization of about 0.1 if the material were ionized out to a radius of 10 pc. The region appears dense enough so that it should be possible for the cloud to cool to the order of 100 K in less than 10^4 yr, well under the probable age of the SNR (using the cooling curves of Cox and Daltabiut, 1971).

In other cases, the remnants 3C391, W49B and CTB37A show a very significant decline in their emission at low frequencies (Caswell *et al.*, 1971; Holden and Caswell, 1969; and Lockhart, 1971, respectively). That of W49B (G43.3−0.2) is shown in Figure 1. A number of sources have been listed by Dulk and Slee (1972) as possible SNR with low frequency absorption but several of these are definite H II regions (e.g. 337.8 −0.1, Dickel and Milne, 1972) or close associations of non-thermal and thermal sources (e.g. 3C397, Dickel and Seaquist (1972) reported in Willis (1973). A few objects require further investigation; one of these is G11.2-0.4 for which, as shown in Figure 1, two different measurements both using the radioheliograph of CSIRO at Culgoora, N.S.W. at the same frequency (80 MHz) differ by almost a factor of 2 (Dulk and Slee, 1972; Dickel, 1973). We can find nothing wrong with either measurement; the most likely culprit is irregularities in the terrestrial ionosphere. This case illustrates the great difficulty in trying to evaluate the low frequency absorption in SNR and emphasizes the extreme caution with which such determinations should be regarded. In any case, the phenomenon does not appear all pervasive (for example the SNR CTB37A shows definite absorption whereas CTB37B, an SNR more distant from the Sun but only 20′ away in angular position, shows no such effect – Lockhart, 1971). Therefore the ionization is likely to be in the immediate vicinity of the supernova, as suggested for HB21 above.

Recently, Cesarsky and Cesarsky (1973) have compared their observations of the H92α recombination line emission in the direction of 3C391 with the low frequency absorption measured for that source. The two independent measurements have a slightly different temperature dependence which allows a separation of the temperature and emission measure. Their best fit gives a temperature for the gas of 20 K and an emission measure of 4 cm^{-6} pc. Because both measurements are extremely difficult, the results are necessarily uncertain but as our knowledge of more sources improves we should be able to evaluate the cause of the ionization and, if associated with the supernova, to place quantitative limits on the ionizing flux which can, in turn, be related to the supernova energy.

3. High Frequency Breaks in SNR Spectra

Kundu and Becker (1972) have reported an integrated flux density at 5000 MHz for the Cygnus Loop which is considerably below that which would be expected from an extrapolation of the power law spectrum from lower frequencies. A similar phenomenon has been recorded for HB9 (G160.4+2.8) as shown in Figure 1. These spectra appear to mimic that of the general galactic background (Purton, 1966) but with the whole spectrum of the SNR shifted to higher frequencies than that of the galaxy. Such a frequency shift can be caused by an adiabatic compression of a radiating medium (Scheuer and Williams, 1968) which has led Kundu and Becker (1972) and DeNoyer (1972) to suggest that the expanding shell of the Cygnus Loop has compressed the surrounding interstellar medium to cause the observed emission. (This mechanism was first put forth by van der Laan in 1962.) The observed shell of radio emission in the Cygnus Loop appears thin enough to have been sufficiently compressed to give both the observed spectrum and the total intensity found. For HB9, however, an analysis of the source emissivity (Willis, 1973) has shown that the emis-radio emission in the Cygnus loop appears thin enough to have been sufficiently compressed to produce the observed shift in the spectrum. As mentioned above, HB9 is probably too young (certainly less than 10^5 yr) for synchrotron losses to have seriously altered the spectrum so, most likely, the observations reflect a non-power-law distribution of the energies of the relativistic particles responsible for the emission.

In summary, a number of interesting features in the continuum radio spectra of SNR have been identified and several of these remain to be completely explained. Many colleagues have participated in discussion of ideas described herein; in particular, A. G. Willis deserves special recognition.

References

Assousa, G. and Erkes, J. W.: 1973, *Astron. J.* **78**, 885.
Caswell, J. L., Dulk, G. A., Goss, W. M., Radhakrishnan, V., and Green, A. J.: 1971, *Astron. Astrophys.* **12**, 271.
Cesarsky, D. A. and Cesarsky, C. J.: 1973, *Astrophys. J.* **184**, 83.
Cox, D. P. and Daltabiut, E.: 1971, *Astrophys. J.* **167**, 113.
DeNoyer, L. K.: 1972, Ph.D. thesis, Cornell Univ.
Dickel, J. R.: 1973, *Austr. J. Phys.* **26**, 369.
Dickel, J. R. and Milne, D. K.: 1972, *Austr. J. Phys.* **25**, 539.
Dulk, G. A. and Slee, O. B.: 1972, *Austr. J. Phys.* **25**, 429.
Erkes, J. W. and Dickel, J. R.: 1969, *Astron. J.* **74**, 840.
Holden, D. J. and Caswell, J. L.: 1969, *Monthly Notices Roy. Astron. Soc.* **143**, 407.
Kardashev, N. S.: 1962, *Sov. Astron. J.* **6**, 317.
Kundu, M. R. and Becker, R. H.: 1972, *Astron. J.* **77**, 459.
Lockhart, I.: 1971, Ph.D. thesis, Australian National University.
Oster, L.: 1961, *Rev. Mod. Phys.* **33**, 525.
Purton, C. R.: 1966, *Monthly Notices Roy. Astron. Soc.* **133**, 463.
Scheuer, P. A. G. and Williams, P. J. S.: 1968, *Ann. Rev. Astron. Astrophys.* **6**, 321.
Willis, A. G.: 1973, *Astron. Astrophys.* **26**, 237.

THE EVOLUTION OF YOUNG SUPERNOVA REMNANTS

S. F. GULL
Cavendish Laboratory, Cambridge, England

1. Radio Emission from Young Supernova Remnants

It has been shown (e.g. Woltjer, 1970) that many of the properties of a supernova remnant can be predicted from general considerations of the expansion of a massive gas cloud into the interstellar medium. However, this simple picture has been unable to explain the very powerful non-thermal radio emission from the young shell supernova remnants such as Cas A and 3C 10. In order to account for this it is necessary to invoke other mechanisms to produce magnetic field and relativistic particles.

Shklovsky (1969) proposed that both the field and particles were expelled in the initial explosion and were contained inside the remnant as it expanded. This has the drawback that the energy is rapidly lost through adiabatic expansion and cannot be preserved until the dimensions are of the order of 1 pc. The expansion problem can be overcome if field and particles are *continuously* injected from a central pulsar, but there is no evidence for the existence of pulsars in these young shell remnants. Further, their thin shell structures are in complete contrast to that of the Crab Nebula, where the central pulsar is known to supply the energy of the remnant.

An alternative mechanism, suggested by van der Laan (1962), is that Galactic magnetic fields and cosmic ray electrons are responsible for the radio emission after compression at a shock front. In the case of the old remnants, where the compression is large (Kaplan, 1966, see also Mansfield, this volume), this mechanism may be sufficient to explain the observed emission (e.g. Cygnus Loop; Moffat, 1971)); but since there is only compression by a factor of 4 at an adiabatic shock front, there is no hope of explaining the radio emission from young remnants in this way. To get the high magnetic fields of $\sim 10^{-7}$ T required in Cas A (Rosenberg, 1970) an alternative means of producing magnetic field is required.

2. Hydrodynamical Models of Supernova Remnants

Since the brightness and polarization distributions across these remnants have now been clearly resolved, it seemed important to try to match these observational developments by a correspondingly more detailed theoretical understanding of the hydrodynamics involved. However, the situation is immediately complicated by the lack of a well-defined model of the first stages of a supernova explosion. It was thus necessary to compute how each of several possible models of the ejected material interacted with the interstellar medium, in the hope that certain features would emerge that were in some way independent of the model chosen.

The evolution of the simplest possible model, in which a solid piston drives out

C. B. Cosmovici (ed.), Supernovae and Supernova Remnants, 337–341. *All Rights Reserved*

into a cold external density, has been described by Rosenberg and Scheuer (1973). More realistic modes of the ejecta have now also been studied (Gull, 1973a) and this paper summarizes the results of these computations. From these models it appears that once the ratio of mass swept up to mass ejected (the mass ratio) is greater than about 0.5, the structure and dynamics indeed become independent of the initial conditions. As the remnant expands, the ejecta cools to form a thin shell, separated from the hotter, shocked interstellar medium by a contact discontinuity. Figure 1 shows the distributions of pressure (p), density (ϱ) and velocity (v) as functions of radius when the mass ratio ~ 2 (see also Gull, 1973a, Figure 2).

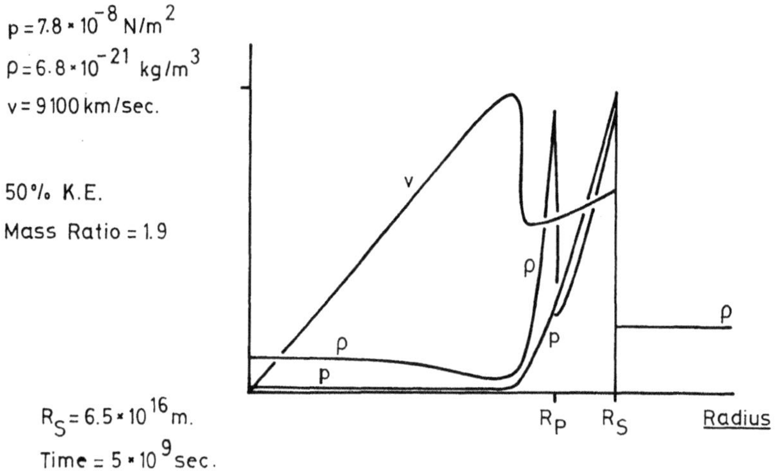

Fig. 1. Computer model of a supernova remnant without convection, showing a Rayleigh-Taylor instability at the outer edge of the ejecta. The parameters of the explosion are: Energy = 10^{44} J: Mass ejected = 10^{30} kg; External density = 10^6 atoms m^{-3}.

Phenomena occurring at this boundary may be important in determining the radio emission. The contact discontinuity is decelerating, due to the momentum of the matter passing through the shock front. In the frame of reference moving with local fluid velocity, this acceleration is equivalent to a gravitational field. Thus, with respect to this field, the denser shell of ejecta lies 'above' the lighter gas and there is a Rayleigh-Taylor instability. Mixing across the boundary then becomes energetically advantageous, random motions appear and the discontinuity is smoothed out to form an unstable convection zone. In this zone it seems likely that the random motions will tangle any pre-existing, frozen-in magnetic field until the magnetic and turbulent pressures are approximately equal. Because such unstable distributions were expected to arise in the course of the evolution, the program was modified to incorporate a simple model of the convective process which could relieve these instabilities by a transfer of mass and thus provide an estimate of the energy available in turbulent motions. The resulting distributions of pressure, density and velocity are shown in Figure 2.

For the range of models considered this random energy was found to rise to between $\frac{1}{2}$ and 1% of the total energy of the explosion when the mass ratio was ~ 1. The equivalent magnetic field rose to a maximum when the mass ratio was ~ 0.3 and thereafter decreased as the convection zone expanded.

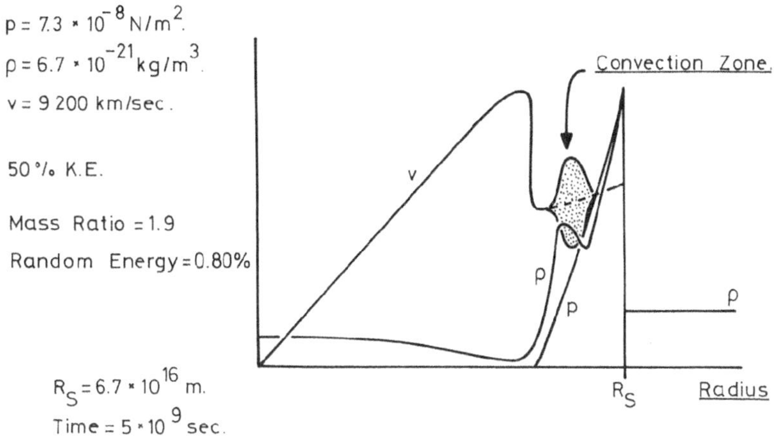

Fig. 2. Computer model of a supernova remnant, incorporating a simple model of convection. The stippled region lies between the curves $(v \pm w)$, where w is the random velocity due to the Rayleigh-Taylor instability. The random energy is defined to be $\int \frac{1}{2}\varrho w^2$ d (volume). The parameters of the explosion are the same as in Figure 1.

3. Properties of the Convection Zone

Any electrons trapped in the magnetic field expected in the convection zone will emit radio synchrotron emission and thus the model is able to predict the thin shell structures of these remnants. The observation that these shells have irregularities on the scale of their thickness (Rosenberg, 1970; Baldwin, 1967) is consistent with the turbulent nature of the field amplification. The radial field indicated by the polarization measurements (e.g. Kundu and Velusamy, 1971; Weiler and Seielstad, 1971) is also consistent with the model since the convective motions are preferentially radial.

The actual flux of the radio emission depends on the distribution of relativistic particle energies. The latter was estimated assuming equipartition between turbulence, magnetic fields and particle energies, an assumption which may be justified if relativistic particles can be accelerated by plasma turbulence (Tsytovich, 1970). If this is the case, the flux rises sharply to a maximum when the mass ratio is ~ 0.3 and then decays as the convection zone expands and dissipates. The model therefore predicts that remnants without central pulsars should not be visible as radio sources until many years after the initial explosion. Further, since the convection zone will quickly dissipate its energy, there should be no middle-aged remnants radiating by this mechanism.

4. Physical Parameters of Cas A and 3C 10

A more detailed comparison of the present model and the observations of Cas A and 3C10 has been made (Gull, 1973b). To do this, an estimate must be made of the mass ratio from the optical and radio data. By identifying the minimum energy in fields and particles with the computed turbulent energy, the physical parameters of these remnants may be calculated. These are summarized in Table I. For 3C 10 a distance of 2β kpc has been used, where β is a numerical factor.

TABLE I

	Cas A	3C 10
Distance (kpc)	3.4	2β
Mass-ratio	1	3
Energy of the explosion (Joule)	5×10^{44}	$1 \times 10^{43}\ \beta^{17/7}$
Ejected mass ($M_\odot$)	2.5	$0.13\ \beta^{3/7}$
External density (atoms cm^{-3})	2.2	$0.18\ \beta^{-19/7}$
Age (yr)	200	400
Predicted secular rate of decrease of flux (per cent per yr)	0.9	0.4

Whilst it must be emphasized that these estimates are sensitive to a breakdown in the assumption of equipartition of energy between fields and particles, it seems that Cas A is the result of a very much larger explosion than Tycho's supernova. Since Tycho's supernova is widely believed to have been of type I (Baade, 1943) and Cas A of type II, these results tend to confirm the idea that type II supernovae originate from much more massive stars than type I supernovae.

5. Conclusions

(1) The shell structure and radio emission from young supernova remnants can be largely explained by a fluid dynamic model in which the radio emission arises in an unstable zone at the outer edge of the ejecta.

(2) The radio emission rises to a maximum when the mass ratio is 0.3 and then decays due to expansion of the unstable region. This model therefore predicts that neither very young nor middle-aged supernova remnants should be powerful radio sources.

(3) Detailed application of the model to individual supernova remnants leads to a prediction of their physical parameters. Using this method it appears that Cas A is the result of a larger explosion than Tycho's supernova.

References

Baade, W.: 1943, *Astrophys. J.* **97**, 119.
Baldwin, J. E.: 1967, *IAU Symp.* **31**, 337, Academic Piess, London and New York.
Gull, S. F.: 1973a, *Monthly Notices Roy. Astron. Soc.* **161**, 47.

Gull, S. F.: 1973b, *Monthly Notices Roy. Astron. Soc.* **162**, 135.
Kaplan, S. A.: 1966, *Interstellar Gas Dynamics*, 2nd. ed., Pergamon Press Ltd., Oxford.
Kundu, M. R. and Velusamy, T.: 1971, *Astrophys. J.* **163**, 231.
Laan, H. van der: 1962, *Monthly Notices Roy. Astron. Soc.* **124**, 127, 179.
Moffat, P. H.: 1971, *Monthly Notices Roy. Astron. Soc.* **153**, 401.
Rosenberg, I.: 1970, *Monthly Notices Roy. Astron. Soc.* **151**, 109.
Rosenberg, I. and Scheuer, P. A. G.: *Monthly Notices Roy. Astron. Soc.* **161**, 27.
Shklovsky, I. S.: 1969, *Supernovae*, John Wiley & Sons, London, New York and Sydney.
Tsytovich, V. N.: 1970, *Nonlinear Effects in Plasma*, Plenum Press, New York and London.
Weiler, K. W. and Seielstad, G. A.: 1971, *Astrophys. J.* **163**, 455.
Woltjer, L.: 1970, in H. J. Habing (ed.), 'Interstellar Gas Dynamics', *IAU Symp.* **39**, 229.

DISCUSSION

L. Woltjer: I can see that you get a Rayleigh-Taylor instability, but this is very different from convection. What exactly drives the convection in such a way that the field is sufficiently tangled up?

S. F. Gull: By 'convection' I merely mean the state produced by the non-linear development of a Rayleigh-Taylor instability. This is certainly very different to the steady-state convection that occurs, for example, in stellar interiors. It seems plausible that vorticity is generated by such an instability, since the smaller scale irregularities have been present for many *e*-folding times, but it is not yet clear in detail how this may lead in turn to the generation of magnetic fields, the acceleration of relativistic particles and radio emission.

L. *Biermann:* From other examples, the equipartition argument is known to be rather dangerous. Did you consider this?

S. F. Gull: The assumption may be valid in this case since it is possible that the magnetic field, relativistic particles and radiation are all closely coupled, but this point needs further theoretical work. For the present, the assumption of equipartion is the best that one can do, but one should remember that this is an upper limit on the efficiency of the process.

V. Mansfield: (1) What fraction of the 0.5 to 1 % of the total energy in the instability do you assume goes into radio emission?

(2) Can you show the radio flux as a function of swept up mass?

S. F. Gull: (1) I assume 50 % goes into the production of particles and fields.

(2) Graph drawn on board (see answer to remark by Biermann). After the maximum radio flux is reached at mass ratio = 0.3, the decay can be approximated by flux $\propto$ (time)$^{-1.7}$; though for an accurate prediction of the flux, one requires a more detailed understanding of the process arising from the instability.

G. A. Tammann: Could you give a typical time for the turn-on of radio emission?

S. F. Gull: The maximum in the radio emission occurs at a mass ratio of 0.3; about 50–80 yr for Cas A parameters. I find that for some models the rise to maximum is quite abrupt.

THE MAGNETIC FIELD DISTRIBUTION IN SUPERNOVA REMNANTS

JOHN R. DICKEL
University of Illinois Observatory, Urbana, Ill. 61801, U.S.A.
and
D. K. MILNE
Division of Radiophysics, CSIRO, Sydney, Australia

1. Introduction

The magnetic fields in SNR can be observed only by polarimetric mapping of their synchrotron emission. The radiation is strongly linearly polarized with the electric vector of the emergent intensity oriented perpendicular to the direction of the magnetic field. Thus, in principal, by measuring the position angle and fractional polarization of the emission we can determine both the direction and the orderlines of the fields, important quantities if we wish to understand the origin of the fields and their relation to the evolution and radio emission of the SNR.

Inherent in the observations are problems of Faraday rotation of the observed vectors, caused by the intervening interstellar medium in front of the source, and smearing of the observed emission both along the line of sight and across the beam of the telescope. The first effect is proportional to ($\lambda^2 \int_{\text{source}}^{\text{observer}} N_e \mathbf{B} \cdot \mathbf{dl}$) where λ is the wavelength, N_e the electron density along the path $\mathbf{l}$, and $\mathbf{B}$ is the magnetic field strength. Because of the λ^2 dependence, it is possible to evaluate the Faraday rotation by observing at two wavelengths. We are continually striving to remove the smearing across the beam by observations with increased resolution, employing both interferometers and higher frequency observations. The final effect of integration of the net emission along the line of sight, can be solved only by 3-dimensional modeling, a procedure still in its infancy.

2. Observational Techniques

For any reasonable geometry, synchrotron sources will have negligible circular polarization, so we need only three measurements to completely specify the partially linearly polarized wave. In general, the radio astronomer measures directly the three stokes parameters I, Q, and U. To specify the polarized signal, Q and U are usually determined by either switching or correlating the signal between two orthogonal feed elements first at one given position angle and then at an angle 45° from the first. An example of such an observation is shown in the paper on 11-cm observations of SNR by Milne (1972). Vector addition of the two output intensities gives the intensity and position angle of the electric vector of the polarized component of the emission. This procedure can be used with either an interferometer or a single dish. To determine

C. B. Cosmovici (ed.), Supernovae and Supernova Remnants, 343–354.

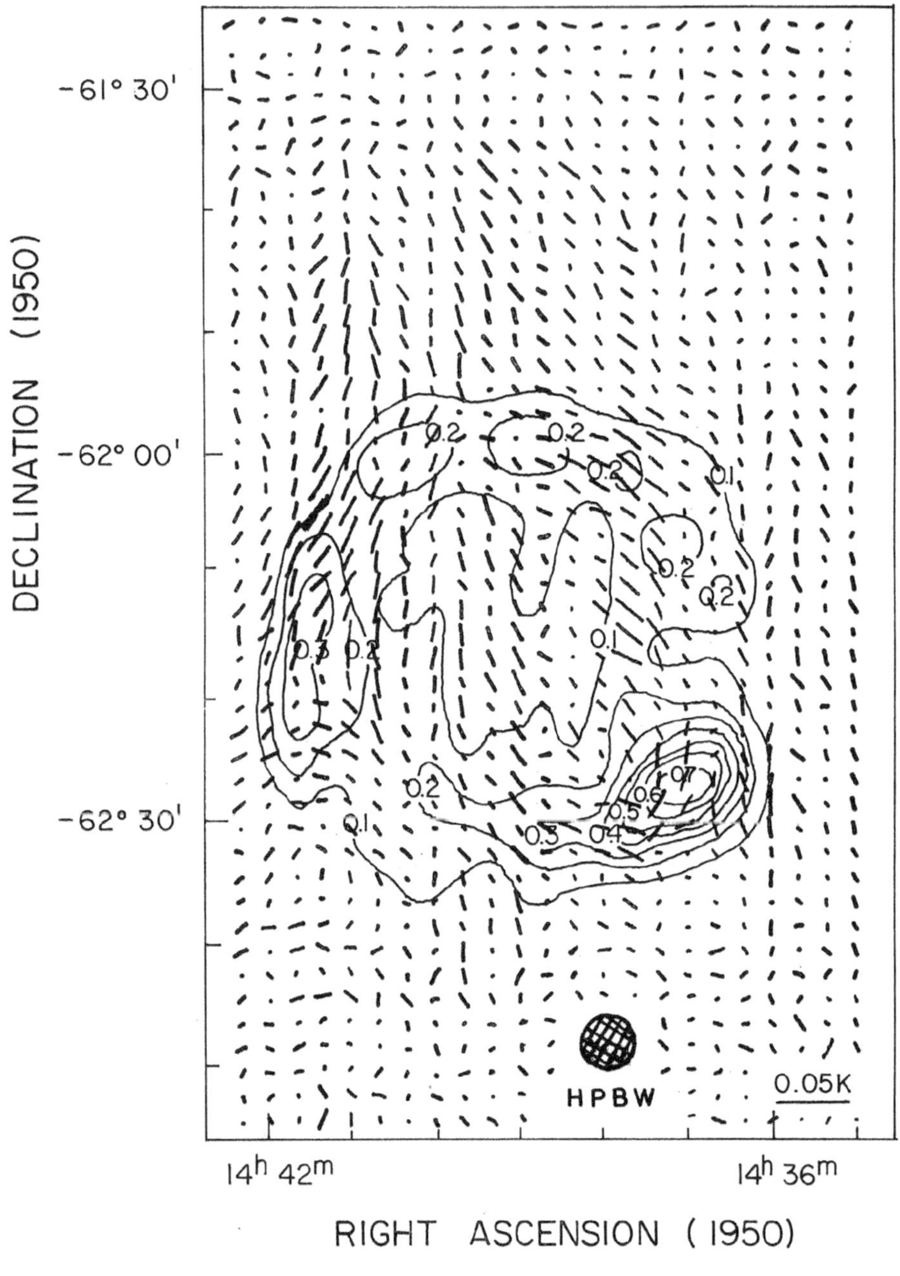

Fig. 1. Total power isotherms and polarization *E*-vectors of the supernova remnant MSH14-63 at 5000 MHz. The scale for the polarization vectors is given in the lower right corner of the figure and the units of the intensity contours are K of brightness temperature. The *E*-vectors are spaced at half beamwidth intervals.

the fractional polarization, we also need to determine the total intensity, I, which can be obtained by measuring the average total power response from the source in the two perpendicular feeds or, if the reciever is not sufficiently stable, by using the output from a Dicke radiometer which is switched between the feed and a stable comparison load.

3. Results

A. FIELD CONFIGURATIONS

Knowledge of orientation of the magnetic fields in SNR can provide valuable insight into the evolution of the shell structure and the energetics of the physical processes establishing the fields. For almost all remnants, the observed polarization distributions show some sort of overall pattern but objects in different stages of their evolution or in interstellar media with varying conditions may show different magnetic field configurations.

An example of the observed polarization of the emission from the old SNR MSH14-

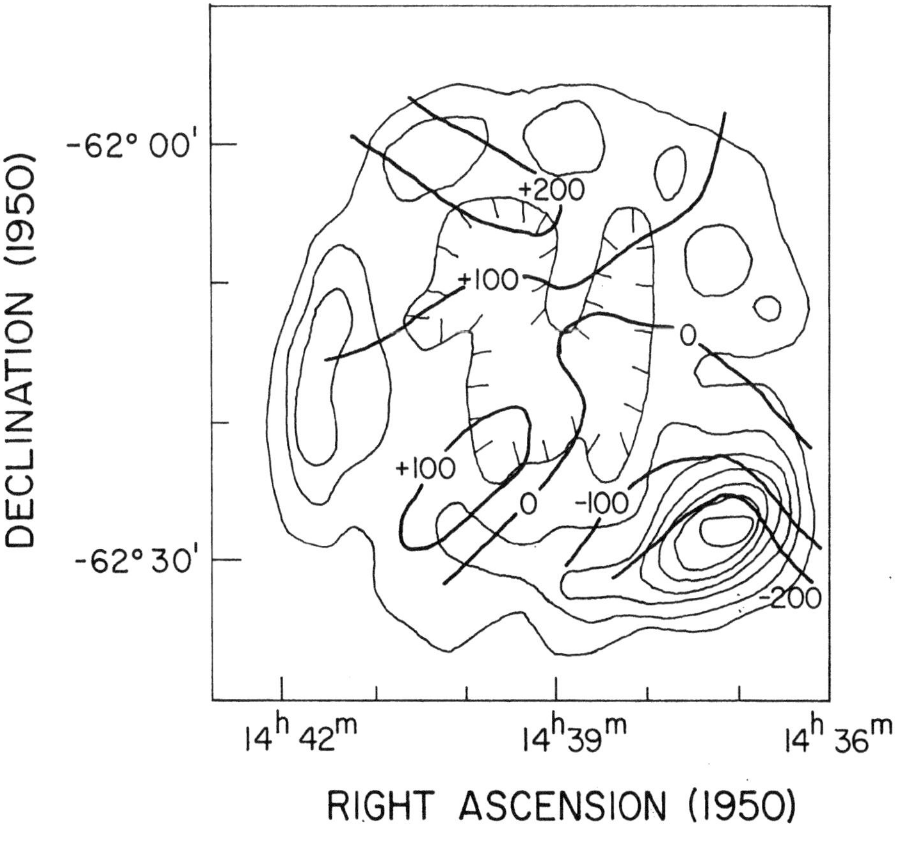

Fig. 2. Contour map of the Faraday rotation of the polarized radio emission in the direction of MSH14-63. The contour units are rad m^{-2} and the total intensity isotherms at 5000 MHz from Figure 1 are sketched by the thin lines.

63 at a wavelength of 6 cm is shown in Figure 1. The contours represent the total intensity from the source and the vectors show the position angle and strength of the *E*-vector of the polarized component. The data was obtained with the 210-ft telescope at Parkes with a resolution of 4′. As mentioned above, the difference in propagation of the ordinary and extraordinary rays in the magnetoionic interstellar medium will cause an apparent rotation of the electric vector and to evaluate this effect we must measure the polarization of the source at two different wavelengths. For MSH14-63,

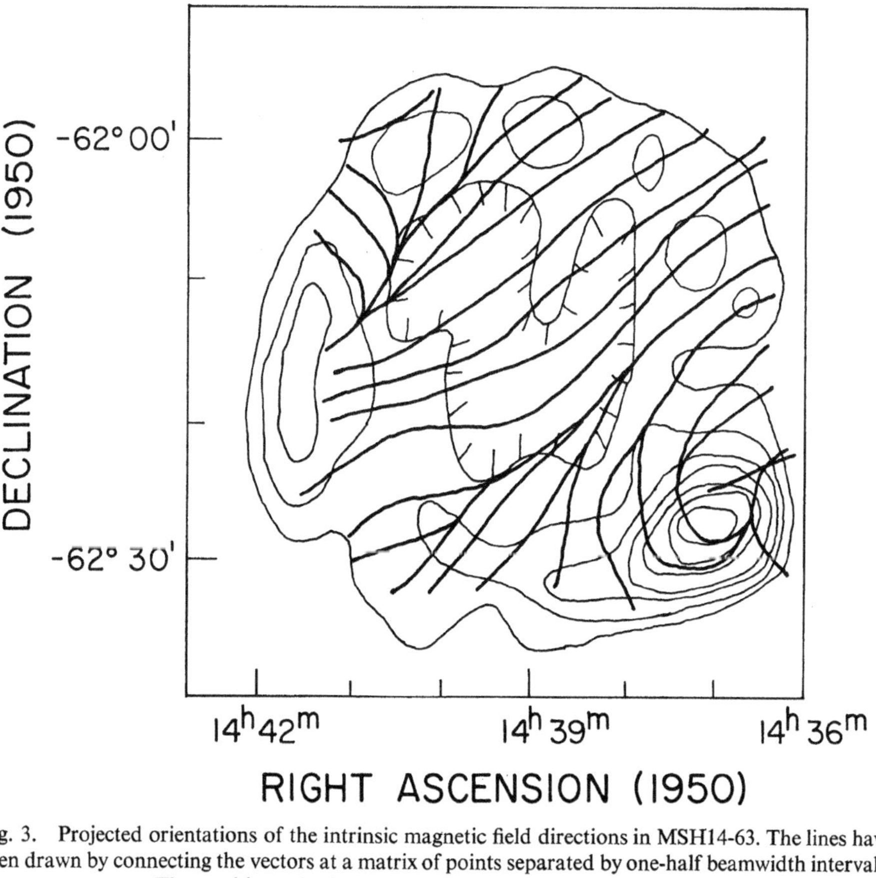

Fig. 3. Projected orientations of the intrinsic magnetic field directions in MSH14-63. The lines have been drawn by connecting the vectors at a matrix of points separated by one-half beamwidth intervals. The total intensity isotherms at 5000 MHz are also sketched.

the 11-cm map obtained by Milne (1971) using the same instrument with a resolution of 8′ was employed for the comparison. The present data were convolved to represent what would have been observed with the 8′ beam. The Faraday rotation was then evaluated from the observed differences in position angle between the two frequencies. The result is shown by the dark contours in Figure 2 (the contour units are rad m^{-2}) and the intrinsic magnetic field directions gotten by extrapolation to zero wavelength are illustrated by the thick lines in Figure 3. Figures 4, 5, and 6 show a similiar sequence for the remnant MSH15-52. Both of these remnants show a pattern which appears to be rather characteristic of many older SNR, namely a quite uniform field across the remnant in one direction with an irregular twisting toward a more-or-less radial structure near the edges of this pattern.

Note that these resultant maps of the magnetic fields have been constructed by connecting the vectors at points in a matrix seperated by half-beamwidth intervals but it should be emphasized that the fractional polarization of the SNR is small (usually 0.05 or less) so that the actual vector at any point may represent only a net orientation

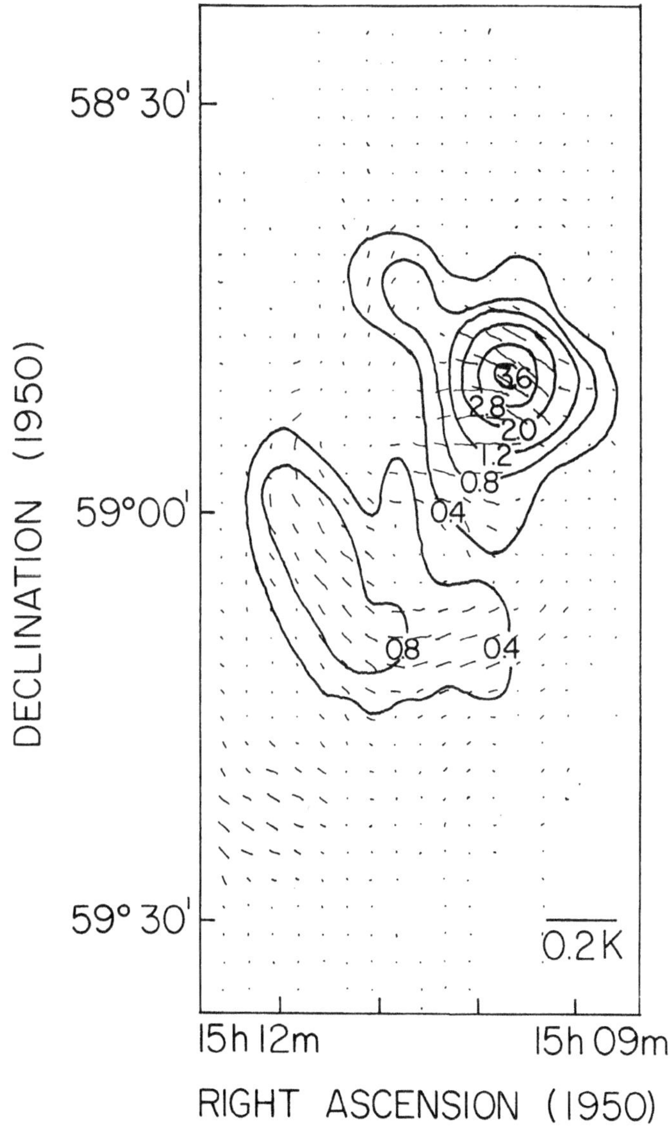

Fig. 4. Total power isotherms and polarization *E*-vectors of the supernova remnant MSH15-52 at 5000 MHz. The scale of the polarization vectors is given in the lower right corner of the figure and the units of the intensity contours are K of brightness temperature.

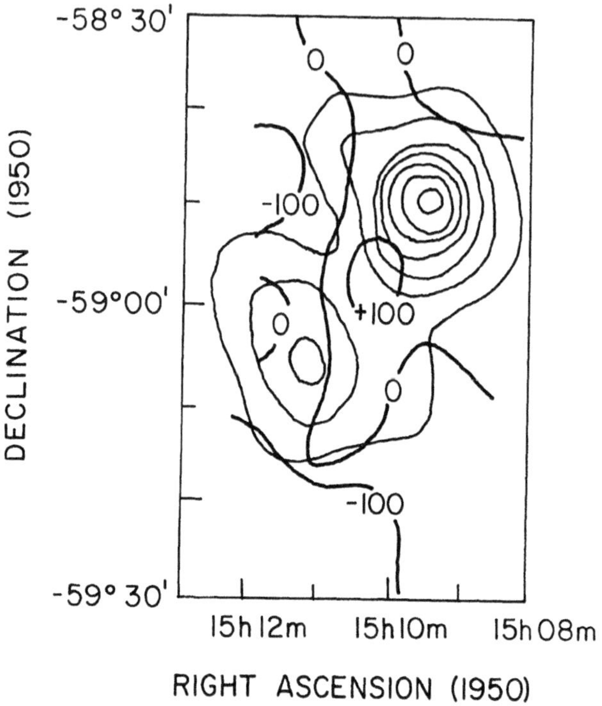

Fig. 5. Countour map of the Faraday rotation of the polarized emission in the direction of the supernova remnant MSH15-52. The contour units are rad m^{-2} and the total intensity isotherms at 2700 MHz are sketched by the thin lines.

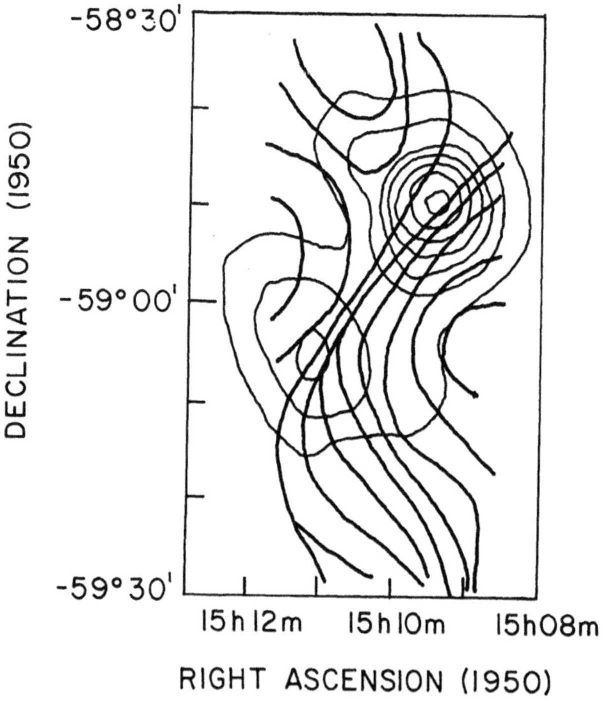

Fig. 6. Projected orientation of the intrinsic magnetic field directions in MSH15-52. The lines have been drawn by connecting the vectors at a matrix of points separated by one-half beamwidth intervals. The total intensity isotherms at 2700 MHz are also sketched.

of many small field loops and not part of a long field line. However, because we see the same net orientation over a larger scale in many cases, there must be some overall ordering mechanism. Also, comparison of polarimetric maps of a given source with different resolution at the same frequency generally show no finer scale structure in the position angle distribution of the polarization vectors.

B. FARADAY ROTATION

An analysis of the Faraday rotation itself can also provide valuable information on the source in many cases. For example, the gradient in the Faraday rotation in MSH14-63 is almost perpendicular to the direction of uniform-field orientation (compare Figures 2 and 3). Another example is shown in Figure 7. The source Puppis A shows the typical pattern of magnetic field lines with a uniform orientation across the source and a more confused pattern near the edges. The Faraday rotation, seen

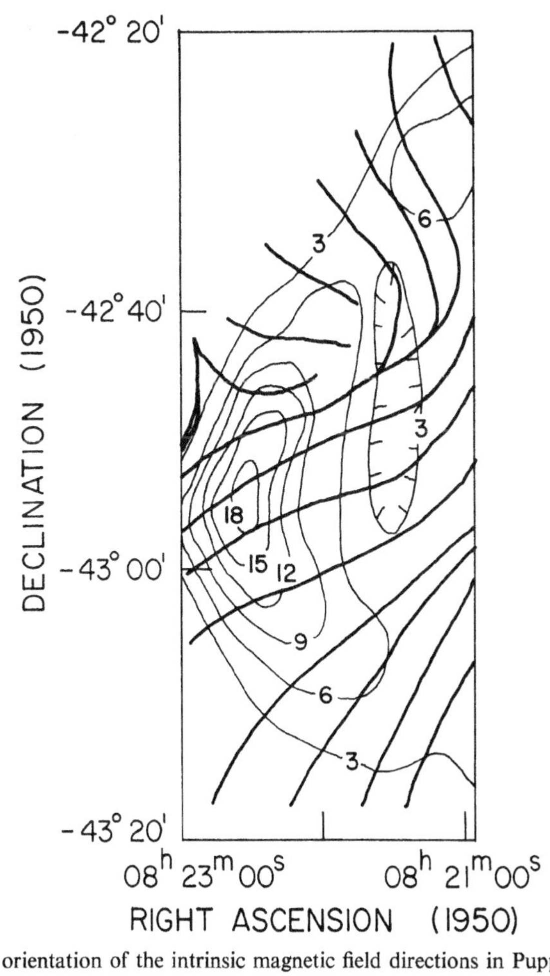

Fig. 7. Projected orientation of the intrinsic magnetic field directions in Puppis A. The lines have been drawn by connecting the vectors at a matrix of points separated by one-half beamwidth intervals. Also sketched by the thin lines are the total power isotherms at 5000 MHz in units of K.

in Figure 8, shows a definite gradient perpendicular to the general orientation of the magnetic field and the maximum value of the rotation (both positive and negative) appear to coincide with the positions of greatest confusion in vector orientation. This suggests that some of the Faraday rotation must be tied up in line-of-sight fields in the confused regions making it difficult to evaluate the orientation of the transverse components. The general relation of the Faraday rotation and the magnetic field structures clearly indicates that the Faraday rotation is internal to the source for these old SNR and the numerical values obtained require internal densities for the thermal electrons of perhaps 1 cm^{-3} and magnetic field strengths of 10^{-4} G, not unreasonable values for SNR. It should be noted here that large internal Faraday rotation may alter the observed λ^2 dependence of the rotation (Burns, 1966) so that the projected field directions in regions of high rotation may be inaccurate. The relation of the rotation to the emissivity of a zone does not appear well established as the regions

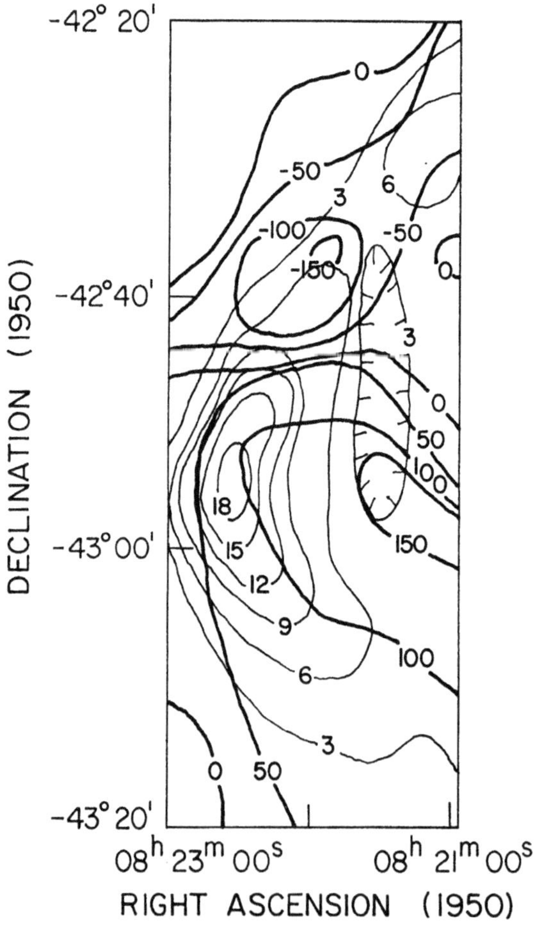

Fig. 8. Faraday rotation map of the supernova remnant Puppis A as determined from data at 2700 and 5000 MHz. The contour units are rad m^{-2}. The total power isotherms at 5000 MHz are also sketched.

with the maximum Faraday rotation in Puppis A lie in areas of weak shell emission whereas for MSH14-63 the largest Faraday rotation fell on the brightest point on the shell. It is interesting to note that the position of the maximum Faraday rotation in Puppis A (Figure 8) is the same as that of the X-ray source as reported by Adams and Zarnecki at this conference (and Zarnecki *et al.* (1973). We are currently working to theoretically evaluate the emissivity generated by the turbulent amplification of the fields and particles in a plasma in order to help determine whether the apparent increase in the line of sight field strength is a large scale structure or a reorientation of small structures.

C. DISTRIBUTION OF THE FRACTIONAL POLARIZATION

Closely related to the arrangement of the magnetic-field orientations in the SNR is the distribution of the degree of polarization which provides information on the uniformity of the fields. As mentioned above the fractional polarization is usually only a few percent of the total intensity and in general there appears to be no variation in the fractional polarization with brightness across a source as illustrated in Figure 9, which is a plot of the fractional polarization of several SNR as a function of their normalized brightness distribution. The errors in these figures are, of course, greater for the lowest intensities, but generally for any given normalized total intensity, the scatter of points about the mean fractional polarization is about the same as the

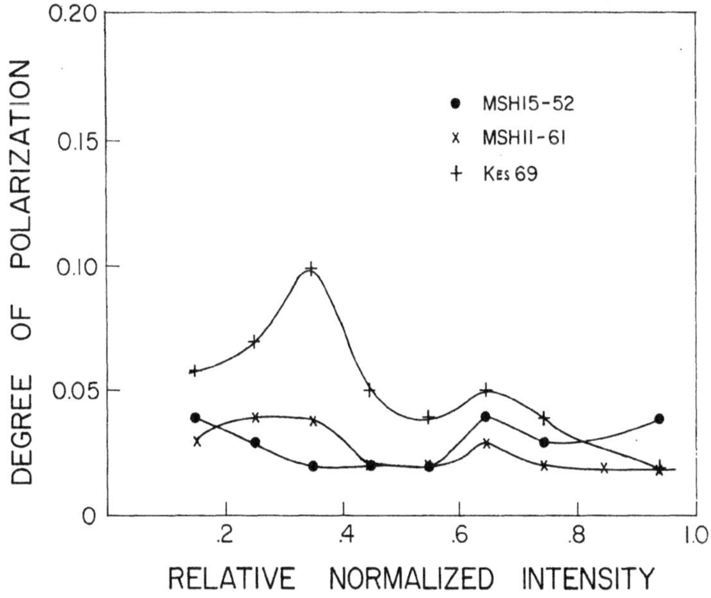

Fig. 9. Distribution of the average fractional polarization at 5000 MHz of three SNR as a function of the normalized total intensity distribution across the source. The fractional polarization was evaluated at each point on the source (with half beamwidth spacing) and the values plotted are the average fractional polarization for all points with total intensity equal to the given fraction of the brightest point on the source.

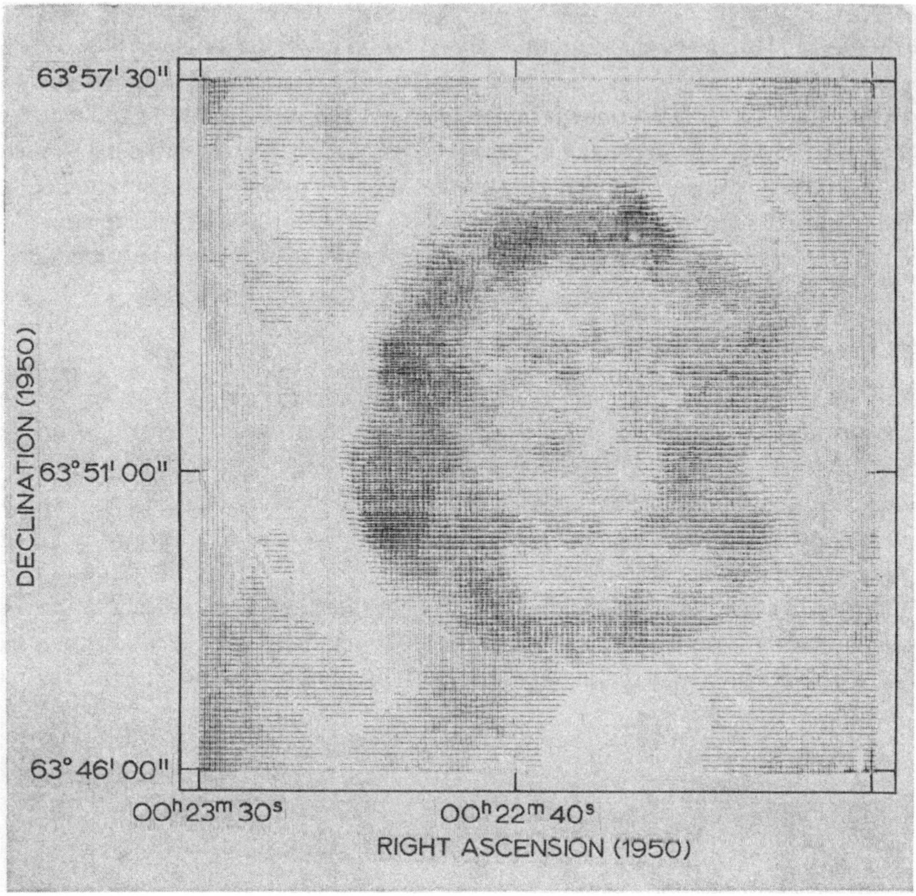

Fig. 10. Distribution of the fractional polarization in Tycho's SNR at a frequency of 2695 MHz. This figure was made by first constructing visual representations of both the polarized and total brightness distributions across the source; then a negative of the total intensity was superimposed upon a clear positive of the polarized intensity and the result photographed.

Zero fractional polarization should appear white and complete polarization black. In this case we observe a nearly uniform gray brightness within the outline of the SNR indicating a small fractional polarization with little variation across the source.

uncertainity so that for MSH15-52 and MSH11-61 there appear to be no significant deviation from a uniform fractional polarization anywhere in the SNR. Therefore, the changes in field direction do not appear to change the fractional polarization. The anomolously high fractional polarization is one area of Kes 69 cannot be explained at present.

In contrast to the older SNR, the younger ones such as Tycho's SNR and Cas A appear to have approximately radial orientation for their net magnetic vectors (Weiler and Seislstad, 1971; Mayer and Hollinger, 1968), but also a quite uniform distribution

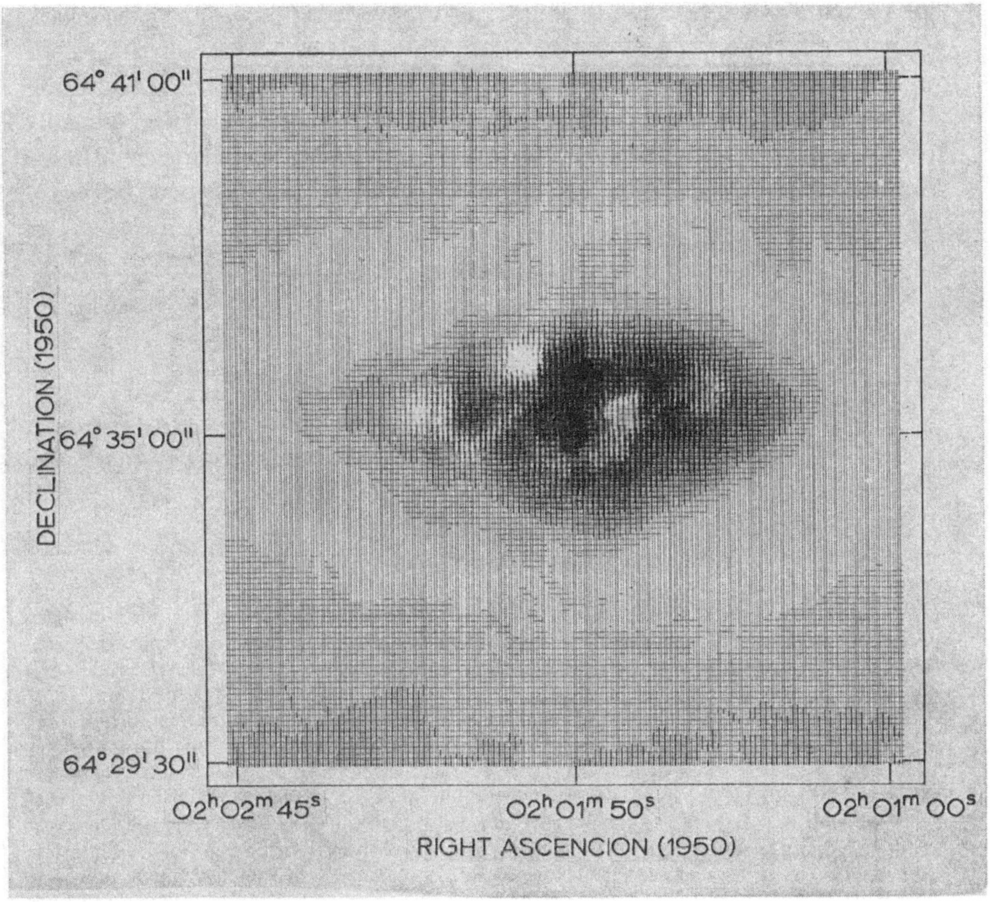

Fig. 11. Distribution of the fractional polarization across 3C58. This figure was constructed in the same manner as Figure 10, but in this case we see large variations in the fractional polarization in different parts of the SNR.

of fractional polarization. To illustrate this for Tycho's SNR, we have constructed visual representations of both the total and polarized intensities of Tycho's SNR (Hermann and Dickel, 1973) and then superimposed a negative of the total intensity upon a clear positive of the polarized intensity and recorded the net intensity through the pair. Regions of small fractional polarization should appear bright and those which are highly polarized should look dark. As can be seen in Figure 10, the result shows a small general polarization on the SNR which follows the outline of the source very well, but only very small irregularities in the distribution of the polarization across the SNR itself. By contrast we can compare this with Figure 11, showing the polarization of 3C58, which, except for the Crab Nebula, is the only SNR with a filled volume of emission rather than a periferal structure. This figure clearly demonstrates how the technique works when there are significant variations in the fractional polarization. In this respect, too, 3C58 is like the Crab Nebula but unlike most other SNR young or old.

The examples of the magnetic field characteristics shown herein are taken primarily from an extensive survey of SNR polarization which we have undertaken with the 63-m radio telescope at Parkes, and which will appear in the *Australian Journal of Physics*.

References

Burn, B. J.: 1966, *Monthly Notices Roy. Astron. Soc.* **133**, 67.
Herman, B. R. and Dickel, J. R.: 1973, *Astron. J.* **78**, 879.
Mayer, C. H. and Hollinger, J. P.: 1968, *Astrophys. J.* **151**, 53.
Milne, D. K.: 1972, *Austr. J. Phys.* **25**, 307.
Weiler, K. W. and Seielstad, G. A.: 1971, *Astrophys. J.* **163**, 455.
Zarnecki, J. C., Culhane, J. L., Fabian, A. C., Rapley, C. G., Silk, R., Parkinson, J. H., and Pounds, K. A.: 1973, *Nature, Phys. Sci.* **243**, 4.

DISCUSSION

A. S. Wilson: I was interested to see that in one of your maps of the magnetic field, in the center you found a uniform magnetic field, but towards the edge the magnetic field lines ran in a circular sense. You said that in these particular regions the rotation measure was high. I want to point out that one sees exactly the same effect in the Crab Nebula. The field is wrapped around the filaments and 'bays' and these regions show strong depolarization. (Wilson, A. S.: 1972, *Monthly Notices Roy. Astron. Soc.* **157**, 229 and paper to be published in *Monthly Notices Roy. Astron. Soc.* later this year). The cause of the 'wrapped around field' is that as the magnetic field presses outwards and accelerate the filamentary shell, it is held up by the denser condensations of gas but can burst through between them. Hence the field wraps around and compresses filaments and bays. Thus in these regions one has both a strong field and a high density of thermal electrons which results in the high rotation measure observed. This interaction between the filaments and the magnetic field I discussed more fully in a paper in *Monthly Notices Roy. Astron. Soc.* **160**.

DISTRIBUTION AND MOTION OF THE RELATIVISTIC ELECTRONS IN THE CRAB NEBULA

A. S. WILSON
Sterrewacht te Leiden, Leiden, The Netherlands

1. Introduction

In this talk this morning, I would like to discuss a model aimed at explaining the variation of structure of the Crab Nebula with frequency and the shape of the integrated nebular spectrum. I shall first summarise the observations of the nebular structure starting at the low frequency end of the spectrum and then go progressively towards the X-rays (Section 2) and after that discuss the model proposed (Section 3). Section 4 considers predictions of the model and Section 5 problems for future investigation.

2. The Relevant Observations

Figure 1 shows a high resolution map of the nebula at 5 GHz made with the Cambridge One-Mile Radio Telescope (Wilson, 1972a). I do not intend to consider the details of this map, but to consider whether there is any change in nebular structure in the radio band. To do this I have made an intercomparison of the available maps (Table I) at different frequencies. After allowing for the differing resolutions by convolution techniques, I can find no reason to believe there is any systematic change of structure over this range. One can calculate the distribution of spectral index over the nebula between any two frequencies, but the small variations of spectral index one does find can probably be completely ascribed to instrumental effects.

Now if one is wanting to work on models of radio sources, it is essential to think in terms of the *emissivity* distribution in the nebula, not the brightness distribution which, after all, involves projection effects through the whole nebula. There are good reasons (Woltjer, 1958; Trimble, 1968) for believing that the object is prolate spheroidal in shape, and if one assumes that and the observed brightness distribution, the distribution of emissivity can be calculated on certain assumptions about symmetry. This distribution (Figure 2) shows the emissivity at 5 GHz is largest in the centre and decreases towards the edge of the nebula.

The next point is that the size to half intensity at radio wavelengths is greater than the corresponding size at optical continuum wavelengths. Using the 5 GHz map and the optical isophotes (Woltjer, 1957), I have made a map of the distribution of spectral index between 5 GHz and optical frequencies. This map (Figure 3) shows the spectral index to be flattest, not surprisingly, in the optically bright region to the north-west of the pulsar and to steepen towards the edge, being particularly steep in the optical 'bays'.

Now lets go on to consider the variation of spectral index over the nebula *at*

C. B. Cosmovici (ed.), Supernovae and Supernova Remnants, 355–364.

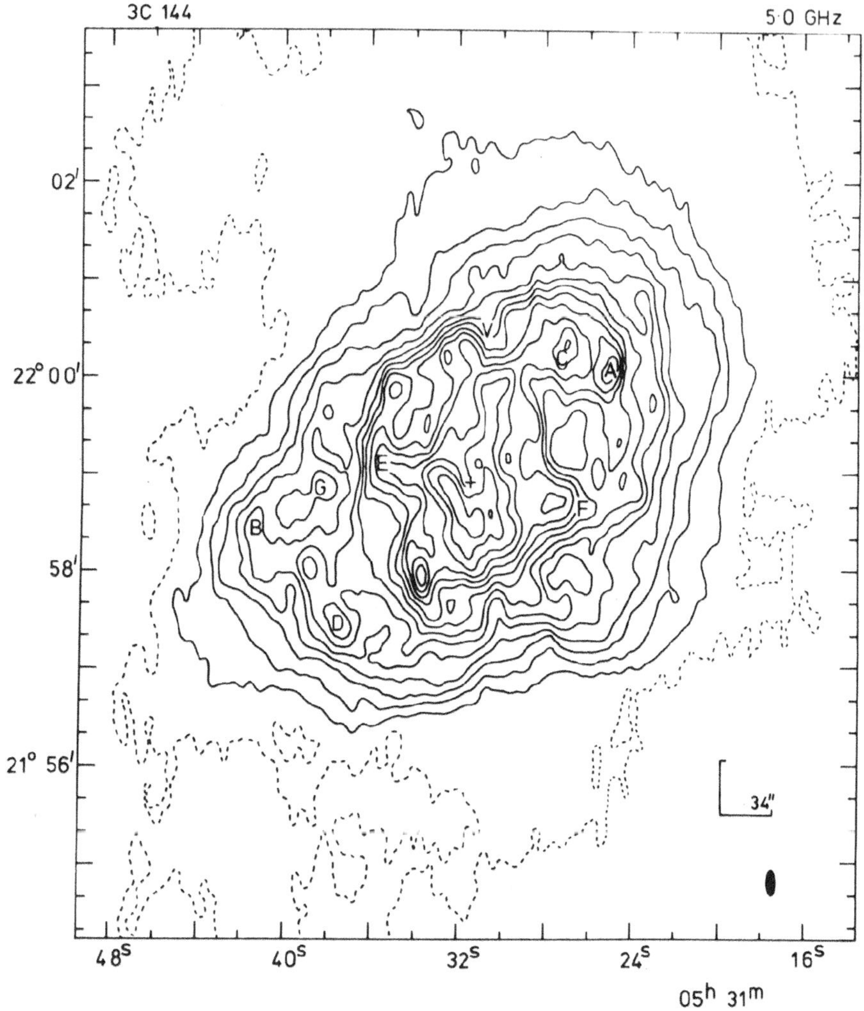

Fig. 1. Map of distribution of total intensity in the Crab Nebula at 5 GHz. The half power beamwidth (6″ × 16″) is shown in the bottom righthand corner. The contour interval is 64 K (0.106×10^{-26} Wm^{-2} Hz^{-1} (beam area)$^{-1}$), the central cross marks the pulsar and the contour of zero brightness is dotted.

optical frequencies. Scargle (1969) has measured the optical spectral index at some points in the nebula and finds that the spectral index is again flattest in the 'wisp' and optically bright region to the north-west of the pulsar and steepens towards the edge of the nebula. This is, of course, simply another way of saying that the nebula is still getting smaller *at* optical frequencies. However, the variations of spectral index *at* optical frequencies are much larger than those between 5 GHz and optical frequencies.

TABLE I

Radio maps of the Crab Nebula used to investigate the change of structure with frequency

Frequency	Resolution (R.A. × Dec.)	Reference
0.4 GHz	79″ × 210″	Branson (private communication)
1.4 GHz	23″ × 61″	Branson (1965)
2.7 GHz	11″ × 29″	Wilson (1972a)
5.0 GHz	6″ × 16″	Wilson (1972a)
19.4 GHz	1.′7 × 1.′7	Mayer and Hollinger (1968)
31.4 GHz	1.′6 × 1.′6	Johnston and Hobbs (1969)
86 GHz	1.′1 × 1.′1	Matveyenko (1971)

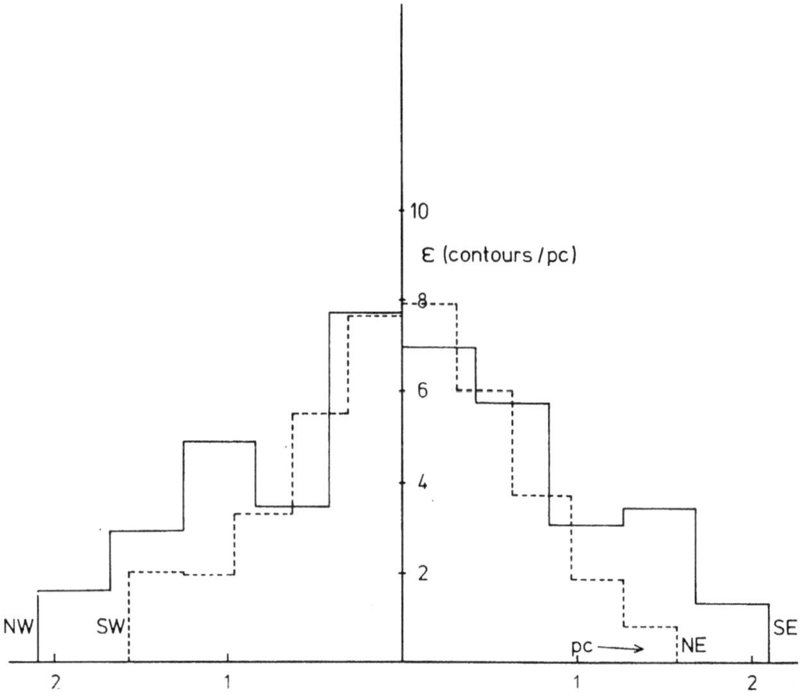

Fig. 2. Distribution of emissivity at 5 GHz versus distance from nebular centre. The dotted line refers to the minor axis, the full one to the major axis. On the emissivity scale, one contour of brightness = 64 K.

It is interesting to note that the data so far are consistent with the hypothesis that the nebular size remains constant from radio frequencies up to some frequency between radio and optical frequencies but above it the size begins to decrease rapidly. The reason that the spectral index variations between 5 GHz and optical are similar to, but less than those at optical frequencies is simply because between radio and optical frequencies there is a large frequency range over which little spectral variations occur.

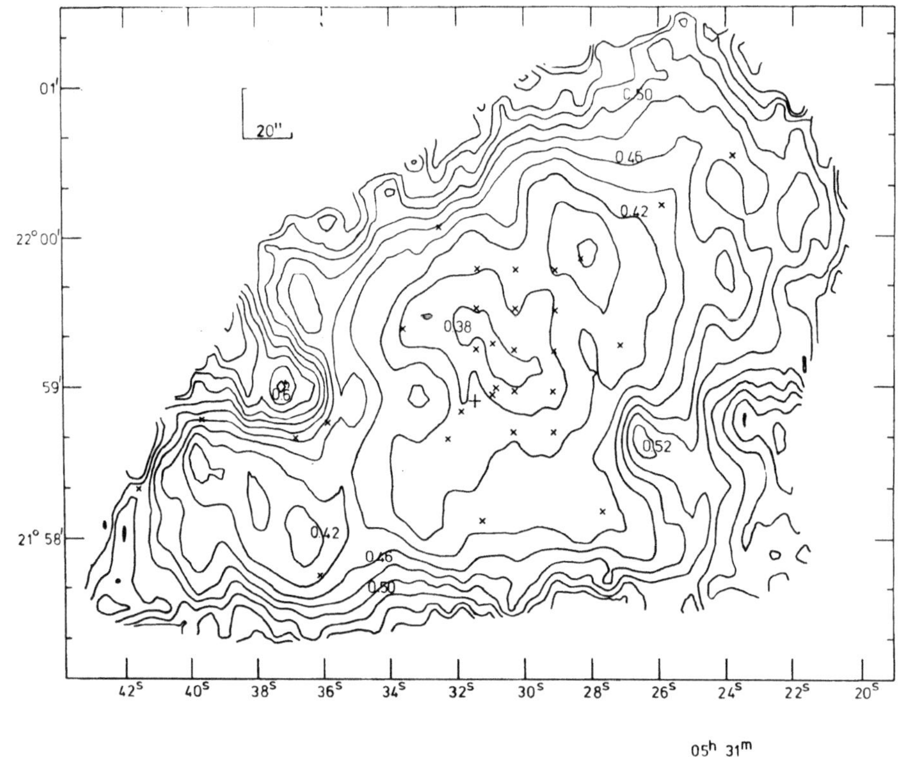

Fig. 3. Contours of constant spectral index between 5 GHz and optical frequencies. The large cross (+) marks the pulsar and the small crosses (×) are places where Scargle (1968, 1969) measured the spectral index of the optical continuum itself.

The last two observational points can be dealt with fairly quickly. Firstly the X-ray size is smaller than the optical (Bowyer *et al.*, 1964; Oda *et al.*, 1967; Floyd, Jr., 1970). Furthermore the X-ray source appears to be localised in the optically bright region to the north-west of the pulsar which has the flattest spectrum and which does not show up on the radio map. The data, therefore, strongly suggest that the X-ray source is simply a continuation to higher energy of the trend shown by radio and optical observations.

Secondly we consider the integrated spectrum (Figure 4). Lets just notice that there appears to be a break in the spectrum round about $\nu = 3 \times 10^{13}$ Hz, with the spectrum being steeper at higher frequencies.

Let me now summarise these observations:

(i) Little variation of spectral index over radio nebula.
(ii) Emissivity at radio frequencies decreases towards the edge of the nebula.
(iii) Half intensity radio size > half intensity optical size.
(iv) Optical spectral index steepens towards edge.
(v) X-ray size < optical.
(vi) Break in integrated spectrum at $\nu \simeq 3 \times 10^{13}$ Hz.

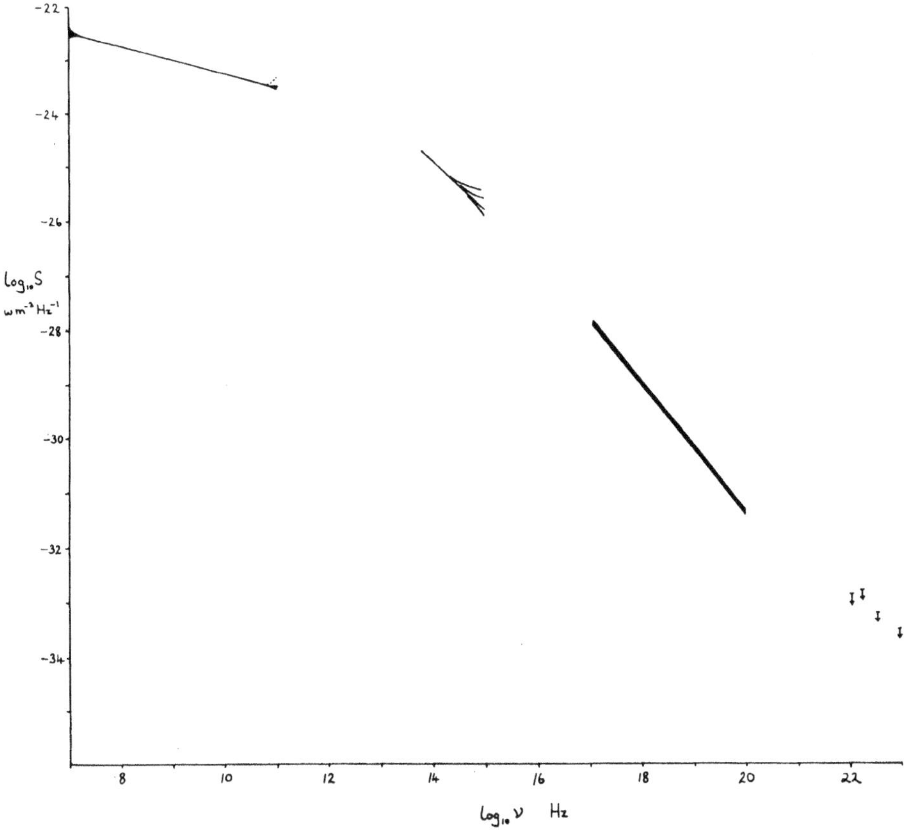

Fig. 4. The electromagnetic spectrum of the total radiation from the Crab Nebula. The thickness of the lines covers the limits of uncertainty at any frequency (from Baldwin, 1971).

3. A Model

The energy source of the Crab Nebula is the rotational energy of the pulsar NP 0532. Ultimately some of this energy is converted into the relativistic particles and magnetic field which give rise to the nebular synchrotron radiation. Ostriker and Gunn (1969) have suggested that these particles are accelerated by the 30 Hz electromagnetic wave emitted by the pulsar, and, if so, the lack of circular polarization in the optical continuum radiation from the nebula (Landstreet and Angel, 1971) strongly suggests that this wave does not propagate more than $\simeq 20''$ from the pulsar. This is supported by the fact that there is a 'valley' in the brightness near the pulsar at both radio and optical frequencies which, it has been suggested (e.g. Scargle, 1971), has been swept clear of thermal gas and static magnetic field by the 30 Hz wave. It is also interesting to note the presence of the 'wisp' activity on the boundary of this region. What I want to make plausible by these arguments is that the relativistic electrons are injected from a source small compared to the total size of the nebula.

Now Ostriker and Gunn (1971) have concluded that not even the electrons presently emitting radio radiation can have been produced in the original explosion because the losses they suffer in adiabatic expansion would have been converted into a kinetic energy of the filaments greater than is observed. Pacini and Salvati (1973) reach a similar conclusion by arguing that the magnetic field of the nebula was probably sufficiently strong in the early phases as to cause even low energy electrons to lose much of their energy to synchrotron radiation. This imples that radio electrons (i.e. those which emit radio radiation in the *present* magnetic field) were (and quite likely still are) being produced long after the original explosion. However, the remarkable straightness of the integrated radio spectrum and the lack of spectral variations over the nebula at radio frequencies both suggest to me that the radio electrons in the nebula *now* have *not* lost appreciable energy by synchrotron losses.

The model, I think, one is led to by these considerations is that relativistic electrons are continuously produced near the centre and move away into the nebula losing energy by synchrotron radiation. The electrons which give rise to the radio radiation have synchrotron half lives $\gg 1000$ yr in the present nebular field so can move to the edge of the nebula without losing an appreciable fraction of their energy. There are thus no variations of spectral index at radio frequencies, the radio nebula essentially delineating the extent of the magnetic field. At some frequency, rather lower than that of optical radiation, the half lives of the appropriate electrons become equal to their diffusion time to the edge. Above this frequency, few electrons can reach the edge, the size of the source decreases with increasing frequency and the spectral index of the integrated emission is then steeper (Ginzburg, 1953) because there is an equilibrium between electron injection and synchrotron energy loss. The gross features of the observations are thus explained.

In order to go further and make a quantitative comparison with observation one must make some simplifying assumptions. I have assumed:

(1) Electrons with a power law spectrum are produced continuously by a point source at $r=0$. The source function is

$$N_s(E)\,\mathrm{d}E = KE^{-\gamma_0}\,\mathrm{d}E$$

where $N_s(E)\,\mathrm{d}E$ is the number of electrons produced per unit time with energies in the range E to $E+\mathrm{d}E$.

(2) The electrons diffuse spherically symmetrically in an infinite, isotropic and homogeneous medium with a diffusion coefficient, D, independent of position r and E.

(3) The electrons lose energy by synchrotron radiation:

$$-\frac{\mathrm{d}E}{\mathrm{d}t} = bE^2$$

where $b=\frac{2}{3}c\,(e^2/mc^2)^2\,H_\perp^2(1/mc^2)^2$ and $H_\perp$ is independent of r and t.

The time independent diffusion equation then becomes:

$$-D\nabla^2 N - \frac{\partial}{\partial E}(bE^2 N) = N_s(E)$$

and its solution is (Webster, 1970; Gratton, 1972):

$$N(r, E) = \frac{KE^{-\gamma_0}}{4\pi Dr} f(\gamma_0, \xi)$$

where

$$\xi = bEr^2/4D$$

and

$$f(\gamma_0, \xi) = \pi^{-1/2}\Gamma(\gamma_0 - 1)\, e^{-\xi}\xi^{1/2}U(\gamma_0 - 1, \tfrac{3}{2}; \xi)$$

where $U(a, b; z)$ is the confluent hypergeometric function of the 2nd kind.

To reproduce the radio spectrum of the Crab Nebula we require $\gamma_0 = 1.52$, but take instead $\gamma_0 = 1.50$ because of the simplification

$$u(\tfrac{1}{2}, \tfrac{3}{2}; \xi) = (\pi/\xi)^{1/2}/\Gamma(\tfrac{1}{2}),$$

and

$$f(\tfrac{3}{2}, \xi) = \exp(-\xi)$$

(Abramowitz and Stegun 1964) so that

$$\log_{10} N(r, E) = -\gamma_0 \log_{10} E + \log_{10}(K/4\pi Dr) - \frac{bEr^2}{4 \times 2.303 \times D} \tag{1}$$

with $\gamma_0 = 1.50$.

Equation (1) has a very simple interpretation. The first term on the right hand side depends only on the index in the electron energy spectrum at injection, the second describes the diffusion while the third takes account of synchrotron losses. It may be seen that $N \propto 1/r$ when synchrotron losses are unimportant. This decrease in particle density is the cause of the decrease in volume emissivity towards the edge of the nebula at radio frequencies (Figure 2). From (1) one may compute the synchrotron radiation spectra for different r and typical curves are shown in Figure 5.

One now has a well defined model – if one assumes a value for the magnetic field, it is a one parameter model, that is, D the diffusion coefficient is the only unknown. The procedure I chose was to determine D from the radio and optical structures (which are well known). The model is then completely defined and one may compare its predictions with:

(a) Variation of spectral index at optical frequencies.

(b) Size of X-ray nebula.

(c) Shape of the integrated spectrum.

The details of this comparison are given in Wilson (1972b). The result is that, for the most part, there is reasonable agreement between this simple model and the observations. Some deviations are, however, found i.e.

(1) The predicted size of the X-ray nebula is too small.

(2) The predicted change in spectral index of the integrated emission between low and high frequencies ($=0.5$) is too small (observed difference $\alpha_{\text{X-ray}} - \alpha_{\text{radio}} \simeq 0.9$).

These difficulties may be resolved if:

(1) Either: the injection region is not a point source but extended over $\simeq 20''$-$30''$ – the size of the active 'wisp' and 'valley' region.

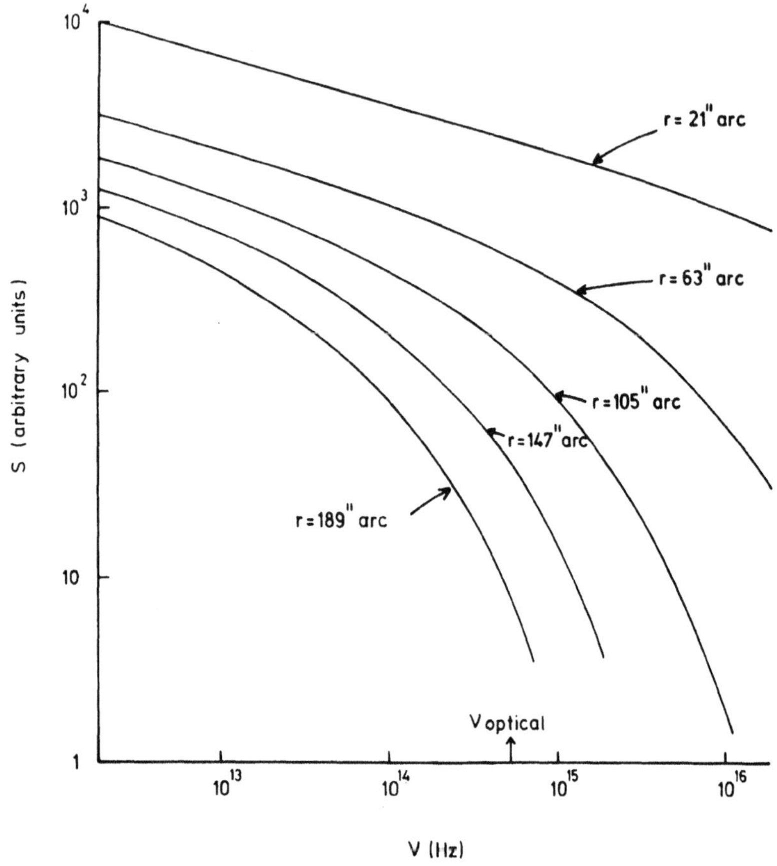

Fig. 5. Synchrotron spectra for different radii from the electron energy spectra of Equation (1). $\gamma_0 = 1.50$, $H = 5 \times 10^{-4}$ G and $D = 7.1 \times 10^{26}$ cm^2 s^{-1}.

Or: D is not constant but increases with energy and (2) the exponent in the energy spectrum at injection, γ_0, is larger for X-ray electrons ($=2.4$) than for radio electrons ($=1.52$).

4. Predictions and Implications of the Model

(1) The model predicts that the size of the nebula is constant at frequencies where synchrotron losses are unimportant. Thus, when maps eventually become available at millimetre and far infra-red ($\sim 100\ \mu$) wavelengths, the source should have essentially the same size as at 5 GHz.

(2) By assuming H one can calculate D from the observations. If one takes $H = 2 \times 10^{-4}$ G, the value of D corresponds to a time of diffusion to the edge of the nebula, T_D, of ~ 1000 yr, if $H = 5 \times 10^{-4}$ G, $T_D \sim 300$ yr. One can reverse the argument – if the diffusion time is known – the magnetic field can be calculated. T_D cannot be longer than the age of the nebula (1000 yr) nor shorter than the light travel time from centre to edge ($\simeq 6$ yr). The corresponding limits on H are $2 \times 10^{-4} < H < 7 \times 10^{-3}$ G.

This shows that the field is not far from the equipartition value of $\sim 10^{-3}$ G. Other field estimates have been 5×10^{-4} G from the acceleration of the filamentary shell (Trimble and Rees, 1970) and 5×10^{-4} G from stability considerations (Scargle, 1968). Although sufficient energy has been lost by the pulsar over the last ~ 300 yr to account for the present energy in the nebula, we cannot say whether all the relativistic electrons which have been injected in the last 1000 yr are still in the nebula or, perhaps, only those injected in the last ~ 300 yr (the rest having escaped into interstellar space).

5. Problems for Future Investigation

(1) It is not clear that the outward motion of the relativistic electron is, in fact, diffusive in nature.

(2) The origin of the magnetic field is not treated. For example, the elliptical shape of the nebula may be intimately related to pulsar electrodynamics.

(3) Scargle's (1969) measurements of the spectral index variation over the nebula at optical wavelengths should be confirmed and extended. An X-ray map is also urgently required.

(4) The origin of the power-law energy spectrum of the electrons at injection is not understood.

Acknowledgements

This work was done while I was a research student at the Mullard Radio Astronomy Observatory, Cavendish Laboratory, Cambridge. The diagrams are reproduced by courtesy of the Royal Astronomical Society.

References

Abramowitz, M. and Stegun, I. A.: 1964, *Handbook of Mathematical Functions*, N.B.S. Applied Mathematics Series.
Baldwin, J. E.: 1971, in R. D. Davies and F. G. Smith (eds.) 'The Crab Nebula', *IAU Symp.* **46**, 22.
Bowyer, S., Byram, E. T., Chubb, T. A., and Friedman, H.: 1964, *Science* **146**, 912.
Branson, N. J. B. A.: 1965, *Observatory* **85**, 250.
Floyd, Jr., F. W.: 1970, *Nature* **226**, 733.
Ginzburg, V. L.: 1953, *Usp. Fiz. Nauk* **51**, 343.
Gratton, L.: 1972, *Astrophys. Space Sci.* **16**, 81.
Johnston, K. J. and Hobbs, R. W.: 1969, *Astrophys. J.* **158**, 145.
Landstreet, J. D. and Angel, J. R. P.: 1971, *Nature* **230**, 103.
Matveyenko, L. I.: 1971, *Astrophys. Letters* **8**, 165.
Mayer, C. H. and Hollinger, J. P.: 1968, *Astrophys. J.* **151**, 53.
Oda, M., Bradt, H., Garmire, G., Spada, G., Sreekantan, B. V., Gursky, H., Giacconi, R., Gorenstein, P., and Waters, J. R.: 1967, *Astrophys. J.* **148**, L5.
Ostriker, J. P. and Gunn, J. E.: 1969, *Astrophys. J.* **157**, 1395.
Ostriker, J. P. and Gunn, J. E.: 1971, *Astrophys. J.* **164**, L95.
Pacini, F. and Salvati, M.: 1973, *Astrophys. Letters* **13**, 103.
Scargle, J. D.: 1968, Ph.D. thesis, California Institute of Technology.
Scargle, J. D.: 1969, *Astrophys. J.* **156**, 401.
Scargle, J. D.: 1971, *Nature Phys. Sci.* **230**, 37.
Trimble, V.: 1968, *Astron. J.* **73**, 535.

Trimble, V. and Rees, M. J.: 1970, *Astrophys. Letters* **5**, 93.
Webster, A. S.: 1970, *Astrophys. Letters* **5**, 189.
Wilson, A. S.: 1972a, *Monthly Notices Roy Astron Soc.* **157**, 229.
Wilson, A. S.: 1972b, *Monthly Notices Roy Astron Soc.* **160**, 355.
Woltjer, L.: 1957, *Bull Astron Inst. Neth.* **13**, 301.
Woltjer, L.: 1958, *Bull Astron. Inst. Neth.* **14**, 39.

DISCUSSION

A. Treves: An injection region of 20″ agrees with the observed delay between spin-ups and 10^{12} eV γ-rays increase observed by Fazio *et al.*

A. S. Wilson: Thank you, but are the 10^{12} eV γ-rays really synchrotron? The electron half-lives against synchrotron losses would be very short indeed.

L. Woltjer: Should one not also include the expansion looses of the electrons?

A. S. Wilson: Yes one should really but note that:

(1) The X-rays and optical electrons are hardly affected by expansion losses because they were injected only very recently.

(2) It is energetically feasible that all the radio electrons were injected in the last ≈ 300 yr, those in it previously having escaped from the nebula. If that is the case, the radio electrons, as well, have suffered little expansion losses.

(3) The radio electrons in any case cannot, as shown by Pacini and Salvati, be remnants of the original explosion and must be injected some time since then. Again, this may mean expansion losses are not so important for the radio electrons presently in the nebula.

M. Reeves: It seems to me that the assumption of a diffusion coefficient D and magnetic field H which are independent of the radius is difficult to accept in your context. I wonder if your conclusion that you require an extended source of electron emission is not simply due to the artificiality of these assumptions.

A. S. Wilson: As I said, the size of the X-ray source can be accounted for by either having an extended injection region or a larger diffusion coefficient for the X-ray electrons. A diffusion coefficient increasing with energy is quite plausible. The philosophy of this work was to see how far you can get with the very simplest possible model. In the model I described there is only one variable parameter, D the diffusion coefficient. If one introduces variation of B with r and D with E and r one could clearly fit all the data because there are then many flexible parameters.

DUST ENVELOPE IN YOUNG SUPERNOVA REMNANTS

S. AIELLO, A. BONETTI, and F. MENCARAGLIA
Space Physics Chair, Univ. of Florence, Florence, Italy

and

E. MASSARO
Physical Institute, University of Palermo, Palermo, Italy

1. Introduction

As stated by Hoyle and Wickramasinghe (1970), hereafter referred to as HW, a significant fraction of the mass ejected by exploding supernovae may condense into solid particles during the expansion phase following explosion. Hence, observable effects on supernova light curves are to be expected.

Absorption by dust should affect the ultraviolet and, to a lesser extent, the visible spectral region, while emission from dust should contribute to the infrared, in the wavelength interval from a few microns to a few hundred microns.

In the following an attempt is made to a semiquantitative evaluation of the modulation induced by dust on young supernova UV and IR luminosities.

2. Experimental Evidence

Present experimental evidence is scarce. From observations in the Crab Nebula, Aitken and Polden (1971), and Furniss *et al.* (1972) report an infrared excess at 10 μm, and an upper limit in the 40–300 μm interval, which are consistent with the assumption of a graphite (and perhaps silicate) dust cloud with a mass of a few percent of one solar mass. This evaluation is close to the suggestion of HW.

Henry *et al.* (1972) analyse in detail the absorption of Crab Nebula soft X-rays. They find that the amount of the equivalent hydrogen density to the Crab is larger by a factor of 2 than the amount of hydrogen obtained from 21 cm absorption measurements. The authors suggest, as a possible mechanism for this extra absorption, the existence within the X-ray generating region of about 1 solar mass of a mixture of elements of atomic weight similar to carbon, amounting to a uniform distribution of about 4 atoms per cm^3. This assumption may be reconciled with the previous results and with the HW model of dust formation.

Kirshner *et al.* (1974) made detailed observations in the visible region from a number of supernovae in nearby Galaxies during the very first stages after the explosion.* In particular they derive a value for the expansion velocity which is close to 2×10^9 cm s^{-1}.

* We are indebted to dr. Searle for a very useful discussion on their results and for the permission of quoting them.

Also they obtain details on the structure of the exploded system and on the thermal behaviour of the inner part and of the expanding envelope. We will refer to these results in the following.

3. The Supernova Dust Envelope

3.1. The Expanding Shell

A model for grain formation is described in HW under the assumption that the material produced by C and Si burning expands adiabatically after the explosion, following a polytropic law of index 3. The following relationships hold among nucleon number density n, temperature T, and radius of the expanding shell R:

$$n \simeq n_0 (T/T_0)^3 \text{ cm}^{-3},$$
$$n \simeq n_0 (R_0/R)^3 \text{ cm}^{-3},$$
$$T \simeq T_0 R_0/R \text{ K},$$
$$R \simeq vt + R_0 \text{ cm},$$

where $n_0 = 10^{31}$ cm^{-3}, $T_0 = 10^{10}$ K, $R_0 = 10^9$ cm are the initial values of n, T, R, and $v = 5 \times 10^8$ cm s^{-1} is the velocity of expansion taken as constant. The mass of the shell is taken of the order of one tenth of the mass of the exploding star (order of 1 $M_\odot$). Following Woltjer (1970), the assumption of constant velocity is plausible as long as the interaction of the shell with the interstellar medium is not important. This is no more true when the mass collected by the expanding shell is comparable to the mass of the shell itself. With the velocity adopted by HW this appears after about 400 yr, at an R of 6×10^{18} cm. With the velocity obtained by Kirshner *et al*, the constant velocity expansion would last about 100 yr.

3.2. Formation of the Dust Envelope

Following HW, the formation of solid particles takes place when the temperature becomes lower than 2000 K, with the precipitation of the most refractory species. Taking into account the abundances of Arnett (1969) and the evidence from interstellar light extinction (Bless and Savage, 1972), the expected composition is mainly graphite, forming at the higher temperatures, with silicates and iron forming at lower temperatures (Wikramasinghe, 1971 and Hoyle and Wikramasinghe, 1968).

At $T = 2000$ K, the radius of the expanding shell is about 5×10^{15} cm, at a time after the onset of the expansion

$$t_{i_{\rm HW}} \simeq 10^7 \text{ s} = 120 \text{ days}.$$

With the velocity of Kirshner *et al.* $t_{i\,\rm Searle} \simeq 30$ days.

At this stage, the number density of nucleons is about 10^{10} cm^{-3}. It decreases by about one order of magnitude as the radius of the shell doubles. After a time $t_0 \simeq 3t_i$, the temperature is down to less than 1000 K, and the number density is less than 10^9 cm^{-3} The rate of grain formation and grain growth becomes negligible.

HW suggest that $\frac{1}{10}$ of the expanding mass may convert to grains during a time interval comparable to $\Delta t = t_0 - t_i$. Using an average mass density of 2.4 gm cm^{-3} for the condensed matter, one obtains ($\Delta t \simeq 2 \times 10^7$ s) for the diameter of grains (mainly graphite) $a_g \simeq 10^{-5}$ cm, and for their mass $m_g \simeq 10^{-14}$ g.

The number density of grains turns out to be $n_g \simeq 2 \times 10^{-2}$ cm^{-3}.

The formation and growth of grains makes the expanding shell very opaque. With the above data, the optical thickness at time t_0 is:

$$\tau_{\max} \simeq \tau_0 = n_g \pi a_g^2 Q_{\mathrm{ext}} R_0 \sim 10^5 .$$

Q_{ext}, the extinction coefficient, varies slowly with wavelength and grain radius. Making use of the expansion velocity measured by Kirshner *et al.*, one may assume in first approximation that the thermodynamical conditions remain the same, except that temperature and shell radius evolve four times as fast. So Δt may be only 5×10^6 s, and the grains radius may reach a value of only 2.5×10^{-6} cm, while the grain number density may remain much the same, 2×10^{-2} cm^{-3}. This means that the mass transformed into solid particles would be only 2×10^{-3} times the ejected mass. But since Q_{ext} is not much sensitive to the dimensions of the grains in a large spectral region from UV to visible, $\tau_{\max} \simeq \tau_0$ decreases essentially with a^2, that is by a factor of 16, down to 6×10^3: the opacity is still very high.

We evaluate now the time after which the expanding cloud becomes optically thin, that is the time t_f after which τ is again about 1.

From $\tau_f = n_{gf} a^2 Q_{\mathrm{ext}} R_f = n_{gf} a^2 Q_{\mathrm{ext}} v t_f$, one obtains (from HW):

$$t_f \simeq 150 \text{ yr}$$

(t_f would be 16 times less in the case of Kirshner *et al.*: $\simeq 10$ yr).

The radius of the expanding shell is now

$$R_f \simeq 2.3 \times 10^{18} \text{ cm}$$

and the number density of the grains

$$n_f \simeq 6 \times 10^{-9} \text{ cm}^{-3} .$$

The shell is still expected to expand at constant velocity. In the case of Kirshner *et al.*: one would obtain $R_f = 10^{17}$ cm with a corresponding larger value of n_f. However, at the time when the expansion velocity is expected to change substantially to lower values (see Section 3.1), the grain densities would be much the same in both cases, and rather near to 10^{-9} cm^{-3}.

3.3. Comparison with Observations

It may be interesting to compare these results with those reported above (Henry *et al.*, 1972) from the absorption of soft X-rays in the Crab Nebula. If one tenth of the suggested density of 4 C atoms per cm^3 gives rise to graphite grains of mass about 10^{-14} gm, the number density of the grains in the 1′ region interesting for X-ray generation would be 10^{-9} cm^{-3}, in agreement with the HW model. (The rest of carbon atoms would be present in that region in molecular compounds, see Hoyle and Wikramasinghe, 1968).

Comparison with the infrared measurements, as reported in Section 2, gives again a fair agreement with the assumption of a (predominantly) graphite dust envelope of the expected total mass. The optical depth due to a grain density of 10^{-9} cm^{-3} over a geometrical path of 1 parsec gives 3×10^9 grains on the line of sight. This is much the same as that observed on the line of sight along 1 kpc in a clear region of the sky on the galactic plane, giving rise to the average stellar light extinction of 1 mag. per kpc. The extinction due to the dust envelope of the Crab Nebula should be of the same order of magnitude, not in contradiction with the available observations (Mayall, 1937).

4. Modulation by Dust of UV and IR Luminosity

4.1. The Heating Sources

Grains absorb electromagnetic radiation, mainly in the UV, to a minor extent in the visible. Grains are heated also by interaction with particles. We examine here briefly the heating mechanisms.

4.1.1. *Heating from Interaction with Particles*

The interaction with particles within the expanding shell takes place in two ways:

(a) encounters with atoms having thermal energies;

(b) interaction with accelerated nuclei (CR).

In case (a), the energy received per unit time by a grain is given by

$$\left(\frac{dE}{dt}\right)_n \propto n\pi a^2 \frac{(KT)^{3/2}}{m^{1/2}};$$

since

$$n \propto t^{-3}, \qquad T \propto R^{-1} \propto t^{-1},$$

one has

$$\left(\frac{dE}{dt}\right)_n \propto t^{-9/2}.$$

It is seen that the amount of energy transmitted to the grains decreases rapidly with time, so that the contribution to heating becomes soon negligible. In case (b), Wickramasinghe (1971) states that the low energy section (10 MeV) of the Cosmic Ray flux accelerated within the expanding supernova shell could contribute substantially to the energy input of grains. Too little is known about the flux and intensity of such accelerated particles to make here an attempt to quantitative conclusions.

We point out that this effect would not depend on the optical properties of grains.

4.1.2. *Heating by Electromagnetic Radiation*

As shown by the results on stellar light extinction, a mixture of graphite and silicate grains of about 10^{-5} cm radius is a powerful absorber of em radiation from far UV down to the visible. So the supernova dust envelope is expected to black out very effectively a source embedded into it during the time $\Delta t = t_0 - t_i$ of grain formation as defined above. From t_0 to t_f the envelope becomes again transparent to the radiation from inner sources. At the same time grains are heated up and reemit in the IR with a quasi-planckian distribution depending on their temperature.

We take into consideration two kinds of sources:

(a) a central thermal source;

(b) an extended, non-thermal, synchrotron source expanding with the shell.

In the various cases we will make a qualitative evaluation of the time dependence of both the UV and the IR luminosities.

4.3. Central thermal source

We make the assumption that a central source exists, and that it keeps substantially constant during a time not less than the optical thinning time t_f.

(a) UV luminosity. One has $I_{UV}=I_0e^{-\tau_{UV}}$, with $\tau_{UV}=n_g\pi a^2 Q_{ext}R$, where $n_g \propto t^{-3}$ and $R \propto t$; so, for a time $t_0<t<t_f$, $I_{UV} \propto e^{-t^{-2}}$. For our purposes Q_{ext} is independent from wavelength in the UV region. The (constant) luminosity at $t \leqslant t_i$ goes down to black-out from t_i to t_0; then it increases again to the initial value at the end of time t_f with the above time dependence.

(b) IR luminosity. We consider three phases:

$t_i<t\leqslant t_0$: grains form, UV absorption takes place, and so does IR reemission;

$t_0<t\leqslant t_f$: grains stop forming, the envelope thins down to transparency. The IR luminosity is calculated for the volume V whose radius is such that $\tau_{UV}=1$.

So:

$$I_{IR} \propto n_g V_{\tau=1} = n_g \tfrac{4}{3}\pi r_{\tau=1}^3$$

$$n_g = n_{g_i}\left(\frac{t_i}{t}\right)^3 \propto t^{-3}; \qquad r_{\tau=1} = \frac{1}{Q_{ext}\pi a_g^2 n_g} \propto t^3$$

$$I_{IR} \propto t^6 .$$

$t>t_f$: the energy rate input into the grains is

$$\frac{dE}{dt} = \int_0^R \frac{I_0}{4\pi r^2} Q_{abs}\pi a^2 n_g 4\pi r^2 \, dr = I_0 Q_{abs}\pi a_g^2 n_g R$$

where I_0 is the source luminosity. Q_{abs} is the absorption coefficient valid in the UV region of the source spectrum. Again:

$$n_g \propto t^{-3}; \qquad R \propto t; \qquad dE/dt \propto t^{-2}.$$

I_{IR} has the same time dependence: $I_{IR} \propto t^{-2}$.

4.4. Extended synchrotron source

Relativistic electrons trapped in a magnetic field embedded in the expanding shell emit synchrotron radiation, which can be absorbed by grains in the appropriate spectral region. The intensity of the reemitted radiation is proportional to the energy received per unit time, which in turn is proportional to the emissivity J_ν of the source.

We write

$$\left(\frac{\mathrm{d}E}{\mathrm{d}t}\right)_{\text{syncr}} \propto \frac{1-e^{-1}}{n_g Q_{\text{ext}} \pi a^2} J_\nu \propto \frac{J_\nu}{n_g}$$

which means that the radiation falling on a grain comes essentially from a region of radius such that $\tau \leqslant 1$.

The time dependence of n_g is $\propto t^{-3}$.

The time dependence of J_ν is related to the time dependence of the magnetic field and of the relativistic electron source. In the case of the existence of a pulsar at the centre of the shell, Pacini and Salvati (1973) give a calculation of the emissivity as a function of time of the synchrotron radiation from the electron cloud. γ being the spectral index of the electrons, they find that the emissivity per unit volume J_ν is given by:

$$(\text{a}_1) \qquad J_\nu \propto \nu^{(1-\gamma)/2} t^{(1-\gamma)/2} t^{-3} \quad \text{for} \quad \nu < \nu_b$$

$$(\text{a}_2) \qquad J_\nu \propto \nu^{-\gamma/2} t^{1-\gamma/2} t^{-3} \quad \text{for} \quad \nu > \nu_b .$$

ν_b corresponds to a break in the frequency distribution, associated to a break in the electron energy distribution: ν_b shifts linearly with time towards higher energies.

In the case of the UV luminosity curve (and the associated IR luminosity from dust reemission), it is important to know at what time after the explosion the break reaches the UV region of the frequency distribution. Calling t_{UV} this time, two extreme cases are possible:

(b$_1$) $t_{\text{UV}} < t_i$ (or t_0): the break frequency ν_b is already beyond the UV region toward larger frequencies when grains form. The time dependency of J_ν is given by (a$_1$);

(b$_2$) $t_{\text{UV}} > t_f$: the break frequency reaches the UV region only after the time when the dust envelope has become transparent. The time dependence of J_ν is given by (a$_2$);

(b$_3$) an intermediate case is that in which t_{UV} falls between t_0 and t_f. The time dependence of J_ν changes at time t_{UV} from (a$_2$) to (a$_1$), with a corresponding break in the luminosity curves.

In the time interval $t_0 < t < t_f$ the UV luminosity increases with the volume of that part of the expanding shell for which $\tau \simeq 1$: this volume varies with time as t^5. After

TABLE I
The UV luminosity, I_{UV}

	$t<t_i$	$t_i<t<t_f$	$t_0<t<t_f$	$t>t_f$
$t_{\text{UV}}<t_i$ $\nu_b>\nu_{\text{UV}}$	$t^{(1-\gamma)/2}=t^{-\alpha}$	?	$t^{(5-\gamma)/2}=t^{2-\alpha}$	$t^{(1-\gamma)/2}=t^{-\alpha}$
$t_{\text{UV}}>t_f$ $\nu_b<\nu_{\text{UV}}$	$t^{1-(\gamma/2)}=t^{1/2-\alpha}$	?	$t^{3-(\gamma/2)}=t^{5/2-\alpha}$	$t^{1-(\gamma/2)}=t^{1/2-\alpha}$
$t_0<t_{\text{UV}}<t_f$	$t^{1-(\gamma/2)}=t^{1/2-\alpha}$	?	$t_0<t<t_{\text{UV}}$, $\nu_b<\nu_{\text{UV}}$: $t^{3-(\gamma/2)}=t^{5/2-\alpha}$ $t_{\text{UV}}<t<t_f$, $\nu_b>\nu_{\text{UV}}$: $t^{(5-\gamma)/2}=t^{2-\alpha}$	$t^{(1-\gamma)/2}=t^{-\alpha}$

TABLE II
The IR luminosity, I_{IR}

	$t<t_i$	$t_i<t<t_0$	$t_0<t<t_f$		$t>t_f$
$t_{UV}<t_i$ $\nu_b>\nu_{UV}$	–	?	$t^{-\alpha}$		$t^{-\alpha}$
$t_{UV}>t_f$ $\nu_b<\nu_{UV}$	–	?	$t^{1/2-\alpha}$		$t^{1/2-\alpha}$
			$t_0<t<t_{UV}$ $\nu_b<\nu_{UV}$	$t_{UV}<t<t_f$ $\nu_b>\nu_{UV}$	
$t_0<t_{UV}<t_f$	–	?	$t^{1/2-\alpha}$	$t^{-\alpha}$	$t^{-\alpha}$

t_f, the UV luminosity varies with the whole volume, that is as t^3. The IR luminosity (neglecting reabsorption) varies for any $t>t_0$ as t^3.

The various cases are summarized in Table I for I_{UV} and in Table II for I_{IR}. As usual, $\gamma=2\alpha+1$, where α is the spectral index of the synchrotron radiation.

4.5. Experimental evidence

No direct experimental evidence exists at present on UV and IR light curves in the early stages of the expansion of a supernova shell.

It seems interesting to refer again to the data described by Kirshner *et al.*

Their evidence in visible light seems to give support to the assumption of the existence of a central 'core' with temperature going from 10000 to 5000 K in a time interval of a few weeks after the explosion, surrounded by a cooler envelope, where strong absorption of a number of prominent lines is observed. Also the luminosity seems to fade out as if obscured by an increasingly opaque screen.

5. Conclusions

The HW model of dust envelope associated with the expanding shell of supernovae gives plausible results for the modulation of UV and IR luminosities after the explosion. Present evidence from IR and soft X-ray observations in the Crab nebula, and from visible observations in a few recent supernovae in nearby galaxies, seems to lend some support to the dust envelope model.

References

Aitken, D. K. and Polden, P. G.: 1971, *Nature (Phys. Sci.)* **233**, 45 and **234**, 73.
Arnett, W. D.: 1969, *Astrophys. J.* **157**, 1369.
Bless, R. C. and Savage, B. D.: 1972, *Astrophys. J.* **171**, 293.
Furniss, I., Jennings, R. E., and Moorwood, A. F. M.: 1972, *Nature (Phys. Sci.)* **236**, 7.
Henry, R. C., Frits, C., Meekins, J. F., Chubb, T. A., and Friedmann, M.: 1972, *Astrophys. J.* **174**, 389.
Hoyle, F. and Wikramasinghe, N. C.: 1968, *Nature* **218**, 1126.
Hoyle, F. and Wikramasinghe, N. C.: 1970, *Nature (Phys. Sci.)* **226**, 62.

Mayall, N. V.: 1937, *Publ. Astron. Soc. Pacific* **49**, 101.
Pacini, F. and Salvati, M.: 1973, NAIC, Cornell University; 1973, *Astrophys. J.* **186**, 249.
Kirshner, R. P., Oke, J. B., Penston, M. V., and Searle, L.: 1974, this volume.
Wikramasinghe, N. C. and Nandy, K.: 1970, *Nature (Phys. Sci.)* **227**, 51.
Wikramasinghe, N. C.: 1971, *Nature (Phys. Sci.)* **230**, 168.
Woltjer, L.: 1970 in H. J. Habring (ed.), 'Interstellar Gas Dynamics', *IAU Symp.* **39**, 208.

DISCUSSION

B. Rossi: What is the actual value of the time when the maximum of the infrared emission is reached?

A. Bonetti: It is the 'clearing' time, when the dust shell becomes transparent to its thermal radiation.

V. Mansfield: What are the main destruction processes for the grains?

A. Bonetti: Destruction processes have not been included in the present analysis.
This should be done in a complete treatment of the problem.

PRE-DETONATION LIFETIME AND MASS OF SUPERNOVAE FROM GALAXY DENSITY WAVE THEORY

E. P. MOORE

New Mexico Institute of Mining and Technology, Socorro, N.M. 87801, U.S.A.

Abstract. Stars in a normal spiral galaxy are formed in a shock front on the inner edge of a spiral arm, according to non-linear density wave theory, then drift across the arm as they age. Ten of the nineteen supernovae measured occur at the innermost edge of a spiral arm, implying a lifetime of less than five million years. Assuming a supernova is a single star at the end of its evolutionary life, the mass implied by this lifetime is greater than 35 solar masses. (Full paper in press, *Publ. Astron. Soc. Pacific*).

DISCUSSION

R. M. Duin: Can you also, the other way around, use the Remnants of type II SN to trace the spiral density waves in our own Galaxy?

E. Moore: Is indeed possible in principle. There may be some difficulty in assigning good distances, but the SN II Remnants should be good spiral arm tracers.

M. F. McCarthy: How do you define 'Sufficiently face on' for the purposes of your paper.

E. Moore: In this first survey, if I can clearly see the spiral arm structure and place the Supernova in that structure, the Galaxy is included in the tabulation. It would be a better statistical practice to set a limiting inclination. Only a couple of high inclination Galaxies were discarded, so any bias introduced must be small. A much more serious potential source of error is the presence of many Supernovae entirely outside all visible traces of spiral structure.

C. B. Cosmovici (ed.), Supernovae and Supernova Remnants, 373.

ON THE FORMATION OF WISPS IN THE CRAB NEBULA

ATTILIO FERRARI
Laboratorio di Cosmo-Geofisica del CNR, Istituto di Fisica Generale dell'Università, Torino, Italy

We discuss here a sequence of physical processes leading to a possible explanation of the formation of the 'wisps', well-known microfeatures visible inside the Crab Nebula. More generally we will be interested in the energetic correlation between the young pulsar NP0532 and the surrounding supernova remnant.

Actually the wisps are centres of dynamical and radiative activity at the inner boundary of the nebula where a rather definite hole encircles the pulsating star. Their activity consists of irregular oscillatory motions alternatively toward and away from the pulsar, over time scales of years or even months. This phenomenon is reported in old plates and has been known since Lampland's observations in 1921. More recently Scargle and Harlan (1970) have given evidence of a sudden enhancement of the activity in connection with the occurrence of a pulsar speed-up.

All these observations follow a general trend suggesting an energetic dependence of the nebula from the pulsar. Other theoretical and experimental arguments are not discussed here, but are well-known already (for a general review see Davies and Smith, 1971). Let us now study how this connection pulsar-nebula can take place.

The usually accepted model for a young pulsar is a neutron star of about 1 $M_{\odot}$ with rapid spinning, $\omega \approx 200$ rad s^{-1}, and high surface magnetic field, $B \approx 10^{12}$ G. The structure of this field is generally assumed dipolar (in relation with models for pulsed radiation) with the symmetry axis at any angle with the spin axis. Even though thermal effects are negligible in extracting particles from a collapsed star (the eventual density scale height should be ≈ 1 cm), a full magnetosphere has been shown to exist as created by strong electric fields induced around the star by the rotation and surface electrodynamic conditions. We refer for this to the works by Goldreich and Julian (1969) and Mestel (1971). This magnetosphere corotates quasi-rigidly with the star up to $\approx R_c = c/\omega$, the speed-of-light radius; beyond this limit a sort of relativistic stellar wind arises with continuous leakage of particles of both sign. This picture has been analyzed by Goldreich and Julian (1969), Michel (1969), Henriksen and Rayburn (1972) in different approximations; from these works one derives a mass flux:

$$10^{9} \leqslant \phi \leqslant 10^{16} \text{ g s}^{-1}$$

corresponding to a particle density (both for protons and electrons in the case of a two-component plasma) at R_c:

$$10^{6} \leqslant n_c \leqslant 10^{13} \text{ part cm}^{-3}.$$

This mass flux is not important with respect to the angular momentum loss of the pulsar related to the slowing-down of the period: the angular momentum is mainly lost by electromagnetic stresses. The above models of the magnetosphere up to R_c

C. B. Cosmovici (ed.), Supernovae and Supernova Remnants, 375–379.

are valid in principle for any magnetic axis orientations. On the other hand, if the observed pulsations are related to an asymmetric magnetic configuration, we know that the em fields outside R_c are time-dependent and a wave propagation arises (Deutsch, 1955). This is a large amplitude wave with two main effects: (i) energy and angular momentum loss from the star, in quantitative agreement with observational data; (ii) acceleration of leaking particles up to very high energies. In this last process we have already a way by which the pulsar affects the nebula, i.e. by acting as a long term source of ultrarelativistic particles (see, e.g., the paper by Woltjer in these proceedings). In addition, if we accept a low mass flux from magnetosphere, $n_c \to 10^6$ part cm^{-3}, so that the em wave has enough power to push all particles out to the maximum attainable energies ($\gamma \propto (eB/mc\omega)^{2/3}$, $\gamma_{pr} \approx 10^5$, $\gamma_{el} \approx 10^7$ for standard data), an asymptotic plasma wind is formed, expanding at relativistic speed. The difference in velocities between protons and electrons cannot affect the expansion since electrostatic forces are never relevant inside a distance of the order of $(c/\omega)\,\gamma_{pr}^2$, as we shall see later.

The interaction between the expanding plasma and the em wave which produces it is a completely non-linear problem. In an ultrarelativistic limit it can be shown (Ferrari, 1972) that the scale length for attenuation of the wave amplitude during the propagation is always higher than the scale length for radiation field decay:

$$\langle l \rangle \approx 10^{10}\,(r/R_c).$$

Thus the wave is not effectively attenuated by the interaction with the wind, even though $\omega^2 \ll \omega_{pe}^2$, classical criterium for reflection. The modifications to this limit are of relativistic origin.

Rees (1971) has proposed that the wave does in fact cross the whole nebula and be responsible for its radiation over the entire em spectrum through the synchro-Compton process. On the other hand observations of the wisps activity seem to suggest that a fraction at least of the wave energy is caught by the nebular plasma at the edges of the central supernova cavity. This effect occurs where the wave pull $P_w = B_c^2/8\pi\,(R_c/r)^2$ becomes comparable to the nebular thermal pressure, which in the case of energy equipartition in the plasma becomes $P_{\mathrm{neb}} \approx B_{\mathrm{neb}}^2/8\pi$; eventually assuming $B_c \approx 10^6$ G and $B_{\mathrm{neb}} \approx 3 \times 10^{-4}$ G (Fazio *et al.*, 1971), the critical point is $r \lesssim 4.5 \times \times 10^{17}$ cm and this agrees with the position of the wisps. Nevertheless the usual absorption processes are shown to be not effective because of the relatively low density in the nebula (Max and Perkins, 1971).

An interesting possibility of resonant absorption can indeed be found in a non-linear interaction of the pulsar wave and relativistic wind 'riding together'. In particular, referring to a frame in which the protons of the wind are at rest (moving at $\gamma \approx 10^5$ with respect to the stellar inertial frame), we notice that the formal plasma frequency ω'_{pe} of the wind falls down $\propto (R_c/r)$ because of the isotropic expansion. Close to R_c, $\omega'_{pe} \gg \omega'$ ($\omega' \approx \omega/\gamma$ is the wave frequency in the wind frame) as we discussed above. Assuming a steady expansion, a distance exists at which $\omega' \approx 2\omega'_{pe}$, namely $R_w \approx 10^9\ R_c$ ($\approx 10^{17}$ cm) again in good agreement with the wisp location.

Where this condition is fulfilled, an efficient non-linear wave coupling may occur leading to a resonant energy exchange in the sense that the strongest mode feeds the weakest ones. The plasma modes that can be excited in this situation by the em transverse wave are in principle Langmuir and ion-sound turbulent modes, which should rapidly deplete a large fraction of the em supply from the pulsar transforming it in plasma heating. For a quasi-monochromatic wave ($\Delta\omega_{\text{em}} \lesssim \omega_{pe}$) the relevant theory of the instability leading to Langmuir turbulence gives a relativistic growth rate:

$$\sigma \gtrsim (\pi/2)\, \omega_{pe} W_{\text{em}} (n_e m_e c^2)^{-1}$$

(Tsytovich 1970), where W_{em} is the energy density associated with the incident mode; a similar growth rate is found for excitation of ion-sound modes directly by em waves or through a plasma electronic oscillation. On the basis of these formulae the time scale for the development of turbulence (in the inertial frame) is $\tau \lesssim 10^6$ s: this means that the pulsar wave is largely absorbed in less than a light month. The efficiency of this instability, even at low densities as in the present case, is both due to the large energy density in the em wave and to the steepness of its spectrum. Notice that the above estimates are based on the assumption of weak non-linearity, so that the energy in the turbulent plasma modes cannot exceed the thermal plasma energy $W_{pl} \lesssim n_e m_e c^2$ (the same is certainly true for the decaying wave). Therefore we cannot follow the whole process of wave decay. Actually the mode coupling does not stop at this stage because two more effects are likely to occur: (i) spreading-out of the em spectrum with decrease of the instability efficiency; and (ii) coupling of Langmuir and ion-sound modes and then with other turbulent modes. By these effects a quasi-stationary situation is created in which a hierarchy of plasma waves is excited and most likely MHD turbulence is settled in if a static magnetic field component starts to play an essential role.

We stress again that the physical conditions for the occurrence of these effects are fulfilled starting from $R_w \approx 10^{17}$ cm; thus we summarize by saying that the wisps could be condensations of plasma generated in the turbulent and relativistic wind from the pulsar by resonant non-linear interaction with the em wave. The plasma theory assures a relevant transfer on a very short time scale, so that a conspicuous fraction of the em energy does not propagate outside the central part of the Crab Nebula. The strict morphology of the wisps would also be related to the field topology of the emitting pulsar in connection with open lines allowing for the wind formation.

A support to this suggestion comes from the fact that the distance to be travelled by wind particles before seeing a full wave oscillation is $\approx \lambda\gamma^2 \approx 10^{18}$ cm; this means that no plasma interaction could be actually realized inside the wisp region. In addition this is also the distance required to provide a spatial charge between protons and electrons (travelling at different velocities) sufficient to start plasma oscillations inside the wind.

From an evolutionary point of view it can easily be shown that, if $\omega(t)$ is the time progression of the rotation frequency, i.e. a decreasing function of t (apart during glitching), the distance of wisps from the pulsar increases steadily $\propto \omega^{-5/6}$. Namely for $\dot{\omega} \propto \omega^n$ the wisps expand with velocity $V_w \propto \omega^{n-11/6}$ over time scales $\propto \omega^{1-n}$. Assuming $n=3$ as given for slowing-down from pure dipole radiation, the present velocity of expansion is evaluated to be less than 1 km s^{-1}, too difficult indeed to be detected observationally due to proper chaotic motions of wisps themselves.

The final picture of the region between pulsar and wisps is summarized in Figure 1. Inside the light cylinder (and up to) pulsed radiation mechanisms are at work. Outside R_c the low frequency em wave accelerates particles escaping from the magnetosphere to very high energies. This process is highly relativistic and non-linear, and

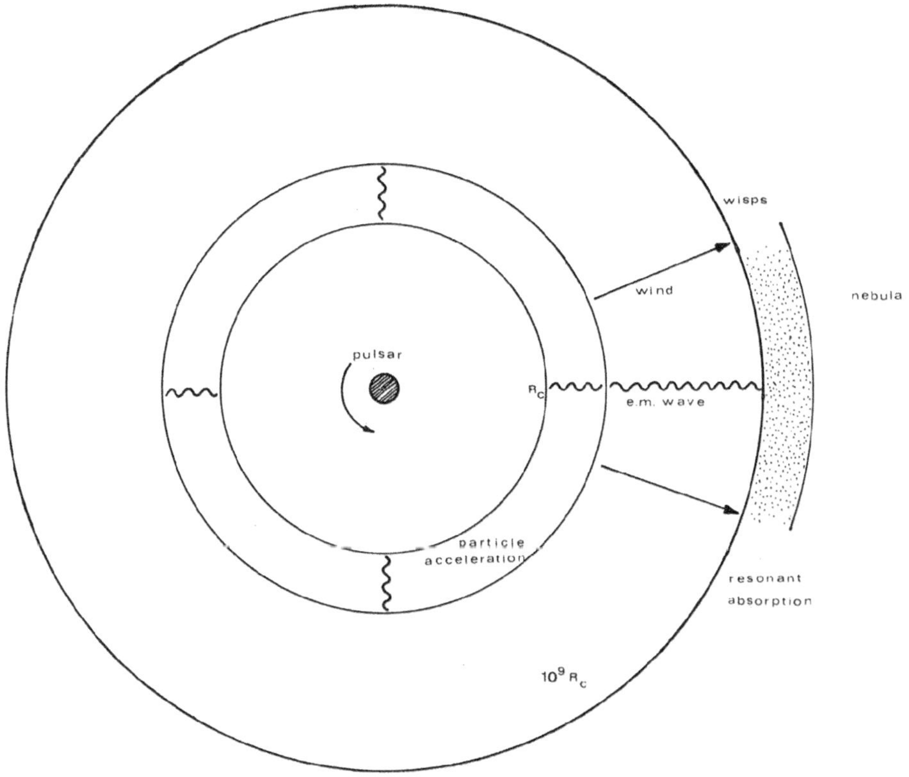

Fig. 1.

radiation emission of synchro-Compton type is of course produced. When the acceleration has been completed at $r \approx 10^9$ cm, charged particles can be considered as a plasma in relativistic expansion. Then collective effects may occur, but the wave attenuation is very weak and requires too long time scales in comparison with the wind dynamics. Thus the wave and the plasma propagate outward without strong interaction and radiation is probably negligible. Then at $r \gtrsim 10^{16}$ cm a resonant coupling of the transverse em wave with Langmuir and ion-sound modes may take place with generation of plasma or MHD turbulence in the pulsar wind that is therefore deviated from free expansion: this entangling of trajectories would be the origin of wisps. Across them a reduced fraction of em energy is propagated, but energetic particles escaping the resonance and turbulent energy are fed into the nebular gas.

The author wishes to thank Prof. Martin Rees for several timely comments; discussions with Dr Edo Trussoni are also acknowledged. This work was prepared in part during a visit at the Institute of Astronomy of Cambridge, England, under joint support of the Accademia Nazionale dei Lincei and Royal Society in the scheme of a European Exchange Fellowship.

References

Davies, R. D. and Smith, F. G. (eds.): 1971, 'The Crab Nebula', *IAU Symp.* **46**.
Deutsch, A.: 1955, *Ann. Astrophys.* **18**, 1.
Fazio, G. G., Helmken, H. F., Rieke, G. H., and Weekes, T. C.: 1971, 'The Crab Nebula', *IAU Symp.* **46**, 160.
Ferrari, A.: 1972, Mem. Soc. Astron. Ital. **43**, 715.
Goldreich, P. and Julian, W. H.: 1969, *Astrophys. J.* **157**, 869.
Henriksen, R. N. and Rayburn, D. R.: 1972, *Astrophys. Letters* **11**, 107.
Max, C. and Perkins, F.: 1971, *Phys. Rev. Letters* **27**, 1342.
Mestel, L.: 1971, *Nature* **233**, 149.
Michel, F. C.: 1969, *Astrophys. J.* **158**, 727.
Rees, M. J.: 1971, *Nature* **230**, 55.
Scargle, J. D. and Harlan, E.: 1970, *Astrophys. J. Letters* **159**, L143.
Tsytovich, V. N.: 1970, *Nonlinear Effects in Plasma*, Plenum Press, New York.

DEUTERIUM AND BORON FROM SUPERNOVAE?

HUBERT REEVES
Centre D'Etudes Nucléaires, Saclay, 91-Gif-Sur-Yvette, France

From an analysis of the various measurements of deuterium abundance in a number of locations (Trauger *et al.*, 1973; Rogerson and York, 1973) it appears that the cosmic value of $n(\mathrm{D})/n(\mathrm{H})$ should be $\sim 1.5 \times 10^{-5}$ (see Reeves, 1974 for a detailed discussion).

The origin of D is generally ascribed to the Big Bang, where the estimated value, corrected for the effect of destruction in stars, is compatible with a rather low density universe ($\varrho \simeq 1$ to 6×10^{-31} g cm^{-3}). Recently the suggestion has been made (Hoyle and Fowler, 1973 and Colgate, 1973), that D could also originate from ^{4}He spallation in shock waves traveling through the atmospheres of exploding celestial objects. Colgate (1973) uses his previous dynamical study of supernovae explosions to evaluate the mass fraction of the star in which high velocities will be reached: if an amount of energy W per nucleon is deposited throughout the object by the detonating mechanism, the fraction F of the exploding mass in which spallation will take place is given by

$$F(E > Q) \simeq 0.6\, W^2/Q^2 \qquad W < Q$$

where Q is the threshold for the reaction $^4\mathrm{He} + \mathrm{p} \to \cdots \mathrm{D}$ ($\simeq 25$ MeV).

The ratio of deuterium to hydrogen in a given object is given by

$$\left.\frac{n(\mathrm{D})}{n(\mathrm{H})}\right|_\sigma = F(E > Q)\,\frac{n(^4\mathrm{He})}{n(\mathrm{H})}\,h_d$$

where $n(^4\mathrm{He})/n(\mathrm{H})$ is the abundance of ^{4}He in the shocked atmosphere ($\simeq 0.1$) and h_d is the number of D formed per initial ^{4}He ($\simeq 1$). Hence

$$\left.\frac{n(\mathrm{D})}{n(\mathrm{H})}\right|_\sigma = 10^{-4}\, W^2$$

where W is the energy deposited in the SN in MeV N^{-1}. An evaluation of W can be obtained for the reports made at this conference on observations and studies of type II supernovae in exterior galaxies or of supernovae remnants in our Galaxy (Kirshner *et al.*, 1973; Gordon, 1972). It appears that the sum of light and gas kinetic energy is ~ 1 to 2×10^{51} erg while the expanding mass is $\simeq 5\, M_\odot$: hence $W \simeq 0.2$ MeV N^{-1}. Hence $n(\mathrm{D})/n(\mathrm{H})|_\sigma \simeq 4 \times 10^{-6}$ in any given object. The mass fraction of our Galaxy f_g which has been involved in exploding objects is at most 0.1; hence the contribution of SN explosions to the 'D' in the galaxy is

$$\left.\frac{n(\mathrm{D})}{n(\mathrm{H})}\right|_g = f_g \left.\frac{n(\mathrm{D})}{n(\mathrm{H})}\right|_\sigma \sim 4 \times 10^{-7}$$

far too low to be of any significance.

C. B. Cosmovici (ed.), Supernovae and Supernova Remnants, 381–382.

A similar analysis can be made for the formation of ^{11}B through the $^{14}N+p\rightarrow\rightarrow\alpha+{}^{11}C\rightarrow{}^{11}B$. Here the effective Q (obtained from the excitation curve (Epherre, 1972)) is $\simeq 7$ MeV, h_d is $\simeq 0.1$ in realistic conditions, $n(^{14}N)/n(H)\simeq 10^{-4}$ yielding $n(^{11}B)/n(H)\simeq 10^{-9}$, again too low to account for the meteoritic boron (Cameron *et al.*, 1973), if the new revised value (10^{-8}) is justified (Adouze *et al.*, 1973).

What about type I supernovae? Although energetically the situation is better ($W\simeq 1$ MeV) their masses are small and they do not appear to have in the presupernova stage the thin and extended envelope required for the shock wave to gain speed and reach several tens of MeV per nucleon. Furthermore, the post-shock temperatures are likely to be very hot, too hot for any D or LiBeB to resist thermonuclear burning and destruction.

References

Audouze, J., Lequeux, J., and Reeves, H.: 1973, *Astron. Astrophys.* **28**, 85.
Cameron, A. G. W., Colgate, S. A., and Grossmann, L.: 1973, *Nature* **243**, 204.
Colgate, S. A.: 1973, *Astrophys. J. Letters* **181**, L 53.
Epherre, M.: 1972, Thesis, Orsay.
Gordon, C.: 1972, *Astron. Astrophys.* **20**, 79.
Hoyle, F. and Fowler, W. A.: 1973, *Nature* **241**, 384–86.
Kirshner, R. P., Oke, J. B., Penston, M. V., and Searle, L.: 1973, *Astrophys. J.* **185**, 303.
Reeves, H.: 1973, to appear in *Ann. Rev. Astron. Astrophys.*
Reeves, H., Audouze, J., Fowler, W. A., and Schramm, D. N.: 1973, *Astrophys. J.* **179**, 909.
Rogerson, J. B. and York, D. G.: 1973, *Astrophys. J.* **186**, L 95.
Trauger, J. T., Roesler, F. L., Carleton, N. P., and Traub, W.: 1973, *Astrophys. J.* **184**, L 137–43.

CONCLUDING REMARKS

L. WOLTJER
Dept. of Astronomy, Columbia University, New York, U.S.A.

Of course in the brief time available it is not possible to give anything like a complete summary and I shall just make a few brief remarks on some of the things that have been discussed here.

Concerning the supernovae I have been struck by the convergence of the different types. We have seen many light curves these days and it is my impression that the differences between the light curves in different colors for the same type are almost larger than the differences in the light curves for different types in the same color. Of course there remain real differences – otherwise the observers could not agree so readily on the type on the basis of either spectra or light curves – but one would be surprised if the basic physical situation were all that different.

With regard to the supernova frequency it is now apparent that obvious differences exist between different types of galaxies. To date this has been shown on the basis of samples of galaxies in which no very systematic search for supernovae has been made – like the Shapley-Ames galaxies. It is to be hoped that future supernova researches will be concentrated on carefully preselected samples of galaxies so that the frequency determination and by inference the populations responsible for the supernovae become more definitely established. Also frequent observation of few fields may well yield more useful data than more incidental observations of many fields. Frequent observation (every few days) is especially important if critically important pre maximum observations and magnitudes of maximum are to be obtained. The latter have gained much in importance because of their potential value as distance indicators. As the discussion of the supernova in Stephan's quintet showed they may well help settle the question as to whether there are non cosmological redshift components in the spectra of galaxies. The radio data for the supernova in M101 look quite intriguing. Observations of the very early phases of supernova remnants may well be of prime importance in elucidating pulsar dynamics and also the question whether all supernova remnants have pulsars. Of course the presence of much ionized gas in the early phases may still cause trouble. The theory of the supernova explosions and the relation with stellar evolution are still imperfectly understood, although progress has been made.

The X-ray data have given us an entirely new view on the supernova remnants, which should be complimentary to that obtained from optical and radio observations. At the moment, however, in some respects these data seem more contradictory than complimentary. To make further progress accurate positions of emission features in X-rays are needed to determine the precise interrelation between the X-ray emitting regions and the regions responsible for radio and optical emission. This requires high spatial resolution and the Copernicus observations on Cas A are a

C. B. Cosmovici (ed.), Supernovae and Supernova Remnants, 383–384.

promising indication of the possibilities: If with an X-ray telescope of effectively 1 cm^2 aperture in a satellite devoted only for 10% of time to X-ray work so much can be learned, one can expect much of a 100 cm^2 telescope in an X-ray satellite. The recent discovery of coronal optical emission lines (Fe XIV) by Angel and Woodgate shows that optical observations of the X-ray emitting region also may be fruitful in further elucidating its kinematics and dynamics.

The great precision and resolution of the newer radio data is very encouraging and as a result very detailed theoretical models now may be constructed and compared with the observations. The observations of the structure of Tycho's remnant at high frequency, however, show that there is more to be done than just finishing off the details. The presence of some diffuse structures in the source with inverted spectra could have profound implications.

The general question of the understanding of the radio spectra of the supernova remnants is still in a very unsatisfactory state. While our theoretical understanding of the basic structure of the pulsar magnetosphere and the mechanisms for generating magnetic fields and energetic particles may still be primitive, there is no theory which explains why the relativistic electrons have a power law type energy spectrum with the kind of slope observed. We possibly do understand the maximum energy to which a pulsar may accelerate electrons, but not the way in which the energy is distributed over many electrons. This is quite serious, because a large body of precise observational information on spectral indices exists; in the absence of some kind of theory this information is completely wasted.

REVIEW OF THE CONFERENCE

HALTON ARP

Hale Observatories, Carnegie Institution of Washington, California Institute of Technology, Pasadena, Calif., U.S.A.

With regard to the gigantic stellar explosions that take place within galaxies, this conference has most impressed me with the variety of ways in which these can take place. For example, I have heard for many years comments to the effect that only type I and type II supernovae were bona fide classes and that not much else was important. It seems to me that not only are the original five types of supernovae still valid, but, as the papers here have shown, an additional very exciting result is emerging. It seems to me that within the types I and II, there are gradations in properties becoming apparent and possibly the existence of subclasses within these larger types is presented.

This possibility appears clearly from the recent work done at Asiago, by Barbon on the fast and slow type I light curves and to a lesser extent from the work done by Ciatti on the type II light curves. Reinforcing this photometric evidence are the spectral peculiarities pointed out by Bertola and several observers in the supernova in NGC 2841 and elsewhere. Remarks by Tamman implied that the type II supernova in M 101 might be similar, in straight light curve decline and relatively high absolute magnitude, to the one studied by Arp in NGC 7331. These two, in turn seem considerably different from some of the type II light curves discussed by de Vaucouleurs and the fainter absolute magnitudes generally attributed to type II supernova.

The establishment of gradations or subclasses within supernovae types would have the important consequence that it would imply that there are differences in luminosity associated with differences in light curve form or spectra. If the luminosity differences are small, their study might be mainly valuable in understanding the mechanisms of the explosion process. If the luminosity differences are large they could affect our present interpretation of the supernovae as distance indicators. Are these differences in types, and possibly within types, related to differences in population types within the galaxies in which they explode? What is the physical reason for differences in supernovae?

Dramatic progress has been made on the study of supernova remnants. We have seen in this conference detailed radio and X-ray maps of the shells produced by the supernovae. With the general identification of pulsars as the remaining kernel of this ancient explosion there is a general picture of how a supernova explodes and developes. But it will be a challenging puzzle, perhaps not impossible, but very difficult, to relate the nature of the fossil supernova to exact type of supernova outburst which we are used to seeing in real time. In this sense it will be exceedingly difficult to understand what physically causes the differences in the different types and sub types

C. B. Cosmovici (ed.), Supernovae and Supernova Remnants, 385–387.

of supernovae which we have observed. Perhaps this project will be aided to some extent by observations of supernova remnants in external galaxies along the lines outlined by De Bruyn and other users of high resolution radio telescopes.

The second problem discussed here which impressed me with its importance, is a problem on which progress has been made more slowly. Nevertheless the distribution of supernovae within galaxies and in different morphological types of galaxies offers the possibility of identifying the types of stellar populations which furnish the pre-supernova candidates. Such work should give some of the basic evidence for the physical explanation of the differences between different kinds of supernovae and possibly fundamental insight into the cause of the explosion as well.

The frequency of supernova occurrence in galaxies is still a vexed question. But as we hear more discussion of this problem I think we begin to realize it is predominantly a problem of terminology. The fact that supernova production varied as a function of the mass of the producing galaxy was known from Zwicky's earlier work. Apparently the statement that one supernova occurs per galaxy per three hundred years is as accurate as the statement that one per thirty years occurs, depending on what sample of galaxies or what specific kinds of galaxies one is talking about. The process at this stage seems to be one of defining accurately the class of objects about which the frequency statements are made. The controversy may in fact be generated by the recent tendency to generalize specific results too quickly to ill defined 'total' categories and 'overall' conclusions.

One extremely important and very specific need which this conference has sharpened in my mind is the need to investigate rigorously the distance moduli for such key supernova host galaxies as IC 4182 as well as NGC 3253, NGC 5236, NGC 5128 and the whole Centaurus chain of galaxies. By doing the best possible derivation of moduli for these galaxies we can begin to study the relationship of the intrinsic properties of a number of well studied, nearby supernovae. We also need to resolve some of the current large discordances in luminosity for some of these nearby supernovae. The provocative results presented by D. Branch on the possible systematic differences between derived luminosities of supernovae in spirals and ellipticals needs to be checked. The supernovae in the various well known clusters of galaxies such as Coma and Virgo need to be more accurately intercompared and also compared to the parameters derived from supernovae from smaller groups of galaxies.

For the answers to all these questions we need more data. We need the discovery of bright supernovae sooner in nearby systems. The earlier observations are needed particularly for the attempted independant predictions of absolute magnitudes from velocities and temperatures of early, uniformly radiating supernova shells. We also need discovery of fainter supernovae to fainter magnitudes and spectra and light curves to fainter magnitudes. Most particularly we need discovery and through observations of supernovae in peculiar galaxies and abnormal systems of galaxies. Fortunately because of the many new telescopes and auxiliary instruments coming into use in different sites in the world today I do have the feeling that there will be a great upsurge in the observational data which we need to solve the many supernova problems of the present time. I hope that this conference will motivate some people to pursue and coordinate these supernova investigations.

Finally, I would like to note that this conference is a tribute to those astronomers who pioneered the investigation of this subject such a short time ago and who carried it through to its present state. It has been a pleasure to summarize the current state by all of us working together with some of the originators in the field. And certainly most of all it is a pleasure to thank our Italian hosts for initiating and conducting such a stimulating and successful conference.

ASTROPHYSICS AND SPACE SCIENCE LIBRARY

Edited by

1. C. de Jager (ed.), *The Solar Spectrum. Proceedings of the Symposium held at the University of Utrecht, 26–31 August, 1963.* XIV + 417 pp.
2. J. Ortner and H. Maseland (eds.), *Introduction to Solar Terrestrial Relations. Proceedings of the Summer School in Space Physics held in Alpbach, Austria, July 15–August 10, 1963 and Organized by the European Preparatory Commission for Space Research.* 1965, IX + 506 pp.
3. C. C. Chang and S. S. Huang (eds.) *Proceedings of the Plasma Space Science Symposium, Held at the Catholic University of America, Washington, D.C., June 11–14, 1963.* 1965, IX + 377 pp.
4. Zdeněk Kopal, *An Introduction to the Study of the Moon.* 1966, XII + 464 pp.
5. Billy M. McCormac (ed.), *Radiation Trapped in the Earth's Magnetic Field. Proceed ingsof the Advanced Study Institute, Held at the Chr. Michelsen Institute, Bergen, Norway, August 16–September 3, 1965.* 1966, XII + 901 pp.
6. A. B. Underhill, *The Early Type Stars.* 1966, XIII + 282 pp.
7. Jean Kovalevsky, *Introduction to Celestial Mechanics*, 1967, VIII + 427 pp.
8. Zdeněk Kopal and Constantine L. Goudas (eds.), *Measure of the Moon. Proceedings of the Second International Conference on Selenodesy and Lunar Topography held in the University of Manchester, England, May 30–June 4, 1966.* 1967, XVIII + 479 pp.
9. J. G. Emming (ed.), *Electromagnetic Radiation in Space. Proceedings of the Third ESRO Summer School in Space Physics, held in Alpbach, Austria from 19 July to 13 August, 1965.* 1968, VIII + 307 pp.
10. R. L. Carovillano, John F. McClay, and Henry R. Radoski (eds.), *Physics of the Magnetosphere. Based upon the Proceedings of the Conference held at Boston College, June 19–28, 1967.* 1968, X + 686 pp.
11. Syun-Ichi Akasofu, *Polar and Magnetospheric Substorms.* 1968, XVIII + 280 pp.
12. Peter M. Millman (ed.), *Meteorite Research. Proceedings of a Symposium on Meteorite Research held in Vienna, Austria, 7–13 August, 1968.* 1969, XV + 941 pp.
13. Margherita Hack (ed.), *Mass Loss from Stars. Proceedings of the Second Trieste Colloquium on Astrophysics, 12–17 September, 1968.* 1969, XII + 345 pp.
14. N. D'Angelo (ed.), *Low-Frequency Waves and Irregularities in the Ionosphere. Proceedings of the 2nd ESRIN-ESLAB Symposium, held in Frascati, Italy, 23–27 September, 1968.* 1969, VII + 218 pp.
15. G. A. Partel (ed.), *Space Engineering. Proceedings of the Second International Conference on Space Engineering, held at the Fondazione Giorgio Cini, Isola di San Giorgio, Venice, Italy, May 7–10, 1969.* 1970, XI + 728 pp.
16. S. Fred Singer (ed.), *Manned Laboratories in Space. Second International Orbital Laboratory Symposium.* 1969, XIII + 133 pp.
17. B. M. McCormac (ed.), *Particles and Fields in the Magnetosphere. Symposium Organized by the Summer Advanced Study Institute, held at the University of California, Santa Barbara, Calif. August 4–15, 1969.* 1970, XI + 450 pp.
18. Jean-Claude Pecker, *Experimental Astronomy.* 1970, X + 105 pp.
19. V. Manno and D. E. Page (eds.), *Intercorrelated Satellite Observations related to Solar Events. Proceedings of the Third ESLAB/ESRIN Symposium held in Noordwijk, The Netherlands, September 16–19, 1969.* 1970, XVI + 627 pp.
20. L. Mansinha, D. E. Smylie, and A. E. Beck, *Earthquake Displacement Fields and the Rotation of the Earth. A NATO Advanced Study Institute Conference Organized by the Department of Geophysics, University of Western Ontario, London, Canada, June 22–28, 1969.* 1970, XI + 308 pp.
21. Jean-Claude Pecker, *Space Observatories.* 1970, XI + 120 pp.
22. L. N. Mavridis (ed.), *Structure and Evolution of the Galaxy. Proceedings of the Nato Advanced Study Institute, held in Athens, September 8–19, 1969.* 1971, VII + 312 pp.

23. A. Muller (ed.), *The Magellanic Clouds. A European Southern Observatory Presentation: Principal Prospects, Current Observational and Theoretical Approaches, and Prospects for Future Research. Based on the Symposium on the Magellanic Clouds, held in Santiago de Chile, March 1969, on the Occasion of the Dedication of the European Southern Observatory.* 1971, XII + 189 pp.
24. B. M. McCormac (ed.), *The Radiating Atmosphere. Proceedings of a Symposium Organized by the Summer Advanced Study Institute, held at Queen's University, Kingston, Ontario, August 3–14, 1970.* 1971, XI + 455 pp.
25. G. Fiocco (ed.), *Mesospheric Models and Related Experiments. Proceedings of the 4th ESRIN-ESLAB Symposium, held at Frascati, Italy, July 6–10, 1970.* 1971, VIII + 298 pp.
26. I. Atanasijević, *Selected Exercises in Galactic Astronomy.* 1971, XII + 144 pp.
27. C. J. Macris (ed.), *Physics of the Solar Corona. Proceedings of the NATO Advanced Study Institute on Physics of the Solar Corona, held at Cavouri-Vouliagmeni, Athens, Greece, 6–17 September 1970.* 1971, XII + 345 pp.
28. F. Delobeau, *The Environment of the Earth.* 1971, IX + 113 pp.
29. E. R. Dyer (general ed.), *Solar-Terrestrial Physics 1970. Proceedings of the International Symposium on Solar-Terrestrial Physics, held in Leningrad, U.S.S.R., 12–19 May 1970.* 1972, VIII + 938 pp.
30. V. Manno and J. Ring (eds.), *Infrared Detection Techniques for Space Research, Proceedings of the Fifth ESLAB-ESRIN Symposium, held in Noordwijk, The Netherlands, June 8–11, 1971.* 1972, XII + 344 pp.
31. M. Lecar (ed.), *Gravitational N-Body Problem, Proceedings of IAU Colloquium No. 10, held in Cambridge, England, August 12–15, 1970.* 1972, XI + 441 pp.
32. B. M. McCormac (ed.), *Earth's Magnetospheric Processes. Proceedings of a Symposium Organized by the Summer Advanced Study Institute and Ninth ESRO Summer School, held in Cortina, Italy, August 30–September 10, 1971.* 1972, VIII + 417 pp.
33. Antonin Rükl, *Maps of Lunar Hemispheres.* 1972, V + 24 pp.
34. V. Kourganoff, *Introduction to the Physics of Stellar Interiors.* 1973, XI + 115 pp.
35. B. M. McCormac (ed.), *Physics and Chemistry of Upper Atmospheres. Proceedings of Symposium Organized by the Summer Advanced Study Institute, held at the University of Orléans, France, July 31–August 11, 1972.* 1973, VIII + 389 pp.
36. J. D. Fernie (ed.), *Variable Stars in Globular Clusters and in Related Systems. Proceedings of the IAU Colloquium No. 21, held at the University of Toronto, Toronto, Canada, August 29–31, 1972.* 1973, IX + 234 pp.
37. R. J. L. Grard (ed.), *Photon and Particle Interaction with Surfaces in Space. Proceedings of the 6th ESLAB Symposium, held at Noordwijk, the Netherlands, 26–29 September, 1972.* 1973, XV + 577 pp.
38. Werner Israel (ed.), *Relativity, Astrophysics and Cosmology. Proceedings of the Summer School Held, 14–26 August, 1972, at the BANFF Centre, BANFF, Alberta, Canada.* 1973, IX + 323.
39. B. D. Tapley and V. Szebehely (eds.), *Recent Advances in Dynamical Astronomy. Proceedings of the NATO Advanced Study Institute in Dynamical Astronomy held in Cortina d'Ampezzo, Italy, August 9–21, 1972.* 1973, XIII + 468 pp.
40. A. G. W. Cameron (ed.), *Cosmochemistry. Proceedings of the Symposium on Cosmochemistry, held at the Smithsonian Astrophysical Observatory, Cambridge, Mass., August 14–16, 1972.* 1973, X + 173 pp.

In preparation:

41. M. Golay, *Introduction to Astronomical Photometry.* 1974, approx. 450 pp.
42. D. E. Page (ed.), *Correlated Interplanetary and Magnetospheric Observations. Proceedings of the Seventh ESLAB Symposium, held at Saulgau, W. Germany, 22–25 May, 1973.* 1974, XIV + 662 pp.

GPSR Compliance

The European Union's (EU) General Product Safety Regulation (GPSR) is a set of rules that requires consumer products to be safe and our obligations to ensure this.

If you have any concerns about our products, you can contact us on ProductSafety@springernature.com

In case Publisher is established outside the EU, the EU authorized representative is:

Springer Nature Customer Service Center GmbH
Europaplatz 3
69115 Heidelberg, Germany

Batch number: 10824484

Printed by Printforce, the Netherlands